Handling Emotions in Human-Computer Dialogues

Johannes Pittermann · Angela Pittermann
Wolfgang Minker

Handling Emotions in Human-Computer Dialogues

Springer

Johannes Pittermann
Universität Ulm
Inst. Informationstechnik
Albert-Einstein-Allee 43
89081 Ulm
Germany
johannes.pittermann@alumni.uni-ulm.de

Angela Pittermann
Universität Ulm
Inst. Informationstechnik
Albert-Einstein-Allee 43
89081 Ulm
Germany
angelapittermann@gmx.de

Wolfgang Minker
Universität Ulm
Fak. Ingenieurwissenschaften
And
Elektrotechnik
Albert-Einstein-Allee 43
89081 Ulm
Germany
wolfgang.minker@uni-ulm.de

ISBN 978-90-481-3128-0 e-ISBN 978-90-481-3129-7
DOI 10.1007/978-90-481-3129-7
Springer Dordrecht Heidelberg London New York

Library of Congress Control Number: 2009931247

Cover design: Boekhorst Design b.v.

Printed on acid-free paper

Springer is part of Springer Science+Business Media (www.springer.com)

Preface

"The finest emotion of which we are capable is the mystic emotion"

(Albert Einstein, 1879–1955)

During the past years the "mystery" of emotions has increasingly attracted interest in research on human–computer interaction. In this work we investigate the problem of how to incorporate the user's emotional state into a spoken language dialogue system. The book describes the recognition and classification of emotions and proposes models integrating emotions into adaptive dialogue management.

In computer and telecommunication technologies the way in how people communicate with each other is changing significantly from a strictly structured and formatted information transfer to a flexible and more natural communication. Spoken language is the most natural way of communication between humans and it also provides an easy and quick way to interact with a computer application. These systems range from information kiosks where travelers can book flights or buy train tickets to handheld devices which show tourists around cities while interactively giving information about points of interest. Generally, spoken language dialogue does not only mean simplicity, comfort and saving of time but moreover contributes to safety aspects in critical environments like in cars, where hands-free operation is indispensible in order to keep the driver's distraction minimal. Within the context of ubiquitous computing in intelligent environments dialogue systems facilitate everyday work, e.g., at home where lights or household appliances can be controlled by voice commands, and provide the possibility, especially in assisted living, to quickly summon help in emergency cases.

In parallel to the progress made in technical development the customer's demands concerning the products have increased. While car owners in the 1920s might have been completely satisfied once they arrived at a destination without any major complications, people in the 1970s would have already tended to become annoyed once their engine refuses to start on the first turn of the ignition key. And nowadays a navigation system showing the wrong way might even cause more anger. For ubiquitous technology like cars this means on the one hand that the driver is literally at the mercy of sophisticated technology on the other hand this does not hinder him/her from building some kind of personal relation to the car, ranging from decorations like car fresheners or fuzzy dice to expensive tuning. Such a relation includes as well the expression of emotions towards the car – just imagine drivers spurring on their cars when climbing a steep hill and being glad having reached the top, or drivers shouting at their non-functioning navigation system, hitting or kicking their cars... A similar behavior can be observed among computer users. Having successfully written a book using a word processing software might arouse happiness, however a sudden hard disc crash destroying all documents will probably drive the author up the wall.

Normally neither the car nor the computer is capable of replying to the user's affect. So why not enable devices to react accordingly? Think of a car that refuses to start and the driver shouting angrily *"Stupid car, I paid more than $40,000 and now it's only causing trouble!"*. Here a car's reply like *"I am sorry that the engine does not run properly. This is due to a defective spark-plug which needs to be replaced."* would certainly defuse the tense situation and it moreover provides useful information on how to solve the problem. This again contributes to safety aspects in the car as the driver can be calmed down, e.g., in the case of a delay due to a traffic jam, whereupon the driver tries to make up the loss of time by speeding. Here the car's computer could try to rearrange the planned meeting and inform the user: *"Due to our delay I have rescheduled your meeting one hour later. So there is no need to hurry."*

To implement a more flexible system, the typical architecture of a spoken language dialogue system needs to be equipped with additional functionality. This includes the recognition of emotions and the detection of situation-based parameters as well as user-state and situation managers which calculate models based on these parameters and influence the course of the dialogue accordingly.

Constituting a hot topic of interest in current research there exist several approaches to classify the user's emotions. These methods include the measurement of physiological values using biosensors, the interpretation of gestures and facial expressions using cameras, natural language processing spotting emotive keywords and fillers in recognized utterances or classification of prosodic features extracted from the speech signal. Concentrating on a monomodal system without video input and trying to reduce inconveniences to the user, this work focuses on the recognition of emotions from the speech signal using Hidden Markov Models (HMMs). Based on a database of emotional speech, a set of prosodic features has been selected and HMMs have been trained and tested for six emotions and ten speakers. Due to variations in model parameters multiple recognizers have been implemented.

According to the output of the emotion recognizer(s) the course of the dialogue is influenced. With the help of a user-state model and a situation model the dialogue strategy is adapted and an appropriate stylistic realization of its prompts is chosen. I.e., if the user is in a neutral mood and speaks clearly, there are no confirmations necessary and the dialogue can be kept relatively short. However if the user is angry and speaks correspondingly unclearly, the system has to try to calm down the user but it also has to ask often for confirmation, which again makes the user turn angry... Principally there exist two methods to model the influence of these so-called control parameters like emotions: a rule-based approach where every eventuality in the user's behavior is covered by a rule which contains a suitable reply, or a stochastic approach which models the probability of a certain reply in dependence of the user's previous utterances and corresponding control parameters.

So how is this book organized? An introduction to the research topic is followed by an overview on emotions – theories and emotions in speech. In the third chapter, dialogue strategy concepts with regard to integrating emotions in spoken dialogue are described. Signal processing and speech-based emotion recognition are discussed in Chapter 4 and improvements to our proposed emotion recognizers as well as the implementation of our adaptive dialogue manager are discussed in Chapter 5. Chapter 6 presents evaluation results of the emotion recognition component and of the end-to-end system with respect to existing spoken language dialogue systems evaluation paradigms. The book concludes with a final discussion and an outlook on future research directions.

Ulm,
May 2009

Johannes & Angela Pittermann
Wolfgang Minker

Contents

Chapter 1
Introduction

"How may I help you?" (cf. Gorin et al. 1997) – Imagine you are calling your travel agency's telephone hotline and you don't even notice that you are talking to a computer. Would you be surprised if your virtual dialogue partner recognized you by means of your voice and if it asked you how you liked your previous trip?

The ongoing trend of computers becoming more powerful, smaller, cheaper and more user-friendly leads to the effect that these devices increasingly gain in importance in everyday life and become "invisible". Within this so-called *ubiquitous computing* there exist a large variety of applications and data structures ranging from information retrieval systems to control tasks and emergency call functionality. In order to handle these applications, a manageable user interface is required which can be realized with the aid of a spoken language dialogue system (SLDS).

In this chapter, we give a brief overview on the functionality of SLDS and their implementation in current dialogue applications. Some of the ideas presented here already apply successfully in state-of-the-art dialogue applications, other ideas are still part of ongoing research. Thus, certain challenges still exist in the development of speech applications (see also Minker et al. 2006b). In this book, we address the user-friendliness and the naturalness of an SLDS. This includes the adaptation of the dialogue to the user's emotional state and, to accomplish that, the recognition of emotions from the speech signal. Therefore we describe the architecture of an SLDS and refer to approaches where regular dialogue systems may be improved and how these improvements can be realized. Here, in Sections 1.2–1.4 especially challenges in the development of adaptive dialogue management are addressed. In Chapters 3–5, we describe our strategies of integrating emotions into adaptive dialogue management and our approach to speech-based emotion recognition and its derivatives like combined speech–emotion recognition and optimization approaches. An evaluation of our methods as well as a summary and a discussion of future perspectives is given in Chapters 6 and 7.

J. Pittermann et al., *Handling Emotions in Human-Computer Dialogues*,
DOI 10.1007/978-90-481-3129-7_1,

1.1 Spoken Language Dialogue Systems

SLDSs enable a user to access a computer application using spoken natural language which is the most natural, convenient and inexpensive way of communication (De Mori 1998; McTear 2004). The general structure of an SLDS is depicted in Fig. 1.1: automatic speech recognition (ASR) for translating an audio signal into sequences of words, natural language understanding (parsing) for extracting a meaning from these words, a dialogue manager for application access and initiating system actions, text generation and speech synthesis for formulating system prompts and transforming them into audio signals, respectively (Minker et al. 2006a, b).

1.1.1 Automatic Speech Recognition

Speech recognition automatically extracts the string of words spoken from the speech signal. Within the last decade of research a significant evolution could be observed from systems supporting isolated word recognition for limited vocabulary to large vocabulary continuous speech recognition (Lefèvre et al. 2001).

The goal of speech recognition can be formulated as a probabilistic problem to find a sequence of words which is most probable given a sequence of acoustic observations (Rabiner 1989; Juang and Rabiner 1991). This problem can be subdivided in two subproblems: finding words the pronunciation of which matches the acoustic properties of the input signal best (acoustic model) and finding the appropriate order of words (language model).

Before classifying the speech signal is preprocessed and relevant features are extracted. The preprocessing includes quantization, preemphasis and windowing, signal enhancement and noise reduction like spectral subtraction (Linhard and Haulick 1999) as well as blind-source separation for use of ASR systems in noisy environments (Bourgeois 2005; Bourgeois et al. 2005). In the short time analysis mainly acoustic features (cepstrum, pitch, formants, energy, etc.) but also acoustic-phonetic features like manner or place of articulation and auditory features like Mel-frequency warping are considered. Temporal aspects are included by calculating delta and acceleration coefficients, i.e., the first and second derivative.

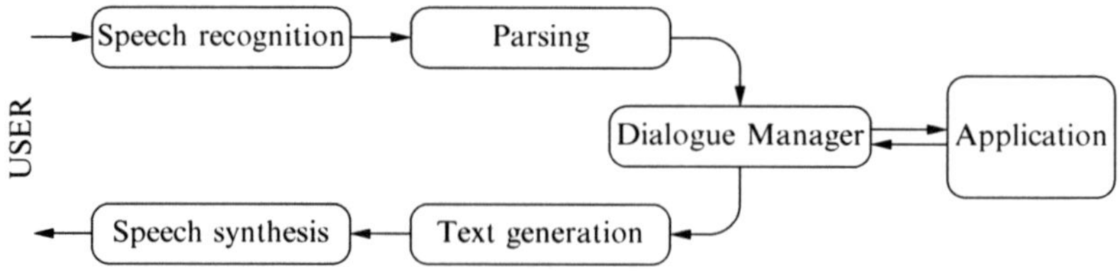

Fig. 1.1 Architecture of a typical SLDS

Each word in the recognizer vocabulary is mapped onto one or multiple sequences of phonemes (depending on how many pronunciations are associated with the word) using a lexicon or dictionary. In simple monophone recognizers, each single phoneme is modeled individually, however, current systems feature more sophisticated sub-word levels such as tied-state triphones (Woodland and Young 1993) which include aspects like co-articulation.

A language model represents the probability of a word sequence which typically includes the probability of a word w_i given the preceding words $w_{i-1}, \ldots, w_1$ (see Young 2001). The simplest approach for a language model is a finite-state network where the allowed word sequences are given explicitly. Other rule-based approaches include context-sensitive grammars in order to approximate natural language. As the complexity of a rule-based language model, and thus the effort to manually create such a model, increases significantly with the vocabulary size, statistical language models are used in large-vocabulary continuous speech recognition and dictation systems. Typically not all preceding words are included in the model but only n-grams, i.e., a word and $n - 1$ preceding words are considered. A unigram ($n = 1$) model only contains the frequency of occurrence of words without any dependencies on adjacent words. A very basic dependency is modeled by bigrams ($n = 2$) which take into account one preceding word (Lesher et al. 1999).

Speech recognition and HMMs constitute a key element in our work on speech-based emotion recognition. Thus, we discuss further details and properties of these models and methods in Chapter 4.

1.1.2 Natural Language Understanding

Having recognized a sequence of words from the speech signal the system now needs to extract the actual meaning from these words (see also Allen 1995). One approach to process a sentence is the syntactic analysis which assigns a certain structure to the sentence describing its different syntactic elements like phrases, verbs, nouns, prepositions, modifiers, etc. Based on this analysis the meanings of the particular elements are derived and merged to an overall meaning. This, however, requires the whole sentence to be analyzed and to be syntactically correct. Optimally, the parsing component should establish the semantic representation of an input word sequence facing various spoken natural language effects (Minker et al. 1999), i.e., the goal of the semantic analysis is to extract the sentence meaning rather than only to check whether the sentence is grammatically correct or incorrect.

The semantic representation is typically determined with the aid of rule-based grammars. This can be semantic grammars (Burton 1976) or case grammars (Fillmore 1968) which are especially suitable for spoken natural language input as they also allow the processing of ungrammatical sentences. The major disadvantage of grammars which are implemented as a set of rules is their lack of adaptability to different applications, domains and languages. Thus data-oriented parsing methods such as grammar inference and stochastic grammars (Jelinek et al.

1992, 1994) or connectionist models (Feldman and Ballard 1982) and hidden understanding models based on Hidden Markov Models (Miller et al. 1994; Levin and Pieraccini 1995) are also employed in spoken language dialogue systems as they are less constraining for modeling and parsing providing a higher coverage of meanings.

Apart from pure semantic meanings, also emotional cues can be extracted from the speech recognizer output with the aid of the semantic analysis. This approach is further specified in Section 4.7, where we use an affective grammar to spot emotional keywords in texts.

1.1.3 Dialogue Management

Being the central component of the dialogue system the dialogue manager handles the user's input in the form of labels provided by the semantic analysis. It interacts with the application and generates a response to the input. By generating suitable responses, the dialogue manager is also in charge of controlling the dialogue flow (Androutsopoulos and Aretoulaki 2003; Cohen et al. 2004; McTear 2004). To accomplish that and to shape the dialogue flow in an appropriate manner the dialogue manager can select between different types of dialogue initiative, employ certain confirmation strategies and access various knowledge sources.

For example, in a simplified travel information system where the user is able to book flights, the system needs to know the destination, the place of departure and the travel date. Such a dialogue, e.g., proceeds as follows:

1 System: *Good morning and welcome to Uta, your virtual travel agency. How may I help you?*
2 User: *I would like to book a flight from Boston to Los Angeles.*

After the user input *departure* = *"Boston", destination* = *"Los Angeles"* is passed from the parser unit to the dialogue manager, the dialogue manager checks if all required fields are filled. As this is not the case, it checks if the missing data is contained in the dialogue history. If, like in this case, the dialogue history is empty or does not provide all of the missing data, the dialogue manager determines the fields which are still missing and requests further details (i.e., the next field):

3 System: *When would you like to depart?*
4 User: *On Wednesday.*

Again the input *date* = *"Wednesday"* is received and converted to a more database-friendly format like *date* = *"2007-10-03"*. Now that all required data is available (departure and destination are obtained from the dialogue history), the dialogue manager is able to access the application and pass the result to the user:

5 System: *The following flights from Boston to Los Angeles are available on Wednesday, October 3rd: ...*

These aspects and further details on dialogue management, particularly with regard to integrating emotions in the dialogue flow, will be addressed in Section 1.3 as well as in Chapter 3.

1.1.4 Text Generation

As soon as the dialogue manager has received a response from the application or external source this information needs to be communicated to the user. Basically there exist two different types of responses: Either the user has not provided all parameters required for a database or an application access, so that a prompt asking for further details needs to be formulated, or the retrieved information, e.g., database records like `''Munich, Paris, 070226:0110''`, needs to be translated into a comprehensible natural language message like *"There is a flight from Munich to Paris on February 26, 2007 at 01:10 am"*.

Depending on the complexity of the application interface current SLDSs employ either a set of predefined (canned) sentences or text templates to generate natural language text. Being the simplest approach to implement, the predefined sentences can already be included in the database rendering the implementation highly inflexible. A higher degree of flexibility is provided by text templates which are used in VoiceXML (cf. Larson 2001) and various other dialogue description languages. Such a text template may be, e.g., *"There are* [number-of-flights] *flights from* [departure] *to* [destination]." where the elements [number-of-flights], [departure] and [destination] are determined by a database query or taken from the dialogue history.

Natural language generation can also be regarded as a planning process transforming communicative goals into comprehensible messages. Such a content planning approach is used in tutoring systems where the users are shown how to accomplish certain tasks with respect to different situations (cf. Wolz 1990). Here, text generation can be subdivided into three processes: Document planning (including content selection), microplanning and surface realization (Reiter and Dale 2000).

1.1.5 Text-to-Speech

Once a suitable response has been generated, it is translated into a speech signal. In analogy to the canned sentences approach for language generation, prerecorded (canned) speech samples may be used. Depending on the complexity of the application, these samples may either contain complete sentences or sentence fragments which are concatenated dependent on the generated text like in voice mailbox user interfaces. Whereas completely prerecorded texts feature a high degree of "naturalness", the concatenation of fragment samples may often lead to a patchy and uneven overall output.

Thus, in order to get a constant output for variable texts typically text-to-speech synthesis is employed (O'Malley 1990; Carlson and Granström 1977). Text-to-speech conversion has evolved to a research field of its own including aspects such as grapheme-to-phoneme conversion, prosodic modeling and speech synthesis. Roughly speaking the text is converted to a sequence of phonemes or diphones using a dictionary like in speech recognition and this sequence is then synthesized to a speech signal. In this process there are certainly more aspects that have to be considered. The analysis of the text does not only comprise the conversion to phonemes but also includes a proper segmentation and interpretation of the text as well as adjustments concerning coarticulation, stress and prosody (Sproat et al. 1992; Sagisaka 2001; Werner and Hoffmann 2007).

The output of the text analysis, typically a phonetic string enriched with prosodic marks denoting stress, pauses and pitch variations, is passed to the actual speech generation component. Based on the prosodic marks synthesizer parameters like pitch (fundamental frequency F_0), speech rate (duration of the speech segments), intensity, timbre, etc. are calculated. The pitch curve, i.e., the contour of the fundamental frequency, is computed with the aid of melody models (d'Alessandro and Mertens 1995) and the speech rate is derived from durational models. These parameters finally serve as the input of the signal generator. In current speech synthesis systems several types of speech generators are employed: Articulatory synthesizers use physical models with respect to the physiology of speech production and the vocal tract (Teixeira et al. 2002). Formant synthesizers apply an acoustic model including speech spectra, formants of the vocal tract and excitation by the glottal flow or noise (Gutiérrez-Arriola et al. 2001). Concatenative synthesizers avail themselves of speech databases containing coded speech segments, e.g., diphones (Isard and Miller 1986), which are concatenated and processed to obtain a smooth signal (O'Brien and Monaghan 2001).

1.2 Enhancing a Spoken Language Dialogue System

In order to provide more functionality and to increase the user-friendliness the architecture of a typical SLDS as shown in Fig. 1.1 can be arbitrarily extended.

Multimodal human–computer interfaces involve the combination of multiple modalities, like audio (speech, dial tone sounds, etc.), haptics (movement, touch, gestures, etc.) or video (mimics, eye/head/body movement, gestures), not only for input but also for output. These systems receive multiple input streams, one per modality, that need to be combined on the recognition or at the parsing level. By that, the dialogue manager is able to, e.g., provide the required information when the user points on a map and says *"show restaurants near this place"* (Johnston et al. 2002). The output of such a system may also be presented multimodally by combining speech with graphics and tables or with the aid of animated presentation agents

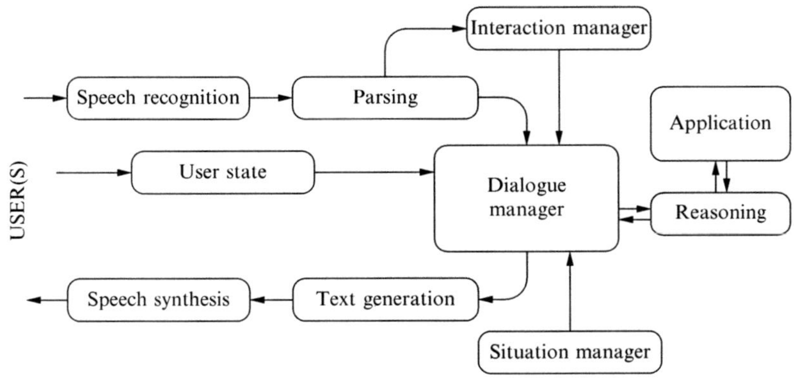

Fig. 1.2 Extended architecture of an SLDS

(André 2000). Integrating multiple input modalities does not only provide a more flexible interaction allowing the users to interact with their preferred modalities (cf. Oviatt 1997), it also helps to reduce recognition errors and misinterpretations (cf. Oviatt 2000). The implementation and evaluation of SmartKom, a multimodal dialogue system is extensively described in Wahlster (2006).

Apart from the involvement of additional modalities there exist numerous approaches to endow SLDSs with certain capabilities, which are provided by separate modules. An exemplary SLDS is shown in Fig. 1.2. Its additional components include a reasoning module which renders the system more efficient by reasoning over the user's input and providing problem solving assistance (Minker et al. 2006b). Furthermore it avails itself of a user-state and situation manager to adapt the dialogue flow to a dynamically changing user model and to the situation in which the dialogue is taking place. An interaction manager enables the system to passively follow a multiparty conversation (between multiple speakers) and to proactively contribute to the conversation when required. Similar to multimodal systems, where the output is communicated using multiple modalities, an extended SLDS may also adapt its output to the current user-state or dialogue situation. Like human dialogue partners who tend to speak louder, slower and more clearly when they have problems understanding each other, an SLDS may proceed accordingly in case of background noise or speech recognition problems. When considering the user's emotional state the system can also apply emotional speech to react accordingly, e.g., to appease an angry user more efficiently. A huge number of state-of-the-art text-to-speech synthesis systems are able to integrate prosodic features like pitch, emphasis, etc., into the output signal (cf., e.g., Nass and Lee 2001; Tesser et al. 2005).

In order to increase the user-friendliness and the efficiency of the dialogue system we focus on rendering dialogue management more adaptive. This is supported by the user-state manager and the situation manager (see also Chapter 3) which adapt the user and situation models on the basis of sensor data, e.g., from emotion recognition.

1.3 Challenges in Dialogue Management Development

Apart from being a data synchronization interface between a user and the application, the dialogue manager is also responsible for maintaining an appropriate dialogue flow so that the interaction ends successfully, i.e., as desired by the user. There exist a variety of parameters and mechanisms to design and also adapt dialogues (cf. McTear 2004), the most important of which are described below.

In contrast to human–human interaction where the initiative shifts perpetually between the dialogue partners, human–computer dialogues are typically maintaining one certain initiative strategy depending on the application area. In an information retrieval or database query application, the dialogue system normally asks a set of questions to obtain all parameters for the database query. Therefore, either system initiative or mixed initiative are applied. In system-initiated dialogue the system asks specific questions like *"Where do you want to travel to?"* restricting the user's input to a very small set of responses. This set may be even reduced by offering options like *"Where do you want to travel to? Paris, London or Frankfurt?"* to increase the speech recognition accuracy. Mixed-initiative dialogues do not only allow the user to ask questions or to request further information but also to provide more information than is asked by the system – here the reply to *"Where do you want to travel to?"* may be *"I want to travel to Munich by train on September 2nd"*. Typically, such a system welcomes the user with, e.g., *"How can I help you?"*, the user provides as much information as possible and the system takes control to ask for further required information. In user-initiative dialogues it is the user who controls the dialogue and who solely asks the questions which the system tries to answer appropriately. Such a natural language interface, however, needs to possess a very accurate large-vocabulary continuous speech recognizer and an adequate parser. The selection of a suitable dialogue strategy is one important issue in dialogue development (see, e.g., Levin et al., 2000a). It is typically defined beforehand and maintained during the whole dialogue. However, in order to fashion the dialogue more natural and user-friendly there also exist approaches to adapt the strategy during the dialogue (Litman and Pan 2002).

Another important issue in dialogue development is how the user's input is processed. In the case of graphical user interfaces where the user needs to select from a limited choice of buttons, list or menu items, the system may assume that the received input is in total accordance with what the user actually intends. As opposed to that, the dialogue manager of an SLDS needs to provide for the possibility that the user's utterance has not been recognized or interpreted correctly by the speech recognizer or the parser, respectively. Moreover, even if the utterance is recognized correctly, it cannot be guaranteed that the user's input can be handled appropriately when accessing the application. This may occur, e.g., due to under- or over-specification when a user just says *"Los Angeles."*, the system is unable to determine whether "Los Angeles" is the departure or the destination. Or when a user says *"I want to go to Los Angeles, no, to San Diego, er, San Francisco."*, the system needs to clarify where the user actually wants to travel to.

In order to cope with wrong input, misunderstandings, speech recognition errors or parsing errors, the system either needs to inform the user that it could not understand or needs to verify what it understood. For the cases of no input (silence) or wrong input (the user's input does not match the parser's grammar), dialogue development tools like VoiceXML (Larson 2002) or the CSLU Toolkit (Sutton et al. 1998) provide predefined prompts or subdialogues which may be adapted to the context of the actual system as shown by means of the following exemplary dialogue with our travel information retrieval system:

1 System:	*Good morning and welcome to Uta, your virtual travel agency. How can I help you?*
2 User:	*(silence)*
3 System:	*Sorry, I did not hear what you said. Please speak loud and clearly. How can I help you?*
4 User:	*Can I buy subway tickets here?*
5 System:	*Sorry, I did not understand what you said. This is an air travel information and booking system. Where do you want to travel to?*
6 User:	*Err, Times Square*
	...

Repeating turns of no-input or not-matching input are also considered in VoiceXML. Here the corresponding prompts and messages may be fashioned according to the idea of incremental or expanded prompts (Yankelovich 1996). E.g., a *"How can I help you?"* prompt is increasingly extended with further help such as *"This is an air travel information and booking system. You can book flights or search our database for flights."* after the first event and *"You can say, e.g., 'show me all flights to Miami'. Please say 'help' if you require further assistance."* after the second event. By that, the understanding problems can be resolved after a certain number of turns depending on the user's experience with spoken language dialogue systems.

Further and even more severe problems may arise from misunderstandings which can not be detected by the system. These may be caused by speech recognition errors or by ambiguities in the parser's grammar. In human–human conversations speakers typically expect their listening partners to indicate whether they understood what has been said. Instinctively being aware of that, listeners automatically conform with that expectation (cf. Brennan and Hulteen 1995) by nodding, shaking their heads or saying *"hm"*, *"OK"*, *"what?"*, *"pardon?"*, etc. In human–computer interaction neither user nor system can apply such an informal method of verification. Especially the user can not expect the system to nod or to mumble *"uh huh"* as the system itself does not actually "know" whether it has understood the user correctly. Instead it is the system's task to verify the recognized user input and to ask the user for confirmation. I.e., it needs to inform the user what it understood and give the user at least one chance to correct eventual errors.

This kind of verification may be accomplished applying an explicit or an implicit confirmation strategy. Using explicit confirmation prompts, the system requires the user to explicitly confirm the recognized input, like *"I understood you wanted to travel to Los Alamos. Is that correct?"*, *"Is it correct that you want to depart on Wednesday?"*, etc., and the user typically replies with *"yes"* or *"no"*. In order to still keep the number of turns at a reasonable level, so that the user does not get bored or annoyed, the confirmation prompts for two or more items may be combined, e.g., *"I understood you wanted to travel to Los Alamos on Wednesday. Is that correct?"*. Especially in cases where understanding errors may have severe consequences, like in monetary transactions or the purchase of expensive goods, the system typically outputs a message summarizing all items and asks for confirmation before it proceeds with the transaction:

1 System:	*Good morning and welcome to Uta, your virtual travel agency. How can I help you?*
2 User:	*I would like to book a flight from Boston to Los Angeles.*
3 System:	*When would you like to depart?*
4 User:	*On Wednesday.*
5 System:	*OK, I understood you want to travel from Boston to Los Angeles on Wednesday. Is that correct?*
6 User:	*Yes.*
7 System:	*The following flights from Boston to Los Angeles are available on Wednesday, October 3rd: ...*

This procedure is similar to the dialogue boxes in graphical user interfaces which ask the user for confirmation, e.g., before deleting data.

When applying an implicit confirmation strategy, the system integrates the confirmation into the question for the next item. Such an implicit confirmation may be integrated into the above dialogue as follows:

3 System:	*When would you like to depart from Boston to Los Angeles?*
4 User:	*On Wednesday.*

In both above examples the user was lucky that the system recognized all utterances correctly. In cases of recognition errors, it strongly depends on the dialogue manager's capabilities how quickly the misunderstanding is resolved. A very simple approach for explicit confirmations is the reset of all items which need confirmation:

5 System:	*OK, I understood you want to travel from Boston to Los Alamos on Wednesday. Is that correct?*
6 User:	*No.*
7 System:	*Where do you want to travel to?*
8 User:	*Los Angeles.*
9 System:	*Where do you want to depart from?*
	...

As it seems to be quite annoying for the user to go through all questions just because one item was misinterpreted, the system may apply a more flexible grammar:

5 System: *OK, I understood you want to travel from Houston to Los Alamos on Wednesday. Is that correct?*
6 User: *From Boston.*
7 System: *OK, I understood you want to travel from Boston to Los Alamos on Wednesday. Is that correct?*
8 User: *To Los Angeles.*
9 System: *OK, I understood you want to travel from Boston to Los Angeles on Wednesday. Is that correct?*
10 User: *Yes.*

However, the use of a more flexible grammar typically involves a lower speech recognition accuracy. Alternatively, the system may change to a subdialogue, first asking which item needs to be corrected, resetting this item and then repeating only the question for the respective item. If the system does not recognize the user's input after a certain number of turns, it should switch to a more robust input strategy, such as asking the user to spell certain words. In cases where errors occur in implicit confirmations, the system can be designed similarly:

3 System: *When would you like to depart from Houston to Los Alamos?*
4 User: *From Boston to Los Angeles.*
5 System: *When would you like to depart from Boston to Los Angeles?*
6 User: *On Wednesday.*

Alternatively, the system may then apply explicit confirmations or include subdialogues as described above.

For the inclusion of confirmations in a dialogue system, the developer needs to find a compromise between an increased robustness (requires more confirmations) and a decreased user annoyance level (requires fewer confirmations). E.g., an approach to dynamically include confirmations has been described in Litman and Pan (2002).

1.4 Issues in User Modeling

As described above, robustness is an important issue in dialogue design. Applying suitable confirmation strategies during the dialogue the dialogue success rate may be increased significantly. The success rate is an objective evaluation metric approximating the probability that the system accomplishes its task successfully according to the user's input. E.g., the virtual travel agent of the previous sections would be successful if it returned all flights from Boston to Los Angeles on Wednesday, whereas it would fail if it returned flights from Houston to Los Alamos. However, both failure of the dialogue or an extremely large number of dialogue turns, even if the system is successful then, typically lead to a lower user satisfaction or user acceptance. To avoid that and to keep the costs of a dialogue, i.e., the number of turns, at a reasonable level, user models may be integrated into dialogue management.

A very simple user-model, implemented as a database record, can contain the user's experience level. This strongly correlates with how often the user has interacted with the system before. E.g., for mobile phone users who are setting up or using their mailbox for the first time it is quite helpful when the mailbox system provides detailed descriptions for all available options:

1 System: *Welcome to your personal XYZ Wireless mailbox system. You don't have any new or saved messages. This is the main menu. You can access existing messages by pressing '1' or saying 'messages'. In the messages submenu you can listen to new or saved messages, delete them or archive these. If you want to configure and personalize your mailbox, press '2' or say 'setup'...*

However, more experienced users will rather prefer short prompts, especially when they have to pay for the time calling their mailbox as higher dialogue costs result in a higher phone bill then. For such a system our simple user model contains the number of times that the user called the system. Once this number is above a certain threshold, e.g., five, the system outputs a shorter message like:

1 System: *You have got four new messages. Press '1' or say 'messages' to listen to the messages. Press '9' or say 'help' if you need any further assistance.*

The duration of the call can then still be reduced if the system supports barge-in, i.e., if the user can press a button or say a command while the system is speaking. An interactive technical help system implementing such a user-model has been described in Peter and Rösner (1994).

Further approaches rendering user interfaces more effective comprise user models including personalization features and user preferences. This technique is common practice for graphical user interfaces like computer applications or websites (Rossi et al. 2001; López-Jaquero et al. 2005) which allow the user to change the interface's appearance (using skins or themes) or behavior (inclusion of macros or user functions) or to access information more easily. E.g., online shops ask their users to create accounts so that they don't need to re-enter their address and payment details every time they order something in the respective shops. Gradually extending their user models, many web applications are walking a tightrope – on the one hand, providing an efficient and powerful user interface by observing the users where they click, which products they are buying or gazing at, etc. to offer suitable products or to assist in information retrieval, on the other hand, however, surpassing the limits of the users' privacy.

SLDSs can avail themselves of similar user model concepts. To implement a very convenient user model based dialogue, the system needs to be capable of recognizing the user. The simplest approach for a telephone-based dialogue would involve assigning login number and personal identification number pairs to the users. When calling the system each user enters the respective numbers using the telephone keypad and the system loads the user's profile. Recognizing users without bothering them to identify themselves could be accomplished by evaluating the caller ID.

This, however, requires that each user can only call from one telephone and this telephone cannot be shared by multiple users. Alternatively the system may extract certain features from the speech signal and apply algorithms to recognize speakers while they are talking to the system. An elaborate overview on speaker recognition involving speaker identification and speaker verification is given in Campbell (1997). Apart from just being used to select the appropriate user model, speaker recognition plays an important role in access control for computer systems. The human voice can not be "forgotten" as opposed to passwords, or lost, like keycards. And state-of-the-art speaker recognition systems are even able to cope with different types of noise and variations in the user's voice. The identification and verification process is comparable to ASR – the extracted features are matched with either template models (e.g., dynamic time warping, cf. Sakoe and Chiba 1978) or stochastic models (e.g. Hidden Markov Models, cf. Rabiner 1989) and a classification is performed.

Having successfully identified the user, the system retrieves the data from the user model and profile and can use that data to optimize the dialogue accordingly. This may be a personal greeting like *"Hello John, ... "* or if a profile shows that the user has booked flights from Boston to Los Angeles almost every week, a dialogue could be as follows:

1 System: *Good morning, John. Would you like to travel to Los Angeles again next week?*
2 User: *Yes.*
3 System: *When would you like to depart?*
...

Apart from objective measures like dialogue success rate, speech recognition accuracy, parsing concept accuracy, duration, etc., the user's subjective impression is an important criterion concerning the acceptance of a dialogue system. This involves short but robust dialogues, an increased user-friendliness and the acceptance of a vast range of input possibilities, i.e., the user's request for a flight to Boston may vary from *"Boston"* or *"to Boston"* to *"I was just wondering whether there are any flights to Boston"* and users tend to become rather disappointed when the system rejects some of their input utterances. Experiments conducted by Weizenbaum (1966) with the ELIZA system or by Reeves and Nass (1996) have shown that users typically do not suppress their feelings towards machines or computers. For humans it is not difficult to observe a dialogue partner's emotional state and to react accordingly, e.g., to appease the dialogue partner in case of anger or to ignore dialogue-irrelevant emotions. However, integrating such a capability into (spoken) human–computer interfaces requires (a) an accurate emotion detection which contributes to the user model as well as (b) a flexible dialogue manager that is able to react on such a dynamically changing user model. The underlying model for dialogue management does not only determine how the system shall react to which emotional state but also includes a pre-selection of "relevant" emotions, i.e., those emotions which actually influence the dialogue flow.

1.5 Evaluation of Dialogue Systems

The development and implementation of a (spoken) human–computer interface can be briefly outlined as an iterative process starting with an idea and approximate specifications what the system shall be capable of. Based on these requirements, concepts and models, e.g., for the course of the dialogue, are developed and approaches to the solution of certain problems are identified, e.g., how to recognize emotions or how to extract semantic meanings from the user input. In the next phase, these methods and approaches are implemented (or purchased from external suppliers) and then integrated into an end-to-end system. However, before the newly developed system is ready to be put into operation, it needs to be evaluated whether it complies with certain criteria and whether potential users are prepared to accept and use such a system. A simplified illustration of this process is shown in Fig. 1.3.

In this illustration, the process contains the development phases as described before plus an evaluation phase in which weaknesses of the system or its components shall be identified. Accordingly, the feedback provided by the evaluation can be taken into account differently in the previous development phases. This is emphasized by the grey arrows – e.g., if the test users involved in the evaluation feel misunderstood by the system, it is very likely that the speech recognizer and/or the linguistic analysis are not working properly (which in turn can be verified in the evaluation protocols). If, however, the test users state that they don't feel comfortable using the system, there could be an integration problem, the concepts might be unsuitable or, in the worst case, the whole idea might be nonsense.

When it comes to measuring the performance of technical systems, the evaluation criteria are typically easy to identify. E.g., in information transmission systems like internet connections, TV or mobile phones, it is desirable to achieve an error-free transmission at maximum speed and at minimum costs (signal power, bandwidth). These criteria can be measured in existing systems or can be simulated for future systems providing a common basis for comparing different transmission schemes.

For human–computer interfaces like SLDSs, there also exist a large variety of objective evaluation measures which, e.g., allow to compare different speech

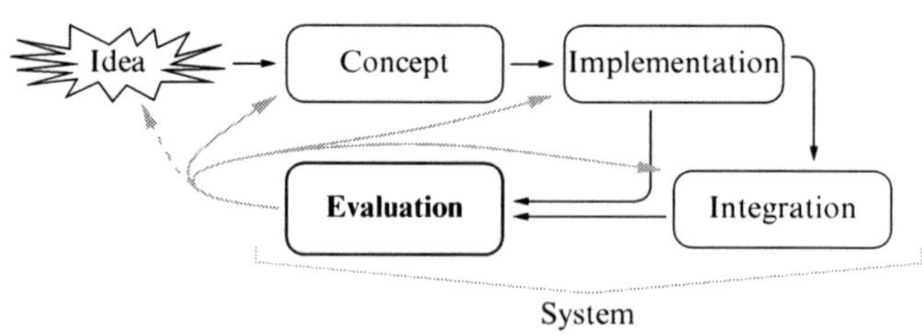

Fig. 1.3 Simplified illustration of the iterative development process of a (spoken) human–computer interface. In the flowchart, black arrows represent the transitions between the development phases and grey arrows represent the feedback from the evaluation triggering a new iteration in the respective phase(s)

recognizers or linguistic analyses. However, even if these measures attest that the system performs "well", the end user may still feel and decide that the system is unusable, unappealing or behaves unacceptably based on this user's *subjective* impression. A list of typical objective measures for end-to-end SLDSs has been compiled by Walker et al. (1998) Among the objective measures are

- The proportion of correct answers compared to reference answers (what did the system actually reply when it was expected to provide certain information or to ask for a certain field?)
- Task completion or transaction success rate (could the system comply with the user's goals?)
- Number of turns, dialogue time or task completion time (how effective is the system in achieving the user's goals?)
- User response time and system response time
- Percentage of diagnostic error messages (how often did the system have to prompt for confirmation, e.g., when the speech recognizer performance was bad)
- Proportion of utterances containing more than one word and the length of these utterances (this indicates how well the system can handle natural input and how much flexibility the users expect from the system)

Concerning the subjective measures, there is no consensus about their application in computer-based dialogue systems. E.g., Grice (1975) proposes four conversational maxims representing guidelines how to communicate successfully. These comprise Quality (Truth: Do not say what you believe to be false or that for which you lack evidence), Quantity (Information: make your contribution as informative as required but not more informative than required), Manner (Clarity: avoid obscurity of expression and ambiguity, be brief and orderly) and Relation (Relevance). Regarding human–computer interaction, the adherence to these maxims can be considered as a measure for cooperativity (Walker et al. 1998) whereas Frederking (1996) claims that these maxims are too vague to be implemented in computational natural language systems.

Being based on human judgments according to qualitative criteria, subjective measures face the problem that they might not be reliable across judges, e.g., in the worst case a user might say that he is very satisfied with the system and on the next day, the same user might reassess the same system as unsatisfactory. Thus, calculating a ratio between different subjective categories, the measures can be made quantitative to a certain degree. Being widely used among the spoken dialogue community, further subjective measures have been collected by Walker et al. (1998):

- Percentage of implicit and explicit recovery utterances where the system tries to recover from errors of partial speech recognition or linguistic analysis
- Proportion of contextually appropriate system utterances
- Proportion of correct and/or partially correct system answers
- Ratio between appropriate and inappropriate system directive and diagnostic utterances
- The whole concept of user satisfaction where a broad spectrum of potential users are asked to assess the system's usability, typically on the basis of questionnaires

These measures are also included in the ITU-T P.supp24 recommendation providing definitions for the above and further parameters which describe the interaction with SLDSs (ITU-T P.supp24 2005).

Individually, each of the objective and subjective measures can be used to compare the behavior or certain features of different dialogue systems within particular limitations. E.g., considering the dialogue time or the number of utterances is only reasonable when comparing systems offering the same features and/or operating in the same domain – contrarily, one may not conclude that a simple train timetable information system requiring approximately three user turns is more effective than a more sophisticated air travel information and reservation system requiring more turns. Also, the use of reference answers is strictly limited to comparing systems applying the same dialogue strategy as for different strategies, the number of "correct" answers is virtually unlimited. Furthermore, on the one hand, there exist correlations between the measures which are difficult to follow or understand, on the other hand, it is not possible to combine different measures or trade off, e.g., whether it is better to feature a large number of dialogue turns but a short system response time or to feature a smaller number of dialogue turns but a longer system response time (see also Walker et al. 1998).

There exist approaches to "replace" the human test persons by computer-based assessment tools in order to reduce the time and effort involved in the user studies. Ito et al. (2006) propose a VoiceXML-based user simulator which is able to evaluate some of the dialogue assessment criteria automatically. Here, it is presumed that the simulator fully knows the structure of the, also VoiceXML-based, dialogue system under test including form layout and task grammar. Based on this knowledge, the system is bombarded with possible user utterances and in combination with the respective system reactions, measures like task success, number of turns, etc., are determined.

1.6 Summary of Contributions

Due to their significant impact on the natural interaction between human dialogue partners, emotions have attracted great interest within the research on adaptive human–computer interaction. This is substantiated by the large number of groups conducting research in this field and reporting progress in the field of emotion recognition from various modalities (speech, gestures, biosignals, etc.) as well as for the integration of emotions in different aspects of human–computer interaction.

Despite the fact that a good emotion recognition performance is achievable with all the other modalities as well, our decision to consider purely speech-based emotion recognition is basically due to the fact that no extra equipment such as cameras or sensors is required and that the user is not burdened with applying this extra equipment. Instead, the speech signal is captured anyway, either by a microphone like in an information kiosk or via telephone in a call center application. Thus, limiting our considerations on spoken dialogue, our work described in this book subdivides into two major aspects: the efficient recognition of emotions from speech signals with low complexity, on the one hand, and the integration of the recognized emotional cues into adaptive dialogue management, on the other hand. Here, particular attention is paid to our following contributions:

- Implementation of a *plain emotion recognizer* using HMMs to classify different emotional states from prosodic and acoustic features extracted from the speech signal. To increase its robustness we use separate acoustic models for female and male speakers. With a performance ranging around the average of existing recognizers, our approach features a considerably lower complexity with respect to state-of-the-art approaches dealing with emotion recognition.
- Development of a *speech–emotion recognizer* combining speech and emotion recognition into one process by classifying "emophonemes" (phonemes with attached emotional states). The feasibility of this new approach is substantiated in our experiments.
- Optimization of our speech–emotion recognizer by extending it to a *two-step recognizer*. In this approach, the speech–emotion recognizer takes advantage of the utterances' textual content provided by an extra speech recognizer. Considering speech and emotion accuracies individually, our system outperforms most plain emotion recognizers while achieving a reasonable speech recognition performance.
- Adaptation of the ROVER (recognizer output voting error reduction, Fiscus 1997) algorithm which combines the output of *multiple* speech *recognizers* to achieve a better word accuracy. Despite the relatively low performance reported for speech recognition, we modify the algorithm such that it processes the output of *multiple speech–emotion recognizers* by what particularly the emotion recognition performance increases.
- Development of a *semi-stochastic dialogue model* enabling a flexible adaptation of the dialogue flow to the user's emotional state without complex rule sets. In the model, the dialogue designer predefines dialogue states (these may include emotional states and/or other dialogue-influencing parameters) the transitions between which are determined by probabilities derived from training data. Due to its small number of internal parameters, the model is less complex than comparable approaches while featuring a higher consistency by introducing tri-turn transitions (from the penultimate state to the previous to the current state) in addition to the commonly used bi-turn transition (from the previous state to the current state).
- Implementation of an *adaptive dialogue manager* which integrates the flexibility of our dialogue model into the convenient programming interface of the VoiceXML framework.

With respect to these contributions, the remainder of this book is structured in the following manner: Constituting the focal point of our considerations, emotions are extensively discussed in Chapter 2. This includes different approaches to define emotions, a variety of emotion theories, the annotation of emotions and an overview on emotional speech databases. These aspects actually influence whether an automatic emotion recognizer performs satisfyingly or not. Especially the quality of the speech samples and the annotation of these have an important impact on the quality of the underlying recognizer models. In Chapter 3, we describe dialogue strategy concepts and how emotions shall influence spoken dialogue. Beginning with existing work on adaptive dialogue management, we advance from rule based approaches to semi-stochastic dialogue and emotion models. We combine these models into an emotional dialogue model and we explicate how such a model can be extended to include further dialogue parameters.

Chapter 4 centers on the automatic recognition of emotions from speech signals. Here, we discuss technical details including signal processing and classification. Having outlined existing work in this field, we derive a plain emotion recognizer from a simple speech recognizer and we describe our speech–emotion recognizer. Furthermore we outline our approach to emotion recognition with the aid of linguistic analysis. Implementation aspects such as improvements to our (speech–)emotion recognizers as well as the realization of our dialogue manager are discussed in Chapter 5, where we also adapt the ROVER idea of combining the results of multiple recognizers to achieve a higher recognition performance. We apply this idea to both emotion and speech–emotion recognition. An evaluation of our proposed concepts and components is given in Chapter 6.

Chapter 2
Human Emotions

Nowadays, computers are more than ever regarded as partners. Users tend to apply social norms to their computer, i.e., typically, they become enraged if the computer makes a mistake or they are delighted if the computer compliments them on a successful work (Reeves and Nass 1996). Moreover, such a relationship is consolidated when users are able to personalize the interface, e.g., by applying themes to their desktops, and, thus, feel more comfortable interacting with the system. In SLDSs, user profiles are employed to provide such a personalized environment relieving the users of repeatedly supplying the same personal data. Complementarily, this feeling or notion of "relationship" between computer and users is intensified when the computer is able to respond to the users situation and/or (emotional) states (Schröder and Cowie 2006; Peter and Beale 2008). In this chapter, we give an overview on the definition of emotions and describe different aspects of emotion theories before discussing the annotation of emotions and emotional speech corpora.

2.1 Definition of Emotion

In order to design adaptive SLDSs which are able to communicate with the user in a more natural way and which are more responsive to the user compared to regular systems, the consideration of social and especially emotional aspects is essential. But what is actually an emotion?

From the etymological point of view, the word "emotion" is a composite built up from the two Latin words *"ex"* in the meaning of "out" or "outward" and *"motio"* in the meaning of "movement" or "action" referring to the spontaneity of emotions. In this section, we discuss further approaches and theories to define and categorize emotions with respect to their use in human–human or human–computer interaction.

Early approaches to the definition of the word "emotion" are described in the 19th century by the Darwinian Perspective and the Jamesian Perspective (Darwin 1872; James 1884). Both views describe emotions as "more or less automatic responses to events in an organism's environment that helped it to survive" (Cornelius 2000). Whereas Darwin concentrates on emotional *expression*, James tries to explain the nature of emotional *experience* by means of the perception of bodily changes.

J. Pittermann et al., *Handling Emotions in Human-Computer Dialogues*,
DOI 10.1007/978-90-481-3129-7_2,

A further approach to explain emotions is the cognitive perspective stating that all emotions imply the appraisal, i.e., the positive or negative judgment of events in the environment (Arnold 1960). The Social Constructivist Perspective defines emotions as products of culture or "social constructions" (Averill 1980), whereas the understanding and the knowledge of social rules play a decisive role.

According to Kleinginna and Kleinginna (1981), emotion "is a complex set of interactions among subjective and objective factors, mediated by neural/hormonal systems, which may

- Give rise to affective experiences such as feelings and arousal, pleasure or displeasure.
- Generate cognitive processes such as emotionally relevant perceptual effects, appraisals, labeling processes.
- Activate widespread physiological adjustments to the arousing conditions.
- Lead to a behavior that is often, but not always, expressive, goal directed, and adaptive".

This complexity concerning emotions is still challenging for computers. E.g., how should a computer identify whether an utterance like *"yes, of course"* is only ironic actually meaning "no" or a happy affirmation, which for human hearers is easy to hear and interpret? Human dialogue partners benefit from the advantage that they are able to combine visual and aural perceptions and they can rely on existing knowledge and experience to determine their partner's emotional state. In order to endow dialogue systems with this kind of social intelligence it is necessary to classify, analyze and recognize emotions.

Due to the subjective factors underlying the term emotion, there is no consistent categorization of emotions forming a common basis for emotional research. Thus, various approaches are made to differentiate emotions and to distinguish emotions from other affective states. E.g., in Scherer (2000) the following affective states are classified:

- Emotions (e.g., angry, sad, joyful, fearful, ashamed, proud, elated, desperate)
- Moods (e.g., cheerful, gloomy, irritable, listless, depressed, buoyant)
- Interpersonal stances (e.g., distant, cold, warm, supportive, contemptuous)
- Preferences/Attitudes (e.g., liking, loving, hating, valuing, desiring)
- Affect dispositions (e.g., nervous, anxious, reckless, morose, hostile)

These states differ from each other in the following design features:

- Intensity
- Duration
- Synchronization
- Event focus
- Appraisal elicitation
- Rapidity of change
- Behavior impact

In contrast to the other affective states, emotions are very intense and shortly lasting. Moreover, different reaction tendencies like physiological responses, motor expression and action tendencies are activated simultaneously. Emotions are highly focused on prior events, elicit a high degree of appraisal, they may rapidly change and have a high impact on the choices of different behaviors (Scherer and Bänziger 2004). Similarly, Cowie (2000) presents the categories "emotions proper/full-blown emotions" and "emotion-related states". Here, however, these emotion-related states, i.e., "arousal" and "attitude", cannot be strictly separated from emotions, as these terms are overlapping. Moreover, in the case of "arousal" it is even questionable whether there is any distinction possible at all.

In order to differentiate among emotional states, a taxonomy of emotional states (see Fig. 2.1) is described in Gmytrasiewicz and Lisetti (2000). As the purpose of the taxonomy is the use in a multi-agent system, emotional states that are not directly measurable are not included in this taxonomy. Furthermore there is uncertainty about which attributes and values have to be used to distinguish between some emotional states.

The aspects described in this taxonomy are, for the most part, also applicable to our approaches to recognizing emotions and to integrating the recognized emotional cues into adaptive SLDSs. On the one hand, for the automatic recognition of emotions from speech signals, we require a set of clearly identifiable emotional states to be classified. This includes differences in the aural perception (how do,

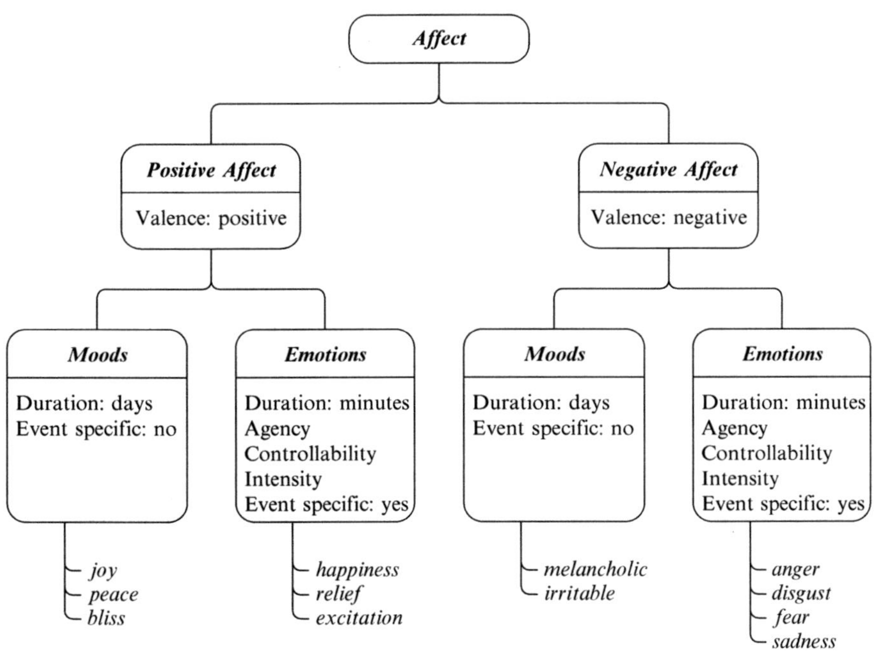

Fig. 2.1 Taxonomy of emotions (Gmytrasiewicz and Lisetti 2000)

e.g., the utterances of a bored person differ from those of a blissful person?) which shall preferably be describable by aural features (e.g., if the speaker is in xy state, the sound intensity increases). On the other hand, these (recognizable) emotional states shall also be meaningful for the adaptation of the dialogue. I.e., we need to cover a wide range of emotional states, but we can actually only include those which require certain actions by the dialogue system (e.g., there exist appeasement strategies to respond to an angry dialogue partner, whereas a blissful dialogue partner typically does not need any special attention). Thus, in our considerations we follow the positive-negative valence distinction described in Gmytrasiewicz and Lisetti's taxonomy but we also include different levels of arousal as represented by the activation–evaluation space or valence-arousal space below.

2.2 Theories of Emotion and Categorization

Several psychological theories try to explain how emotions evolve. According to the James–Lange theory illustrated in Fig. 2.2 actions precede emotions or feelings (James and Lange equate emotions with feelings, see James 1884; Lange 1885; Benyon et al. 2005). The brain interprets a particular situation that has occurred and a corresponding physiological response, i.e., heart rate elevation, is caused by a reflex. Then, as soon as the brain cognitively processes this physiological response, the person becomes aware of the emotion. An example is given by James, one of the theory's two founders: "We are afraid, because we run away from a bear instead of running away from a bear because of being afraid". As for James and Lange, an emotion is the consequence of peripheral physiological changes, this theory is also called the "peripheral" theory.

In contrast, the Cannon–Bard theory (see Fig. 2.3) suggests that after perceiving an emotion-arousing stimulus the action follows from cognitive appraisal (Cannon 1927; Bard 1934; Benyon et al. 2005), i.e., the brain's thalamus simultaneously

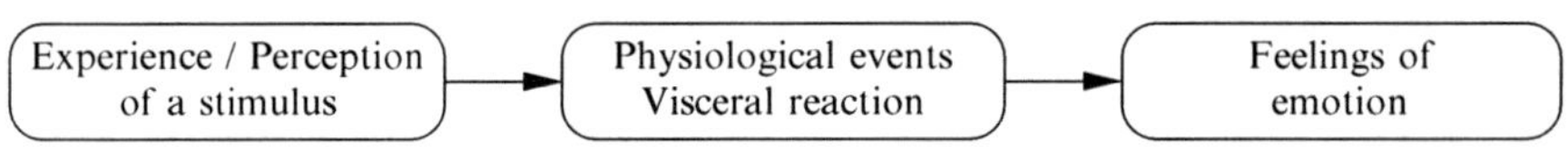

Fig. 2.2 James–Lange emotion theory

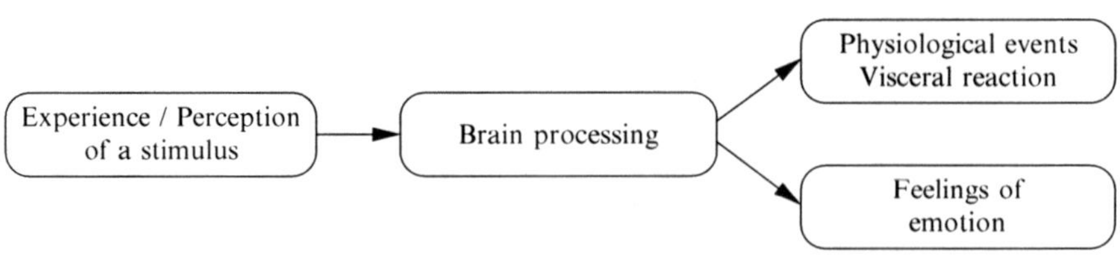

Fig. 2.3 Cannon–Bard emotion theory

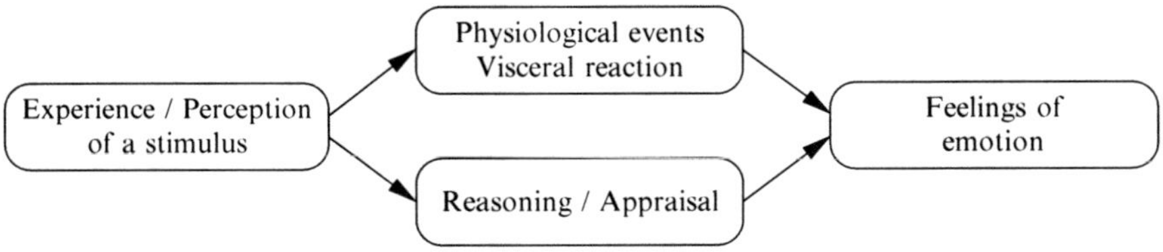

Fig. 2.4 Schachter–Singer emotion theory

sends signals to the autonomic nervous system (ANS), which then regulates bodily functions like heart rate, and to the cerebral cortex, which interprets the situation cognitively.

Similar to the James–Lange theory is the Schachter–Singer theory (see Fig. 2.4), also called "Two Factor" theory (Schachter and Singer 1962; Benyon et al. 2005). The experience of emotions is a consequence of the cognitive labeling of physiological responses to emotion-arousing stimuli. In addition, information is gathered from the situation in order to use it to modify the label of the physiological sensation.

In further cognitive labeling theories it is generally believed that cognitive evaluation is included in the experience of emotion, however, different opinions exist on whether any evaluation precedes the affective reaction like in Lazarus (1982) or whether emotional responses precede any cognitive processing (Zajonc 1984).

In order to categorize and specify emotions themselves, Wundt (1924) proposes three dimensions of emotions taking into consideration all differences among emotional states: pleasure/lust-displeasure/non-lust, strain/tension-relaxation and excitement-calmless. Another widespread and popular representation is the activation–evaluation space integrating possible emotional states into only two dimensions, namely activation and valence. Activation, also called arousal, corresponds to the degree of the emotional intensity, i.e., how intense the emotion is brought out, valence, also called pleasure, refers to the emotional value, i.e., if the emotion is negative like anger or positive like happy. Table 2.1 shows the integration of several exemplary emotional words into the activation–evaluation space according to a study by Whissel (1989) (see also Cowie et al. 2001).

Looking at the values for activation in Table 2.1, a low level of activation is represented by a low number and a high number corresponds to high activation, e.g., terrified (at 6) is a highly active emotional word and disinterested (at 2.1) is an emotional word with low activation. Similar to activation is the classification of the values for evaluation, i.e., guilty (at 1.1) and unfriendly (at 1.6) represented with low numbers of evaluation are negative emotional words and positive emotions like joyful (at 6.1) and delighted (at 6.4) are high-numbered.

Another approach to categorize emotions is the "emotion wheel" by Plutchik (1980b) (see also Cowie et al. 2001): Emotional words are arranged on a circle and their characteristics are reflected by angular measures. The basis for the circle is built by two axes ranging from acceptance with the angular measure of 0° to disgust at 180° and from apathetic at 90° to curious at 270°. On this basis,

Table 2.1 Emotion words (Whissel 1989)

Emotion	Activ.	eval.	Emotion	Activ.	eval.	Emotion	Activ.	eval.
Adventurous	4.2	5.9	Affectionate	4.7	5.4	Afraid	4.9	3.4
Aggressive	5.9	2.9	Agreeable	4.3	5.2	Amazed	5.9	5.5
Ambivalent	3.2	4.2	Amused	4.9	5	Angry	4.2	2.7
Annoyed	4.4	2.5	Antagonistic	5.3	2.5	Anticipatory	3.9	4.7
Anxious	6	2.3	Apathetic	3	4.3	Ashamed	3.2	2.3
Astonished	5.9	4.7	Attentive	5.3	4.3	Bashful	2	2.7
Bewildered	3.1	2.3	Bitter	6.6	4	Boastful	3.7	3
Bored	2.7	3.2	Calm	2.5	5.5	Cautious	3.3	4.9
Cheerful	5.2	5	Confused	4.8	3	Contemptuous	3.8	2.4
Content	4.8	5.5	Contrary	2.9	3.7	Cooperative	3.1	5.1
Critical	4.9	2.8	Curious	5.2	4.2	Daring	5.3	4.4
Defiant	4.4	2.8	Delighted	4.2	6.4	Demanding	5.3	4
Depressed	4.2	3.1	Despairing	4.1	2	Disagreeable	5	3.7
Disappointed	5.2	2.4	Discouraged	4.2	2.9	Disgusted	5	3.2
Disinterested	2.1	2.4	Dissatisfied	4.6	2.7	Distrustful	3.8	2.8
Eager	5	5.1	Ecstatic	5.2	5.5	Embarrassed	4.4	3.1
Empty	3.1	3.8	Enthusiastic	5.1	4.8	Envious	5.3	2
Furious	5.6	3.7	Gleeful	5.3	4.8	Gloomy	2.4	3.2
Greedy	4.9	3.4	Grouchy	4.4	2.9	Guilty	4	1.1
Happy	5.3	5.3	Helpless	3.5	2.8	Hopeful	4.7	5.2
Hopeless	4	3.1	Hostile	4	1.7	Impatient	3.4	3.2
Impulsive	3.1	4.8	Indecisive	3.4	2.7	Intolerant	3.1	2.7
Irritated	5.5	3.3	Jealous	6.1	3.4	Joyful	5.4	6.1
Loathful	3.5	2.9	Lonely	3.9	3.3	Meek	3	4.3
Nervous	5.9	3.1	Obedient	3.1	4.7	Obliging	2.7	3
Outraged	4.3	3.2	Panicky	5.4	3.6	Patient	3.3	3.8
Pensive	3.2	5	Pleased	5.3	5.1	Possessive	4.7	2.8
Proud	4.7	5.3	Puzzled	2.6	3.8	Quarrelsome	4.6	2.6
Rebellious	5.2	4	Rejected	5	2.9	Remorseful	3.1	2.2
Resentful	5.1	3	Sad	3.8	2.4	Sarcastic	4.8	2.7
Satisfied	4.1	4.9	Scornful	5.4	4.9	Self-controlled	4.4	5.5
Serene	4.3	4.4	Sociable	4.8	5.3	Sorrowful	4.5	3.1
Stubborn	4.9	3.1	Submissive	3.4	3.1	Surprised	6.5	5.2
Suspicious	4.4	3	Sympathetic	3.6	3.2	Terrified	6.3	3.4
Trusting	3.4	5.2	Unaffectionate	3.6	2.1	Unfriendly	4.3	1.6
Wondering	3.3	5.2	Worried	3.9	2.9			

each emotional word is assigned a certain angular measure. The simplified circle (see Fig. 2.5) only includes the so-called primary or basic emotions.

For the implementation of an emotion recognition system, it is often easier and more convenient to only recognize a limited number of emotions, i.e., a set of primary or basic emotions. Several approaches have been made to define and to determine primary or basic emotions (see Table 2.2). Descartes proposes the idea to distinguish primary and secondary emotions (Anscombe and Geach 1970;

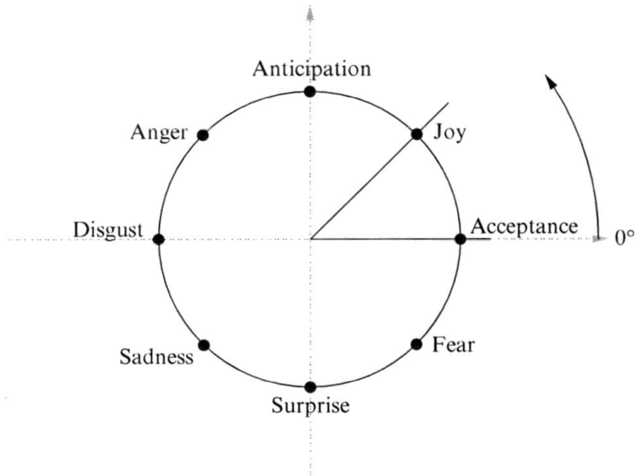

Fig. 2.5 Emotion wheel (Plutchik 1980b; Cowie et al. 2001)

Cowie 2000). First of all, a list of primary emotions has to be set up in order to then analyze how these emotions are expressed in speech. It is assumed that secondary emotions will emanate from this study afterwards. On the one hand, secondary emotions can be derived by mixing the primary ones like colors, which is called the palette theory of emotion. On the other hand, the terms "primary" and "secondary" convey the meaning that there exist a few elementary, primitive and pure emotions as opposed to the secondary emotions. This, however, also implicates that primary emotions are more important than secondary ones. Therefore, in the study by Cowie (2000), the term "second order emotions" is used instead of "secondary emotions" in order to underline that second order emotions are more complex, but not less important.

In the study by Cornelius (2000), primary, basic or fundamental emotions in general "represent survival-related patterns of responses to events in the world that have been selected for over the course of our evolutional history" and all other emotions in some way are derived from them. He proposes the "Big Six" as fundamental, primary or basic emotions, namely happiness, sadness, fear, disgust, anger and surprise, whereas Plutchik (1994) differentiates eight primitive emotions, i.e., fear, anger, joy, sadness, acceptance, disgust, anticipation and surprise. In Nisimura et al. (2006) even 16 basic emotions (including the neutral state) are determined taking into account the basic emotions given in Schlosberg (1954); Russell (1980); and Ekman (1992) and collected in Table 2.3.

Generally, the four primary emotions anger, fear, joy/happiness and sadness mostly appear in literature when characterizing emotional behavior (Devillers et al. 2002). These emotions correspond to relevant problems in life, i.e., anger may be considered as a reaction to competition, fear to danger, happiness to cooperation and sadness to loss (Power and Dalgleish 1997). In Cowie et al. (2001), emotions emerging in almost every list of basic/primary emotions furthermore are called "archetypal" emotions. These are happiness, sadness, fear, anger, surprise and dis-

Table 2.2 A Selection of Lists of "Basic" Emotions (Ortony and Turner 1990)

Reference	Fundamental emotion	Basis for inclusion
Arnold (1960)	Anger, aversion, courage, dejection, desire, despair, fear, hate, hope, love, sadness	Relation to action tendencies
Cornelius (2000)	Happiness, sadness, fear, disgust, anger, surprise	Relation to survival, Responses to events in the world
Descartes (1649)	Admiration, love, hatred, desire, joy, sadness	Derivation of all other emotions from primary emotions
Ekman et al. (1982)	Anger, disgust, fear, joy, sadness, surprise	Universal facial expressions
Frijda (1986)	Desire, happiness, interest, surprise, wonder, sorrow	Forms of action readiness
Gray (1982)	Rage and terror, anxiety, joy	Hardwired
Izard (1971)	Anger, contempt, disgust, distress, fear, guilt, interest, joy, shame, surprise	Hardwired
James (1884)	Fear, grief, love, rage	Bodily involvement
McDougall (1926)	Anger, disgust, elation, fear, subjection, tender-emotion, wonder	Relation to instincts
Mowrer (1960)	Pain, pleasure	Unlearned emotional states
Oatley and Johnson-Laird (1987)	Anger, disgust, anxiety, happiness, sadness	Do not require propositional content
Ortony and Turner (1990)	No "Basic" Emotions	Existence of only basic elements building different emotions
Panksepp (1982)	Expectancy, fear, rage, panic	Hardwired
Plutchik (1980a)	Acceptance, anger, anticipation, disgust, joy, fear, sadness, surprise	Relation to adaptive biological processes
Tomkins (1984)	Anger, interest, contempt, disgust, distress, fear, joy, shame, surprise	Density of neural firing
Watson (1930)	Fear, love, rage	Hardwired
Weiner and Graham (1984)	Happiness, sadness	Attribution independent

Table 2.3 The 16 basic emotions in Nisimura et al. (2006)

Anger	Contempt	Contentment	Depression
Excitement	Fear	Joy	Mirth
Neutral	Pleasure	Pressure	Sadness
Surprise	Tension	Tiredness	Displeasure

gust. Contrariwise, Ortony and Turner (1990) conclude that there do not exist basic emotions at all. Instead, emotions are composed of basic elements, i.e., components of cognitions, feeling states, etc. In this aspect, emotions behave like languages consisting of basic constituents, e.g., phonological properties building up languages but not being languages themselves.

Similarly, Schröder (2004) and Burkhardt et al. (2006b) outline the problem of using so-called "basic" or "fundamental" emotions as these pure emotions rarely appear in natural data. Therefore they propose to encode "emotion-related states" in phrases with strong emotional semantic content, e.g., *"Congratulations, you've just won the lottery!"* as a correspondence to a joyful emotional state. In their experiments, participants had to listen to these sentences and to determine if a sentence was uttered in an appropriate way according to the semantic content or not.

Another possibility to categorize emotions is the subsegmentation of certain emotions as done with anger in Pereira (2000), i.e., anger is segmented in "hot anger" and "cold anger". Moreover, each emotion is gradated on the basis of a six-point Likert scale, e.g., 1 for very unhappy, 2 for unhappy, 3 for slightly unhappy, 4 for slightly happy, 5 for happy and 6 for very happy. The idea of a further categorization of emotions is also picked up in Okada et al. (1999), where 123 subcategories have been defined for Plutchik's eight primitive emotions. E.g., gladness is subdivided into 17 subcategories, i.e., physiological pleasure and psychological pleasure, whereas psychological pleasure in turn consists of 16 subcategories.

A rather simple approach is grouping emotions only with respect to evaluation as follows (Fujisawa and Cook 2004):

- Positive affect (joy, satisfaction, pleasantness)
- Negative affect (sadness, anger, unpleasantness)
- Ambivalence (uncertainty, tension, anxiety)

On the one hand, in this manner, emotions can easily be categorized; on the other hand, emotions like sadness and anger are grouped into the same class even if they are highly different. Similarly, Miwa et al. (2000) defines six emotions assigning them to four major emotion groups:

- Joy (happiness)
- Sadness (disgust, sadness)
- Anger (anger, fear)
- Neutrality (neutrality)

According to Kim et al. (2005), even an infinite number of emotions may be categorized on the basis of these groups, however, this model also faces the problem of regarding highly different emotions as the same group.

In Clavel et al. (2004), the activation–evaluation space is extended by a third dimension named "reactivity" in order to distinguish between different types of a certain emotion. The aim of their study is to detect emotions in abnormal situations like in kidnapping or hostages. Thus, a rather unusual categorization is chosen in particular consideration of fear constituting the predominant emotion in these situations. The exemplary emotion "fear in abnormal situations" is given, i.e., the categories for fear are ranging from a very passive reaction to fear (like inhibition) to a very active reaction (such as anger). In order to depict activation, evaluation and reactivity three axes are defined, two for evaluation and reactivity ranging from wholly negative (–3) to wholly positive (+3) and one axis for intensity providing four levels from neutral (0) to high intensity (3). In total the emotion categories "neutral", "other emotions in normal situations", "fear" and "other negative emotions" are regarded with their activation, evaluation and intensity values.

A further three-dimensional approach by Miwa et al. (2000) defines three levels of the emotional space, namely the activation level, the pleasant level and the certainty level, which is mainly used to describe the mental space of humanoid robots. A modified version of this model is presented in (Kim et al. 2005) removing the certainty level as this level has to be assigned to the cognition category in contrast to the activation and the pleasant level being physiological indices. The result is a model similar to the activation–evaluation space.

A new dimension of emotion research is described by Batliner et al. (2005). The traditional dimensions of valence and arousal are changed to the dimensions of valence and interaction as emotions normally are more private and social interaction should be modeled more intensively. Interaction determines whether the emotion is addressing oneself, e.g., angry and joyful, or the communication partner like motherese or reprimanding. The choice of the suitable dimensions of course depends on the basis of the used data and its emotions.

Similar to this approach is the distinction between cause-type and effect-type description of emotion in Cowie (2000), where the cause-type description refers to the speaker's internal state in opposite to the effect-type description characterizing the effect on the listener.

As negative emotions like fear and anger both are unpleasant and highly active on the basis of the two dimensional activation–evaluation space a third dimension called "potency", "dominance", "power" or "control" is suggested in literature (Osgood et al. 1957; Russell and Mehrabian 1977; Lazarus and Smith 1988) in order to better distinguish between negative emotions. This dimension includes the potential of a person to cope with a particular situation. In Frijda (1970), an additional fourth dimension of certainty, i.e., self assured ↔ insecure, is defined. Even five dimensions are suggested in Roseman (1979), these are need, occurrence of a certain state, probability, type of cause and legitimacy (it should be noted that some corrections to these dimensions have been applied between 1979 and 1996). On the basis of these five dimensions, 48 combinations can be formed, and these combinations in turn correspond with 13 emotions. Another possibility to categorize emotions is described in Abelin and Allwood (2000), where the three dimensions lust ↔ non-lust, active ↔ passive and secure ↔ insecure are used, but not necessarily requiring the use of all dimensions for an emotional term. E.g., happiness may be categorized into only two dimensions, namely +lust and +activity, whereas anger is "three-dimensional" with –lust, +activity and +security.

According to Scherer (1988), five functionally defined subsystems are involved in emotional processes. One of these subsystems, the information-processing subsystem, in turn is based on so-called *stimulus evaluation checks*. Four out of five predominant evaluation checks even possess subchecks. A table detailing which emotion is determined by which combination of these checks and subchecks is given in Scherer (1988).

The approach of the FEELTRACE system proposed by Cowie et al. (2000) follows the idea of Plutchik's emotion wheel as illustrated in Fig. 2.5 (Plutchik 1980b; Cowie et al. 2001). In this system, the activation–evaluation space is transformed into a circle also defined by transversal axes which are ranging from very negative to very positive concerning the evaluation and from very active to very passive regarding activation. Within this circle (on the computer screen) a cursor can be moved with the aid of a mouse, so that a person is able to indicate the current emotional state of a speech sample according to the visual and/or aural impression. According to its position within the circle the cursor in the form of a disc takes different colors ranging from pure red for the most negative evaluation to pure green for the most positive one and from pure yellow for the most active emotional state to pure blue for the most passive state. Furthermore the sizes of the discs of previous cursor positions are decreasing gradually over time (with the current disc featuring the biggest diameter) in order to reconstruct ("trace") how the emotional state or its interpretations ("feel") change over time.

In order to achieve a more realistic model the axes of the activation–evaluation space may also be expanded with action tendencies and appraisals. According to cognitive theories described in Ekman (1977, 1999) and Lazarus (1991), there exist certain mechanisms which attend to certain key elements of a situation and which trigger the respective emotions. These mechanisms called appraisals can be considered as a model identifying these key elements as positive or negative. Ekman and Lazarus distinguish between automatic appraisal quickly attending to some stimuli and deliberate and conscious appraisal slowly adapting to complex events or situations. In Roseman et al. (1990) and Ortony et al. (1988), the correspondence between appraisals and emotions is defined by certain distinctions, e.g., the intrinsic (positive or negative evaluation) and contextual (goals may be achieved or not) value of key elements.

Oatley and Johnson-Laird (1995) consider emotions as an important factor determining how cognitive processes are organized. Their "communicative theory of emotions" assumes that emotion signals control quasi-autonomous processes in the nervous system. Here, certain milestones of a plan or task are communicated and interpreted. E.g., a system is happy when a subgoal is achieved but it is sad when a plan fails (see also Oatley and Jenkins 1996).

Frijda (1986) assumes that an emotion cannot refer to a certain class of phenomena that can be distinguished from other events. Instead, there exist *concerns* which produce preferences and goals for a system. As soon as these emerging goals may not be achieved, the system develops emotions. In order to obtain a functioning emotional system, it needs to include six substantial characteristics:

- Ability to obtain and interpret information/stimuli from itself and the environment (concern relevance detection)
- Appraisal of the influence of the stimulus on the system's concerns
- Change of the system's behavior and priorities according to the intensity of the relevance stimulus
- Changes of the action readiness, i.e., the system tends to certain preferred actions and adapts its attention and processing of events

Table 2.4 Basic Action Tendencies (Frijda 1986; Cowie et al. 2001)

Action tendency	Emotion
Approach	Desire
Avoidance	Fear
Being-with	Enjoyment, confidence
Attending	Interest
Rejecting	Disgust
Nonattending	Indifference
Agonistic (attack/threat)	Anger
Interrupting	Shock, surprise
Dominating	Arrogance
Submitting	Humility, resignation

- Monitoring of all processes concerning the action readiness
- Adjustment to the social nature of the environment

In order to expand the activation axis in the activation–evaluation space, action tendencies which are linked to certain emotions (see Table 2.4) have been defined (Frijda 1986; Cowie et al. 2001). However, looking at further emotions like pity or remorse corresponding action tendencies are difficult to explore. Fox (1992) and Cowie et al. (2001) describe a further way to expand the activation axis, where the action tendencies are differentiated on levels according to the development of emotions. The two broad action tendencies "approach" and "withdraw" representing the first level are seen as origin of all emotions. These two action tendencies are subdivided on the second level into, e.g., approach in order to possibly gain pleasure (joy), approach in order to possibly get information (interest) or approach in order to provoke confrontation (anger).

The idea of mixed emotions is picked up in Carofiglio et al. (2002, 2003). The four goal-based emotion categories "Fortune-of-others" (e.g., sorry-for, happy-for, envy, gloating), "Prospect-based" (e.g., fear, hope), "Well being" (e.g., distress, joy) and "Confirmation" (e.g., disappointment, relief) are modeled on the basis of dynamic belief networks taking into consideration the "generative mechanism" (Picard 2000b). Two ways of mixing up emotions are possible according to the generative mechanism: emotions that coexist (i.e., emotions mixed like a "tub of water") and emotions rapidly switching from each other and not having overlapping generative mechanisms (i.e., emotions mixing according to the "microwave oven" metaphor). The generative mechanism of emotions including the intensity with which they are activated and the development of this intensity with time can be represented by dynamic belief networks (DBNs), i.e., beliefs about the achievement or threatening of goals of an agent A during the time instants (T, T + 1, T + 2...).

A DBN tailored to the monitoring of emotions is depicted in Fig. 2.6 comprising the following elements (the arrows denote the influence of one element on the following element):

- A's Mind at time T with its beliefs about the world and its goals (= M(T))
- the event occurred in the time interval (T, T + 1) with its causes and consequences (= Ev(T, T + 1)

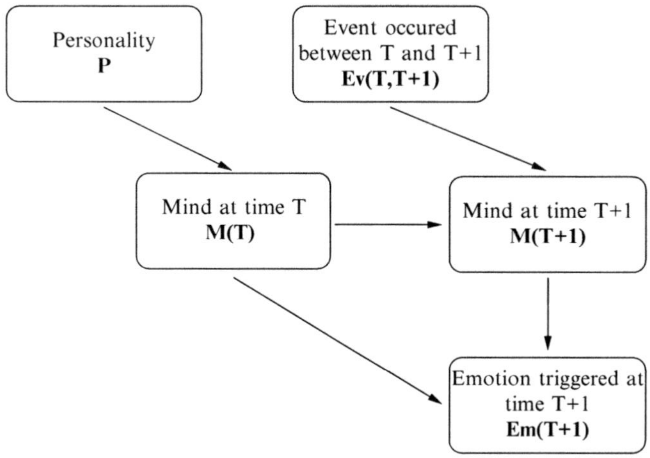

Fig. 2.6 Monitoring emotions with a dynamic belief network (Carofiglio et al. 2002, 2003)

- A's Mind at Time T + 1 (= M(T + 1)) which is dependent on M(T) and Ev(T, T + 1)
- a particular emotion activated in A at time T + 1 (= Em-feel(T + 1)) which is dependent on M(T) and M(T + 1)

As soon as an event occurs, the probability of the belief that a goal in the DBN will be achieved or threatened increases. The variation of this probability influences the intensity of the triggered emotion. Thus the variation of intensity in an emotion ΔIe can be calculated as follows:

$$\Delta Ie = \left[P^{\star}(\text{Bel A Ach}\{G_i\}) - P(\text{Bel A Ach}\{G_i\})\right] \cdot W_{\text{A}}(\text{Ach}\{G_i\}), \qquad (2.1)$$

where $P(\text{Bel A Ach}\{G_i\})$ and $P^{\star}(\text{Bel A Ach}\{G_i\})$ are the probabilities that an agent A attaches to the belief that the goal G_i will be achieved, before and after an event. Here, $W_{\text{A}}(\text{Ach}\{G_i\})$ is the weight that A allocates to achieving G_i. The achievement of a goal (Ach$\{G_i\}$) is replaced by the threat of a goal (Thr$\{G_i\}$) if the valence of the emotion is negative. With the aid of DBNs, also emotionally oriented communication of agents involving sensing, thinking, feeling and acting can be modeled (Carofiglio and de Rosis 2005). In the domain of natural argumentation, the interaction between cognitive and emotional modes constitutes a relevant issue. For agents typically arguing on a rational basis, the emotional persuasion may also be instantiated in the framework of a belief network as described in Miceli et al. (2006).

A specific emotional model with the aim of using it in an artificial intelligence system has been developed by Ortony et al. (1988), also called OCC model (named after Ortony, Clore and Collins, see Fig. 2.7). Emotions appear as reaction to three aspects: *consequences of events*, *actions of agents* and *aspects of objects*, i.e., a person may be pleased or displeased about consequences of events. Actions to agents may be approved or disapproved or a person likes or dislikes certain aspects of

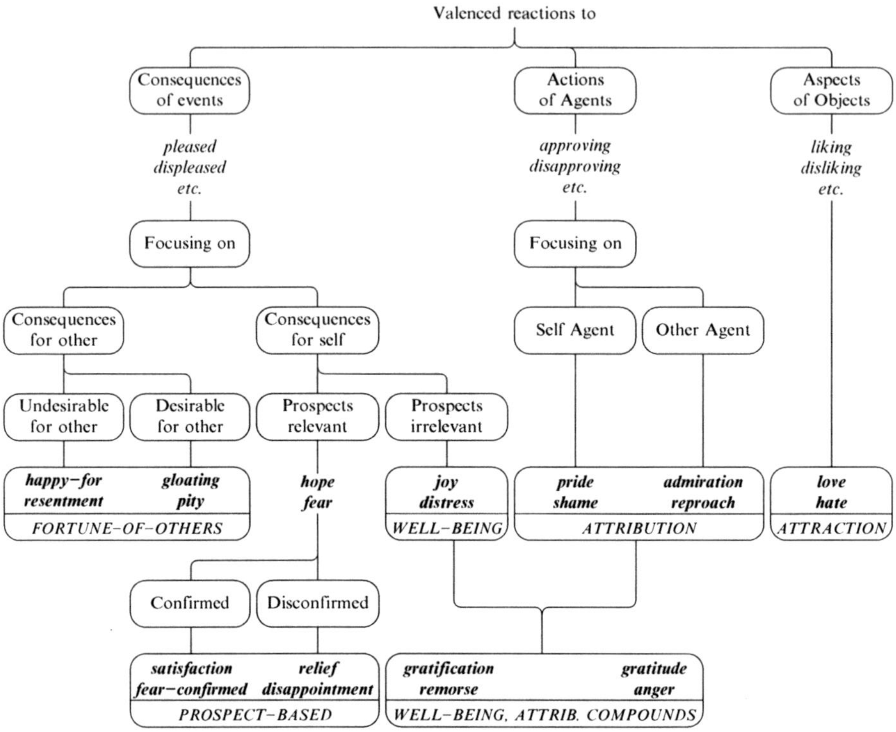

Fig. 2.7 OCC model of emotions (Ortony et al. 1988)

objects. The consequences of events furthermore are subdivided in consequences for others, which may be desirable or undesirable and consequences for oneself, which may be relevant and confirmed, relevant and disconfirmed or irrelevant expectations. Furthermore global and local intensity variables are differentiated, whereas the four global intensity variables are *sense of reality*, *proximity*, *unexpectedness* and *arousal*. The local intensity values are linked to the reactions according to the following three aspects:

- **Events**: *desirability*, *desirability for others*, *deservingness*, *liking*, *likelihood*, *effort*, *realization*
- **Agents**: *praiseworthiness*, *strength of cognitive unit*, *expectation deviation*
- **Objects**: *appealingness*, *familiarity*

To each of these variables a *value* and a *weight* are assigned. Moreover, a *threshold* for each emotion is defined, above which a certain emotion is consciously felt and determined and below which the value "zero", i.e., no emotion, is given. Even if a computer does not and will not have a subjective experience, a computer should be able to analyze these values and give a statement to a person's emotional state. These 22 emotional states are listed in Table 2.5. Surprise and disgust – two important human emotions are not included in this original emotion list, they are therefore

Table 2.5 The 22 emotions defined in the OCC model (Ortony et al. 1988)

Positive Emotions	Negative Emotions
Happy-for	Resentment
Gloating	Pity
Joy	Distress
Pride	Shame
Admiration	Reproach
Love	Hate
Hope	Fear
Satisfaction	Fear-confirmed
Relief	Disappointment
Gratification	Remorse
Gratitude	Anger

added by Kshirsagar and Magnenat-Thalmann (2002). In Andersson et al. (2002), the following emotion attributes furthermore describe the 22 emotions in accordance to the OCC model:

- Class
- Valence
- Subject
- Target
- Intensity
- Time-stamp
- Origin

The *class* corresponds to the emotion type representing related forms, i.e., *concern, fright* and *petrified* are assigned to fear differing from fear in various degrees of intensity. *Valence* refers to whether the value of the reaction is positive or negative. The agent undergoing the emotion is defined by the *subject*, whereas the event, the agent or the action eliciting the emotion is determined by the *target*. The *intensity* is reflected by a scale reaching from zero to ten, whereas zero corresponds to "no emotional feeling". The duration of the emotion-generation is determined by the attribute *time-stamp*. Whether an emotion originates from a physical component or from an affective user modeling component is constituted by the *origin* attribute.

As opposed to most of the existing emotion models which involve a high complexity for their implementation (Picard 2000a) and, thus, are not practical in computer applications, the essential features of the OCC model can be captured with the DETT (disposition, emotion, trigger, tendency) model as used for situated agents (Parunak et al. 2006). This model combines the theoretical richness of the emotional model proposed by Gratch and Marcella (2004) and the efficiency of emotional combat models used in artificial war scenarios. These models implement parts of a combined believe-desire-intention model (cf. Rao and Georgeff 1991) and the OCC model including disposition as illustrated in Fig. 2.8. Assuming the desires remain constant during our considerations, these and the beliefs (influenced by the environment via perception) affect the analysis which produces intentions. These, in turn, elicit actions which have an influence on the environment. This BDI model

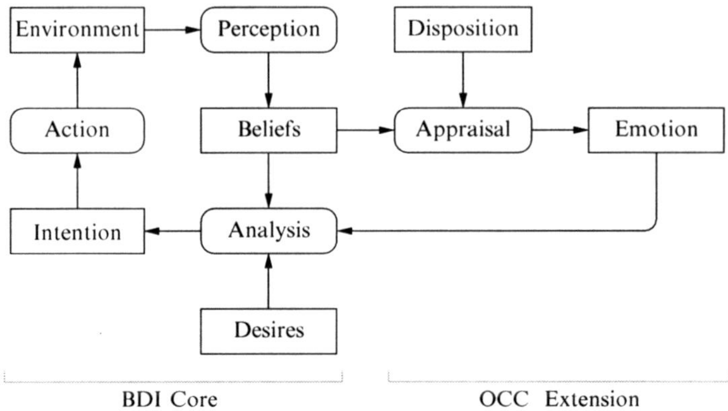

Fig. 2.8 BDI and OCC models of emotions incorporating disposition (Parunak et al. 2006). The rectangular boxes denote states, the boxes with round corners denote processes

is enhanced with an OCC extension where appraisal (driven by beliefs and disposition) generates emotions which also have an effect on the analysis and perception. With disposition modulating appraisal to determine how much the belief influences an emotion and the emotion modulating the analysis to indicate a tendency on the intention, the four key elements of the DETT model are quickly identified.

A comprehensive taxonomy of how to build up and structure an affective lexicon is presented in Ortony et al. (1987). According to this study the best emotion terms for an affective lexicon match the following criteria:

- Reference to internal, mental conditions
- Clear state description
- Focus on affect

If one of these constraints is neglected, more but poorer emotion terms may be added. Furthermore it is pointed out that every term has to be seen in its linguistic context as "being abandoned" is less emotional than "feeling abandoned".

In order to provide a general basis for the categorization of emotions a basic English emotion vocabulary (BEEV) is presented in Cowie et al. (2001). Naive persons are asked to choose 16 words from a list with emotional words that shall form a basic English emotional vocabulary. They furthermore define the positions of these emotional words in the activation–evaluation space and describe the emotional words with respect to appraisal and action tendencies. Table 2.6 contains an excerpt of the BEEV including the emotional words which are chosen by at least 50% of the persons.

The emotional words chosen for the inclusion in a basic English emotional vocabulary (as shown in the first column of Table 2.6) also represent (among others) the so-called basic, primary or fundamental emotions. Nevertheless some other "less common" emotions like "affectionate" or "relaxed" are included. The second column shows the location in the activation–emotion/evaluation Space. In the third column, the emotional strength is given in a scale ranging from zero as minimum

Table 2.6 Excerpt of the basic English emotion vocabulary (Cowie et al. 2001)

Emotional word	Emotional orientation	Strength of emotion	Disposed to en-gage or withdraw	Open- or closed-minded	Own perceived power	Orientation to surroundings	Orientation to other time	Orientation elsewhere
Bored	−166	0.46	+Withdr.	++Closed	−Power	+Surr.	−Past −Future	
Disappointed	−133	0.49	Withdr.				+Past	
Sad	−101	0.78	Withdr.		−Power	+Surr.		
Worried	−80	0.65	Unpred.	+Open	−Power		+Future	
Afraid	−52	0.84	+Withdr.	+Closed	−Power		+Future	
Angry	−48	0.95	+Unpred.			+Surr.		
Interested	17	0.70	Eng.	+Open				
Excited	24	0.95	Eng.	+Open			Future	
Loving	25	0.84	+Eng.					
Affectionate	43	0.72	Eng.					
Pleased	44	0.52	Eng.					
Confident	44	0.75	Eng.		Power			
Happy	47	0.71	Eng.					
Amused	53	0.71	Eng.			+Surr.	−Past −Future	
Content	136	0.66	Eng.					+Else.
Relaxed	153	0.68	Eng.	+Open			−Past	+Else.

to one as maximum. Furthermore, negative oriented emotions are more likely to be regarded as withdrawing. Analogously, positive oriented emotions are more likely to be seen as engaging (see fourth column). The fifth column presents whether the emotions are judged as seeking for information, i.e., open-minded, or not. The last columns specify the individual's own perceived power in the present situation, the orientation to situations in the present, past and future, and elsewhere.

Another type of emotion dictionary for Japanese is proposed by Matsumoto et al. (2005). Emotion-related nouns are extracted from a Japanese lexicon (Ikehara et al. 1999). These nouns are mapped to one or multiple emotion attributes building an emotion dictionary with nouns and their attribute(s). Nineteen emotion attributes are defined (see Table 2.7). Having assigned the emotion attributes to the nouns in the speaker's sentences one out of 12 emotions is derived from these attributes. However, for classifying emotions cultural differences also have to be taken into consideration, i.e., emotions are differently expressed and categorized in Japanese and in English as well as in other languages. This cultural difference is extensively discussed in Abelin and Allwood (2000), where Swedish, Englishmen, Finns and Spaniards had to categorize utterances from a Swedish speaker expressing different emotions according to their aural perception. Whereas 89% of the Swedish classified happy utterances as happiness, only 22% of the Spaniards had the impression that these utterances were happy. By contrast, these utterances were considered as sad by

Table 2.7 The 19 emotion attributes (Matsumoto et al. 2005)

Anger	Appreciation	Approbation
Contempt	Dislike	Excitement
Equilibrium	Fear	Hope
Joy	Like	Pride
Reception	Regret	Relief
Respect	Sadness	Shame
Surprise		

35% of the Spaniards. The emotional categorization is further complicated as, semantically seen, there often are no one-to-one correspondences of emotional words in different languages. E.g., no Swedish word out of "förvånad", "överraskad" and "häpen" is exactly equivalent to the Spanish word "espantado" expressing both meanings "surprise" and "fear". Moreover, specific emotional words from other cultures, e.g., "chagrin", "ennui" or "hubris" are hard to verbalize and accordingly the classification of these emotional words is difficult (Cowie 2000).

In our work, particularly in the field of automatic emotion recognition, we limit our considerations to a set of six emotions resembling the fundamental emotions described by Cornelius (2000): anger, boredom, disgust, fear, happiness and sadness plus neutral. Our choice of emotions is also based on practical considerations such as how to make further use of these cues in adaptive dialogue management, where pain and pleasure as propose by Mowrer (1960) or fear, love and rage (Watson 1930) are rather unpractical. From this point of view, we furthermore consider the use of smaller emotion sets, e.g., without boredom and disgust, reducing the quantization steps of the emotional scale. For the further use of emotional cues in adaptive dialogue management, we aim at a numerical representation of emotional states based on their position in the activation–evaluation space as proposed by, e.g., André et al. (2004).

Our approach to adaptive dialogue modeling claims to be straightforward and quickly implementable. Thus, we set complex underlying models such as the OCC model aside and we rather incorporate temporal dependencies as described by dynamic belief networks (Carofiglio et al. 2003). We argue that the emotional state at time T is strongly dependent on the previous state(s) at $T-1$, $T-2$, ... This is particularly taken into account in the semi-stochastic model which we propose in Chapter 3.

2.3 Emotional Labeling

The emotion-related signs of the utterances in a corpus have to be assigned to valid labels that identify the emotions (Cowie and Schröder 2004). As systematic evaluations and generally applicable regulations are lacking, a subjective evaluation has to be carried out. The difficulty in annotating emotions is the uncertainty factor, i.e., emotions are difficult to be rated for sure (Rigoll et al. 2005). A reasonable criterion therefore is an annotation by human performance. However, an annotation will never be able to reflect the exact emotional state of a speaker, but it should be adapted to the corresponding application such that a recognizer can be trained appropriately and a kind of empathy with the user can be elicited. An overview on existing approaches to emotional labeling and to known problems is given in Craggs (2004).

Emotions are either labeled by persons who are not involved in the process of the recordings, or, in case of acted data, by the actors themselves as done in Rigoll et al. (2005). E.g., the actors of the recordings described there label both their own utterances and the utterances from the other actors, so that a human reclassification error rate can be calculated. Alternatively, in Razak et al. (2003), a confusion matrix table shows which emotions are confused most often and, consequently, which data needs to be recollected again. In Navas et al. (2006), two different groups of persons who are not involved in the recordings, are selected to evaluate two acted databases in Basque, i.e., one database with semantically neutral texts and another database containing texts with emotional content: The first group consists of Basque speaking persons who are also able to understand the emotional content of the texts of the second database, the second group is made up of Spanish speaking people not understanding Basque so that their decisions (based on a list with seven given emotions) are not influenced by semantic information. By labeling these corpora containing artificial data the actor's ability to simulate emotions furthermore is proved. In Alexandris and Fotinea (2004), the actor's ability is validated even with separate labels like "naturally-sounding" or "not naturally-sounding".

A corpus should generally present an association that is accepted as valid, therefore appropriate labels and methods have to be used. The following two approaches to emotion annotation by human performance can be differentiated (Stibbard 2001):

- Free choice, i.e., the labels for the utterances are freely chosen by the annotators
- Forced choice, i.e., a predefined set of labels is given, among which the annotators have to choose from

The free choice normally results in a large number of categories, which are difficult to handle in emotion recognition as due to a lot of similar labels certain emotions cannot be differentiated and recognized clearly. Usually the label set therefore has to be reduced, e.g. similar labels are put together, even if information gets lost. As described in Batliner et al. (2004b) for the SYMPAFLY database, it may be advantageous to let the labelers decide on which user states to annotate and then find a consensus with a limited number of emotions. The consensus labeling approach is typically applied when the agreement between different labelers is too low to be used for a reliable emotion database (Ang et al. 2002). Such a kind of consensus may also be found by majority voting, considering only a certain number of most common emotions (Batliner et al. 2005). Predefined labels enable a better and easier emotion recognition, however, the annotators are forced to assign a label to an utterance even if a label does not exactly match the utterance, i.e., the transition of the emotions are blurred. The predefined label set should be adapted to the application and to the context, in order to keep the label set as small and clear as possible and to omit emotions which do not occur in the application anyway as done in Devillers et al. (2002). A very small label set has been used in Swerts and Krahmer (2000) as a selection of the word "no" had to be annotated with either the label "positive" or "negative". According to Craggs and McGee Wood (2003), a special differentiation has to be made in a database containing dialogue acts: Not only the emotional state of the speakers should be indicated, but also the emotional state of the dialogue itself, i.e., if a serious dialogue is interrupted by a joke the emotional state of the dialogue remains serious whereas the speakers' emotional states may change.

In Douglas-Cowie et al. (2000), two types of annotation are described similar to the two systems for emotion annotation:

- **Dimensional**: Each utterance is given an emotional content, i.e., an activation as well as an evaluation. The activation indicates the intensity of an emotion and the evaluation describes the emotional value by indicating if the utterance has a positive emotional content like happiness or a negative one like despair see Section 2.2.
- **Categorical**: Predefined labels are provided, i.e., a forced choice or logical descriptions are given.

Dimensional labeling is, e.g., accomplished with the aid of a computer program called FEELTRACE as described in Section 2.2. The utterances are annotated continuously and changes of the emotional content can be reconstructed with regard to time. However, emotions cannot be distinguished clearly from each other (Cowie and Schröder 2004). In Douglas-Cowie et al. (2000), the dimensional labeling furthermore is combined with a categorical labeling where annotators had to choose between labels on a list. If a label does not match exactly to the emotional content in the annotator's opinion he may select more labels from either the first list or a second list. The annotation is completed by an indication of the intensity on a scale from one to three. If ambiguous labels are assigned to sentences a further annotator may be brought in to judge the respective sentences as done in Devillers et al. (2002). Various conditions furthermore may influence the emotion annotation, e.g., if the annotator is allowed to change his decision afterwards or not or if the annotator is allowed to hear the utterance only once or several times. According to Navas et al. (2004b), even more than one label may be assigned to an utterance. Phonemes may be labeled instead of a whole utterance or words (Campbell 2000), if the appropriate phonemes are provided, e.g., by a speech recognizer. However, high annotation error rates may occur because of too small labeling units, so that emotions cannot be differentiated clearly. A rather simple approach to one-dimensional labeling is described in Forbes-Riley and Litman (2004) where the evaluation (valence) axis is quantized into three levels – positive, neutral and negative. To assess how speech recognition problems elicit user emotions, only two states (emotional and non-emotional/neutral) are considered in Rotaru et al. (2005).

In the approach by Campbell et al. (2006) different kinds of resolution are suggested in the form of a three phased annotation:

- Global emotion labeling
- Trace labeling
- Quantal labeling

Global emotion labeling in the first stage represents the coarsest approach in which a whole utterance or passage is assigned one common label chosen from a limited set of attributes. In the following stages, i.e., trace labeling and quantal labeling, the annotation is improved by finer time resolution and more accurate labels are chosen from a larger set of attributes. Moreover, different kinds of labels may be assigned to the utterances, which indicate everyday emotion words, authenticity of the emotion, core affect dimensions like intensity, valence, activation, etc.

A similar approach with a flexible tree-structured three-level hierarchical labeling scheme, is proposed by Fék et al. (2004), where the first level consists of only three labels (positive, negative, neutral) and the following two levels contain more labels refining the labels of the first level. Laskowski and Burger (2006) describe a decision tree for the annotation of emotionally relevant behavior. Based on questions like *"does the speaker express disagreement?"* or *"is there any attempt to amuse the listener?"* in the branches, the annotators reach leaf nodes, each representing a certain state.

In case of a visual database, facial expression or gestures may be annotated in addition. Each annotation furthermore is depending on the annotator, e.g., a female annotator might perceive emotions stronger than a male annotator (Alexandris and Fotinea 2004; Wardhaugh 1992). According to Campbell et al. (2006), the annotator, on the one hand, should be trained to be able to label like an expert, on the other hand, the annotator should give labels in a normal "naïve" manner, which might be difficult for him or her after a long training. Therefore multiple experts should be included in the annotation process and the coherence should be verified in the form of an inter-labeler agreement and annotation label confidences. Depending on the number of labelers, the inter-labeler agreement can be measured with either Cohen's kappa statistics (see Cohen (1960), two labelers) or Fleiss' kappa statistics (see Fleiss (1971), any number of labelers). Both statistics include considerations on the relationship between the actual agreement between the labelers and the agreement that would occur if all labelers labeled the data randomly. Cohen's kappa κ_C is calculated as

$$\kappa_C = \frac{p_A - p_R}{1 - p_R}, \tag{2.2}$$

where p_A is the actual agreement portion and p_R is the agreement which is expected if both annotators labeled the data on a random basis. The values for κ_C range from $-\infty$ (very poor agreement) to 1 (full agreement), and there exist different opinions about the interpretation of the values in between. Typically, $\kappa_C \geq 0.75$ is considered as good agreement. Analogously, Fleiss' kappa κ_F is calculated as

$$\kappa_F = \frac{\bar{P} - \bar{P}_d}{1 - \bar{P}_d}, \tag{2.3}$$

where $\bar{P}$ is the average value of the extent to which the annotators agree on the subjects (in this case utterances) and $\bar{P}_d$ is the sum of all squared probabilities for each labeling category. For the interpretation of Fleiss' kappa values hold the same assumptions as for Cohen's kappa.

Similarly, confusion ratios for certain emotion pairs can be calculated as done in Yuan et al. (2002):

$$CR(\text{joy} > \text{anger}) = \frac{N(\text{joy}, \text{anger})}{N(\text{joy}, \overline{\text{anger}})}, \tag{2.4}$$

where $CR(\text{joy} > \text{anger})$ is the ratio that joy is confused with anger, $N(\text{joy}, \text{anger})$ is the number of utterances that have been annotated at least with one label for joy and one label for anger and $N(\text{joy}, \overline{\text{anger}})$ is the number of utterances that are identified with at least one label for joy, but no label for anger. As opposed to the kappa statistics a higher value indicates a lower agreement.

Furthermore, in practice three annotation techniques can be differentiated:

- Manual labeling
- Semi-automatic labeling
- Automatic labeling

The most time-consuming technique is to label manually, as to each utterance or part of the utterance, a label has to be assigned by the annotator. For the annotation of plain speech there exist numerous software tools which allow to mark certain regions in the waveform and to assign labels to these regions – these tools are typically also used for the transcription of data for speech recognition. For multimodal annotation, e.g., the ANVIL tool allowing a task-dependent annotation strategy may be used (Kipp 2001; Clavel et al. 2004). In multimodal labeling, apart from a holistic view considering all modalities, also "blind" annotation, audio-oriented, video-oriented, context-oriented, etc., is used to mask out influences from the other modalities (Clavel et al. 2006). A task-oriented labeling tool for video sequences in augmented multiparty interaction similar to the FEELTRACE tool is described in Reidsma et al. (2006).

A certain level of facilitation is provided by a semi-automatic labeling technique with the aid of the bootstrapping method. Using a small part of the database which is labeled manually, a preliminary (acoustic) model is trained and used to recognize, i.e., to label new unlabeled utterances. These labels are manually corrected and are included in the training of a new model which is used to label further utterances until the whole corpus is labeled (see also Section 4.5). Automatic labeling using a speech–pause detector may apply in cases where the utterances' emotional content is already known and just one emotion shall be assigned to an utterance. Nevertheless a manual correction may be needed in cases, where the assignment is ambiguous. Alternatively a database can be annotated with an already existing functioning emotion recognizer, if available. In Stibbard (2000), the Reading/Leeds Emotional Speech Corpus is automatically annotated using the standard for labeling English prosody called ToBI (Tones and Break Indices). Experts in speech technology and prosodic analysis developed ToBI in order to provide a standard for prosodic transcription which conforms to the IPA (International Phonetic Alphabet) for phonetic segments (Silverman et al. 1992). As several aspects of prosody have to be included, parallel tiers are used in the system, namely a tonal tier and a break index tier. The tiers indicate prosodic events with the aid of symbols as well as the time slots of the prosodic events in the utterance. Pitch events on the basis of Pierrehumbert's intonational phonology (Pierrehumbert and Hirschberg 1990) are transcribed on the tonal tier. Also local maxima in the pitch contour are specified. The break index tier is transcribed by a seven point scale ranging from 0 to 6, whereas no boundary between two orthographic words is represented by 0 and a full boundary of an intonation phrase is indicated by 6. A further tier, namely the miscellaneous tier, specifies spontaneous speech effects like laughs, breaths, etc. E.g., Stibbard (2000) annotate the Reading/Leeds Emotional Speech Corpus with ToBI elements of the tonal tier, so that six emotions can be classified by considering the differences in the occurrence of the elements.

Another phonetic transcription for the annotation of the Berlin Database of Emotional Speech is applied in Burkhardt et al. (2005). Two label files are created, the first file for a phonetic transcription and the second file for indicating stress including a segmentation into syllables. Beside using symbols of the machine-readable SAMPA phonetic alphabet for the phonetic transcription in the first label file, segment and pause boundaries were annotated in this file, too, as well as settings and diacritics by using German abbreviations. Moreover, additional labels like "whispery voice" specify further (emotional) characteristics of voice. In contrast to the Reading/Leeds Emotional Speech Corpus, the Berlin Database of Emotional Speech is labeled manually by aural perception as well as by visual analysis of the oscillograms, spectrograms and electro-glottograms. The transcriptions of stress are furthermore evaluated by eight trained phoneticians.

In addition to acoustic information as basis for labeling emotional corpora, three characteristics for emotion annotation and recognition are presented in Devillers et al. (2002):

- Acoustic characteristics (extraction of prosodic features, e.g., pitch, energy, speaking rate)
- Linguistic characteristics (extraction of linguistic cues identifying emotions, e.g., on lexical, semantic, dialogic basis)
- Combination of acoustic and linguistic characteristics

A combination of acoustic and linguistic information allows a significant error reduction in emotion recognition. An integration of further parameters, e.g., by including labels for nonverbal events like laughter or throat, enables an additional improvement of recognition results. Similarly, in Batliner et al. (2000), a combination of several indicators like key word spotting or annotation and recognition of meta communication, etc. is proposed in order to receive a better emotion recognition rate. Also in Devillers et al. (2003), emotions are labeled on the auditory and lexical level independent of each other. Here, it shows that depending on the level, the utterances may be assigned different emotions.

The annotation of the speech data used in our plain emotion recognition experiments is accomplished on the word and word group level where the relevant sections are automatically determined with the aid of a speech–pause detector. These sections' labels are then selected from a list of seven emotional states according to the annotators' aural impression. The utterances which we also use for the training and testing of our hybrid speech–emotion recognizer are furthermore labeled on the phoneme level. I.e., for each word in the utterance the appropriate emotion is determined and the word is labeled accordingly, e.g., "PLEASE-NEUTRAL". With the aid of our adapted dictionary, this word–emotion is refined into what we refer to as emophonemes such as "pn ln iyn zn" (derived from "p l iy z" for the non-emotional word "PLEASE"). For our considerations we set aside further labels such as those on the prosodic level. Whereas the data in the Berlin Database of Emotional Speech is labeled according to a forced choice limiting the annotators' input to seven possible emotions, we allow a free choice annotation for our spontaneous speech database described in Section 6.2. In the latter case, however, in order to keep a certain degree of consistency, we include a post-processing limiting the number of possible emotions.

For the integration of emotional cues in adaptive dialogue management, we consider that textual representations are somewhat unhandy for both rule bases or stochastic models. Thus, we also employ a numerical representation of emotional states as described in Section 3.7. The respective numbers are derived from the emotion's position in the valence-arousal space taking into account how the user's emotional state influences the perception of a message (good news vs. bad news). Such a numerical representation also allows a flexible quantization of emotions, e.g., depending on their use in a dialogue model as discussed in Section 3.8.

2.4 Emotional Speech Databases/Corpora

Campbell et al. (2006) in general distinguish two kinds of corpora:

- Artificial data (acted data or induced data)
- Real-life spontaneous data (authentic speech methods)

The emotions acted in the artificial data normally are only full-blown emotions as the context, i.e., internal events such as "headache" influencing the emotional state as well as external events such as "someone helping the sick person", are missing. Consequently emotions or affective states are not mixed like in real-life data, in which also multiple events simultaneously may influence the affective state. Therefore the lacking naturalness of the affective states in artificial data is often criticized in literature. A corpus of spontaneous emotions, i.e., authentic speech, may be built by either recording real-life situations such as call center dialogues or TV interviews, i.e., found speech (Campbell 2000), or by inducing certain situations with expected emotions in an interaction between a speaker and a machine or in a Wizard-of-Oz (WOZ) scenario. The induced data may also be regarded as a third separate category including corpora of elicited or induced speech as done in Navas et al. (2004b) and Campbell (2000). While human–machine interaction is restricted to certain words or commands and often does not enable the speaker to express recordable (spontaneous) emotions, the WOZ scenario provides authentic data of multiple speakers on the same conditions in various languages in a laboratory (Aubergé et al. 2003). In this scenario, the speaker communicates with a human via a machine in order to accomplish a task. As the computer's utterances and behavior are only imitated and remote-controlled by a human (typically referred to as wizard), the speaker believes that he is interacting with a computer. The goal of the wizard is to imitate the machine's behavior such that certain emotions of the speaker are elicited. Furthermore, the phonetic and linguistic content of the speaker's utterances is controlled by the wizard's command language. However, the problems of legality, anonymity and privacy, the ethical acceptance of real-life spontaneous data as well as the copyright of data from TV or radio are often discussed (Campbell 2000; Stibbard 2001).

Corpora furthermore can be characterized according to the following criteria:

- Control of the speech characteristic through the observer (i.e., the person collecting the speech material) or no control
- In vitro method (laboratory corpus) or in vivo method
- Professional actors or non-professional speakers
- Utterances linguistically and phonetically predefined or undefined
- Utterances with emotional content or semantically neutral

A very common method to build a corpus is to collect authentic data without any control, e.g., to record everyday speech or talk-shows. Even in such real situations, the speech characteristic may be controlled with the aid of (anonymous) partner or non professional actors who participate in the interactions. A corpus furthermore can be produced in a laboratory or in vivo. The challenge in the application of in vivo methods such as recordings of talk shows or everyday speech is to achieve a corpus of good technical quality especially without background noise outside a laboratory. In special applications where emotions have to be recognized in a noisy environment, corpora with certain types of background noise may be useful.

A common method is to employ actors in a laboratory environment, where the utterances are linguistically and phonetically predefined as done in the Berlin Database of Emotional Speech (Burkhardt et al. 2005). In this corpus the actors are asked to utter semantically neutral sentences in different emotions enabling a good comparability between the emotions and their acoustic features and furthermore providing a phonetic balance. As the utterances of actors generally should sound as natural as possible the actors are often confronted with stimuli like pictures before actually starting the recording in order to get them in the appropriate emotion or mood. On the one hand, corpora composed of acted utterances are criticized for exaggerating and for not representing natural emotions as already mentioned with respect to artificial data, on the other hand, even in real situations, e.g., in social interaction, certain emotions are expressed which are not really felt, i.e., perceived and expressed emotion has to be differentiated (Campbell 2000). Another method to build corpora is to use non-professional speakers reading predefined texts with emotional content or telling a story with emotional background as done in the emotional speech corpus by Amir et al. (2000). Ideally, female and male speakers are equally distributed among both actors and non-professional speakers to obtain more representative data. Different age groups and different ethical background of the speakers additionally increase the representativeness of the data (Rigoll et al. 2005). A mixture of semantically neutral texts and texts with emotional content has been used in the Audiovisual Database of Emotional Speech in Basque (Navas et al. 2004a) to ensure a phonetic balance with the aid of the semantically neutral texts and to make it easier for the speakers to express the emotions by using texts with emotional content. Furthermore different lengths and contents of the texts are included for representative reasons. As it is impossible to include all emotions in a corpus most of the corpora contain only few emotions. As the primary or basic emotions are strongly colored and mostly serve as reference for emotional studies, they are also frequently chosen for corpora (Cowie and Schröder 2004).

In Appendix A, several emotional corpora are listed. The list is sorted in alphabetical order of the language or languages, in which the corpora are developed and by the year of the development or accomplishment of the corpora (starting with the oldest corpus in the appropriate language). In cases where several corpora are produced in the same year and in the same language the corpora are sorted in alphabetical order according to the first author's name of the respective publication. Due to the enormous number of different emotional corpora that have been developed until now, not all existing corpora are included in this list and some restrictions apply: As our work concentrates on unimodal SLDSs, only databases including speech or optionally video are considered. As opposed to speech data which may be used for the training of emotional text-to-speech synthesizers, corpora of synthesized speech are not included as these are not suitable for the use in emotion recognition systems for human speech.

The application of reliable and representative emotional speech data is particularly vital for the development and assessment of speech-based emotion recognizers. With respect to the training of the statistical acoustic models, large amounts of accurate data are required to achieve a robust classification. Moreover, our interpretation of representative data includes the assessment how emotional the speech of a corpus actually is: What is the ratio of emotional utterances to neutral utterances? Are the emotional utterances actually recognizable for human interpreters? How many emotions are included? Some of the corpora listed above unfortunately are not documented comprehensively enough to assess their applicability, other corpora such as the SmartKom or SYMPAFLY databases possess a high proportion (80% or more) of neutral utterances. Furthermore, in some corpora, emotional utterances do not differ significantly to neutral utterances regarding the aural perception. A further aspect is the availability and costs of the data. Whereas most of the corpora from the list are not offered on the open market, other corpora are sold at relatively high prices.

For our experiments, we use the Berlin Database of Emotional Speech (see Burkhardt et al. 2005) which is made publicly (and freely) available by the Technical

University of Berlin and which enjoys an overwhelming popularity among the emotion recognition community. Keeping in mind that the emotions are acted and not spontaneous, we use the data for a comparable performance assessment of our approaches described in Chapter 4. A more detailed description of this corpus as well as of our self-recorded material with spontaneous emotional speech is given in Section 6.2.

2.5 Discussion

Emotion constitute an important share in a human person's everyday life. E.g., in a conversation which is strongly influenced by each participant's emotional state, or in environments like a car where the driver's emotional state is influenced by the situation (e.g., traffic jam) and where the emotional state, in turn, has an impact on the driving behavior. In the latter case, emotions even relate to safety aspects.

Due to the problem that emotions cannot be measured by objective means, the actual handling of emotions poses a great challenge for humans as well as for computers. E.g., Cowie and Schröder (2004) point out that it is not possible to distinguish different emotions clearly from each other. This involves ambiguities already in the development phase of a emotion recognition system, where an utterance of the training data may be labeled with different emotions due to the different annotators' perception. By that, from the first, such a system cannot perform better than the annotators of the underlying training data. Moreover, the exact emotional state of a speaker cannot be reflected by such an annotation – it is rather a maximum-likelihood decision which label (from a predefined set of emotions) matches the user's state best. Predefining such a set emotions, in turn, is also not trivial as the developer here is required to determine which emotions actually influence the dialogue.

A large amount of work has already been accomplished in the field of emotion theories, collection of emotional data, signal processing and classification of emotions. This includes the selection of appropriate features and useful classifiers for speech-based emotion recognition. There exists a large variety of speech-based emotion recognizers, each of which has its own kind of unique feature and performs "best" according to a certain criterion. Keeping in mind the problems mentioned above, the recognizers' actual capabilities are difficult to compare.

Chapter 3
Adaptive Human–Computer Dialogue

When two or more persons are talking to each other, there exists a large variety of parameters and stylistic devices that may influence such a dialogue. This can be situation-related parameters, e.g., a person is in a hurry, the environment is noisy, it is raining, etc., or speaker-related parameters, e.g., a person is sad because a close relative has died, a person is nervous because of an upcoming examination, etc. Furthermore, humans tend to use certain stylistic devices to emphasize what they want to express in the interaction, e.g., irony, gestures, facial expressions, etc. Human dialogue partners are typically able to correctly construe the meaning of the paralinguistic parameters and the interrelation between situation and the behavior of dialogue partners. Moreover, humans are able to account for this context in the dialogue. This adaptation typically occurs subliminally, i.e., the speaker doesn't even notice.

In human–computer interaction, computers, however, face the challenge not only to recognize these paralinguistic or only linguistic cues but also to process these appropriately and to adapt the interaction accordingly. This does not only include situation and user-state parameters but also the fact that humans tend to interact differently with computers than with human dialogue partners. In this book, we address these challenges focusing on the user's emotional state in SLDSs. The idea and the implementation of a speech-based emotion recognition system is described in Chapter 4. In this chapter, we present several approaches how the information provided by such an emotion recognizer can be processed to adapt the dialogue flow.

For the handling of emotions in the adaptive dialogue manager we propose an integrated approach combining user-state management and dialogue management in the SLDS as illustrated in Fig. 3.1. For the adaptation of the dialogue flow there exists a large variety of approaches focusing on different parameters individually, e.g., the dialogue system proposed by Litman and Pan (2002) adapts its confirmation strategy to the confidence measures of the speech recognizer or Yankelovich (1996) describes the adaptation of prompts according to the user's experience level.

Having outlined existing approaches to adaptive dialogue management with particular respect to emotions in the following section, we describe the purpose and the functioning of the user-state and situation managers in Section 3.2. In the remainder of this chapter, we propose an approach considering the entirety of all

J. Pittermann et al., *Handling Emotions in Human-Computer Dialogues*,
DOI 10.1007/978-90-481-3129-7_3, © Springer Science+Business Media B.V. 2010

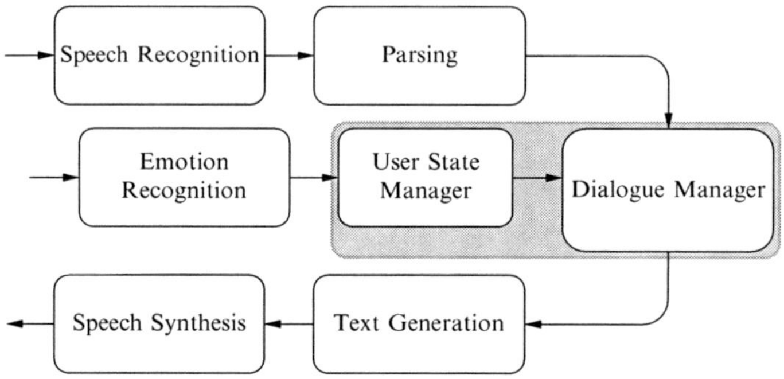

Fig. 3.1 Spoken language dialogue system integrating user-state management into adaptive dialogue management (Pittermann et al. 2007b)

possible dialogue parameters en bloc. As an exemplary situation control parameter we choose noise represented by the speech recognizer's confidence measures based on which we describe the adaptation process in Section 3.4. In Sections 3.6 and 3.7, we introduce semi-stochastic dialogue and emotion models before merging these models into an emotional dialogue model as described in Section 3.8. A perspective on extending the emotional dialogue for multiple dialogue control parameters is then given in Section 3.8.

3.1 Background and Related Research

A comprehensive overview on adaptive dialogue systems in general, also addressing dialogue management, is given in McTear (2004). As for the actual definition and implementation of adaptivity and flexibility, the views and opinions are quite divergent. Whereas flexibility is commonly associated with enhancing a standard system-initiative dialogue system by extending the speech recognizer vocabulary and integrating mixed initiative, adaptivity typically implies the existence of dialogue-influencing parameters based on which the flow and style of the dialogue is adapted.

3.1.1 Adaptive Dialogue Management

The idea of rendering human–computer interaction more efficient by adapting the dialogue flow is actually not new and is not restricted to SLDSs. Ukelson and Rodeh (1989) propose a dialogue manager for screen-based interaction which aims at efficient interaction in terms of a minimum number of questions. Their approach includes the appropriate selection of items to be shown on the screen, exploitation of relations between the dialogue fields and the handling of incorrect user input. In the representation, the dialogue fields are structured in a tree-like graph taking into account their interdependencies (e.g., if an employee's ID number is known, it is not necessary to ask for the name). Apart from simple constraint checks (e.g., the start date must be before the end date), the actual adaptivity of the dialogue is based on rules which allow to infer dialogue field values from other known field values by what the tree structure is simplified and the field selection is adapted.

In order to improve the quality of a spoken dialogue in general, Krahmer et al. (1997) propose what they refer to as seven commandments for spoken language dialogues constituting guidelines for dialogue development. Among these guidelines are the demand for consistency, comprehensibility, error proofness, adaptability and translucency as well as the plea to the developer not to underestimate the design phase and to choose sensible prompts which should be well formulated and which should fit in the ongoing dialogue. They present an SLDS for car-drivers the dialogue manager of which implements these commandments as well as possible.

A step towards adaptivity is presented by Duff et al. (1996) who describe the architecture of a task-based human–computer SLDS. In addition to the five standard components, a discourse processing module is included between linguistic analysis and dialogue management as illustrated in Fig. 3.2. This discourse processor avails itself of the current user input as well as a discourse state and knowledge base containing a static domain model, a dynamic backend model and a user model. Based on a military battlefield simulation environment, the authors propose a four-step algorithm which the discourse processor applies to recover from miscommunication problems such as interpretation failures.

The same problem is addressed in Martinovski and Traum (2003) from the aspect of how strong the dialogue deviates from the user's expectations if too many interpretation errors occur or if certain dialogue strategies inhibit a fluent dialogue. They propose a dialogue system model of conversational partners based on competence and conventions that potential users have learned from their interactions

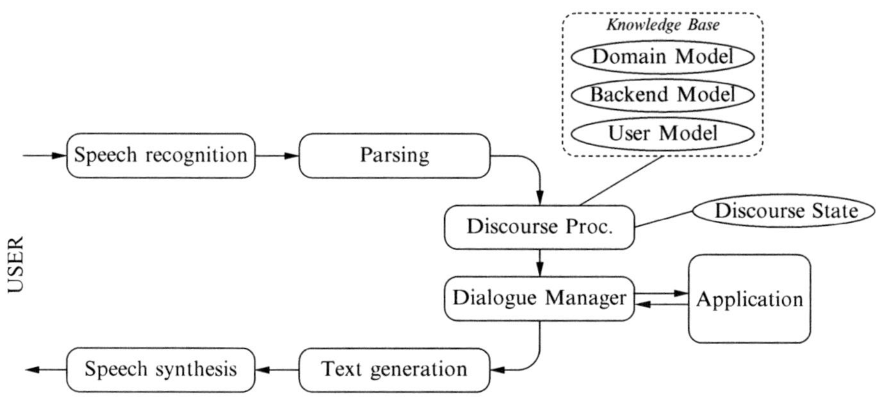

Fig. 3.2 Architecture of an SLDS discourse processing (Duff et al. 1996)

with other humans. Based on the example of another military dialogue application and the TOOT train timetable information system (see also Kamm et al. 1997; Litman and Pan 2002), the authors analyze aspects such as fluency and cooperation as well as causes of communication breakdown. They observe that, apart from speech recognition problems, also the systems' inability to adapt to the users' state and the inappropriateness of repair requests such as confirmation prompts may lead to a breakdown. Signals which indicate such a breakdown can be found in the intonation, articulation, extra-linguistic sounds, attention calls or elliptic speech. They conclude that, as it is not possible to eradicate all errors, a dialogue system shall be either able to handle possible non-fluencies or to respond sensitively to the users' emotional state. Veldhuijzen van Zanten (1999) proposes a user model for the use in adaptive mixed-initiative dialogue management as described in Veldhuijzen van Zanten (1998). The user model is designed for the prevention and handling of misunderstandings in the dialogue and consists of flags which are maintained in a hierarchical slot structure.

For a dialogue system, concepts need to be defined about what the system is capable to understand and to process. A formal description of the concepts and rules defining the relationship between these concepts can be summarized in ontologies. Such ontologies allow the description of generic dialogue models as well as concepts for domain-specific applications.

In order to render dialogue systems in automotive environments more efficient, Hassel and Hagen (2005) adapt the dialogue to the user's experience level. They propose a user model which is updated after each turn containing parameters such as the number of help or option requests, the mean response time and speech recognizer confidence measures. Based on heuristics including a forgetting curve, the user is classified as either expert or novice. Accordingly, the system prompts are adapted, e.g., *"Pardon me?"* for experts or *"Sorry, I could not understand you. Please repeat."* for novice users.

One approach to adaptive dialogue strategies with respect to recovering from speech recognition and interpretation errors is proposed by Litman and Pan (2002). They describe an adaptive version of the TOOT system which is able to consecutively restrict the initiative and the confirmation strategy with decreasing speech recognizer performance. The system is initialized with user initiative without confirmations and, if necessary, the algorithm switches to more conservative strategies using mixed initiative with implicit confirmations and even system initiative with explicit confirmations. The adaptation occurs with respect to the expected misrecognition rate calculated from the confidence measures provided by the speech recognizer. A usability study is also conducted to verify improvements in the system performance compared to a non-adaptive version of TOOT. Maintaining mixed initiative during the whole dialogues, Komatani and Kawahara (2000) present a dialogue strategy involving two levels of confidence measures to achieve a more robust but still flexible dialogue flow. In their approach, confidence measures are calculated for each content word. Two thresholds define whether a word is accepted or rejected or whether confirmation is required. If the word is rejected or if the confirmation is denied also semantic-attribute confidence measures are determined to estimate which (semantic) category the user refers to. Depending on these measures, the system decides whether the user needs guidance (e.g., in a travel information system, if the system asks for a destination and the user replies with information about the travel date) or whether the user shall just rephrase or repeat the previous input.

Hagen and Popowich (2000) propose a flexible approach to dialogue management based on speech acts. Their dialogue engine calculates dialogue primitives to determine system and user utterances allowing simple question–answer dialogues, overanswering as well as complex mixed-initiative dialogues including different confirmation strategies if required. The engine includes a reasoning module and a knowledge base with application description, grammar, dialogue history, etc. In their approach, the reasoner is mainly used to determine the appropriate system reaction to the primitives determined from the user input. E.g., if the system requests a certain information, the reasoner allocates the user's informative answer accordingly. A different use of reasoning is presented in Bühler and Riegler (2005), where a domain reasoning component is operating outside the dialogue manager. Based on atoms received from the user or given by the situation it provides inferences or solutions which are then integrated into mixed-initiative planning dialogues, in this case in the TRAINS-93 domain where different combinations of engines and boxcars are used to transport items between five cities. To transform the solutions into dialogue acts, the dialogue manager is extended by a solution evaluation module filtering the reasoner's output and an interface communicating new information to the reasoner. The interaction manager ensures the collaboration between user and reasoner deciding in which direction information shall be communicated and which information shall be passed to the user. Reasoning is also involved in intelligent tutoring systems which need to infer the students' knowledge and proficiency from their actions as described by Conati et al. (2002). To overcome the uncertainty in modeling the students' learning and reasoning, the authors propose the integration of Bayesian models into these systems. A flexible framework for dialogue management modeling domain knowledge and planning is presented in Delorme and Lehuen (2003) where tasks and methods from problem-solving are used on the dialogue control and the domain level.

The general terms of adaptivity and flexibility can also be limited to a single dialogue scenario or domain which requires a specific behavior or a certain degree of intelligence. For the medical domain, Azzini et al. (2001) propose a telemedicine framework which shall be able to render the (multimodal) human–computer dialogues as "natural" as possible by dynamically changing the content to be presented and adapting to misunderstandings or argumentation. To accomplish that, their proposed architecture includes a large knowledge base consisting of domain knowledge, dialogue history, patient record and a medical unit which are controlled by a system manager passing relevant information to the dialogue engine. A flexible dialogue system in the travel information and booking domain is presented in Stallard (2002). The system includes a tree-shape dialogue control model where the leaf nodes represent system actions and the interior nodes order the relevant actions depending on the goals. As opposed to, e.g., VoiceXML, their approach is asynchronous (by using separate threads) and event-driven minimizing the system response time after user input. Seneff et al. (2004) propose a mixed-initiative SLDS for restaurant information retrieval based on the ideas presented in Glass and Seneff (2003). Their system's flexibility is improved by a statistical language model which is trained with the aid of a user simulator generating multiple paraphrases of a sentence, a generic dialogue model, an automatically updating vocabulary and more flexible response generation.

Generic dialogue strategies for domain-independent dialogue management are described by Polifroni and Chung (2002). Their approach enables a dialogue developer to use self-contained dialogue flow functions and to adapt these to the required domain. The top-level strategy coincides with the VoiceXML idea of filling the fields of a form, the implementation, however, differs significantly from the VoiceXML form interpretation algorithm. Its dialogue control is constituted by a set of rules like, e.g., in the air travel domain if destination known and departure city unknown → ask for departure city or if all fields filled → retrieve flight information from database. On the basis of these rules, input and output are canonicalized with the aid of semantic frames. A different approach to dialogue management for the use in multi-purpose dialogue systems like intelligent secretary agents is proposed in Sugimoto et al. (2002). They describe a linguistic resource database called semiotic base following the systemic functional linguistic theory which is included in the linguistic analysis of the user input. Using the analysis' results the dialogue manager maintains a tree-like plan structure with domain and interaction plans which are accordingly selected. The interaction plans range from simple action plans where *"I want to write a report."* opens a word processor program to more complex plans in which the system requests information from the user. A plan-based structure also including trees is the topic forest approach described by Wu et al. (2001). This structure consists of topic trees as shown in Fig. 3.3 which involve a topic node indicating the type, mid nodes describing logical relations between their sub nodes and leaf nodes relating to a dialogue field storing

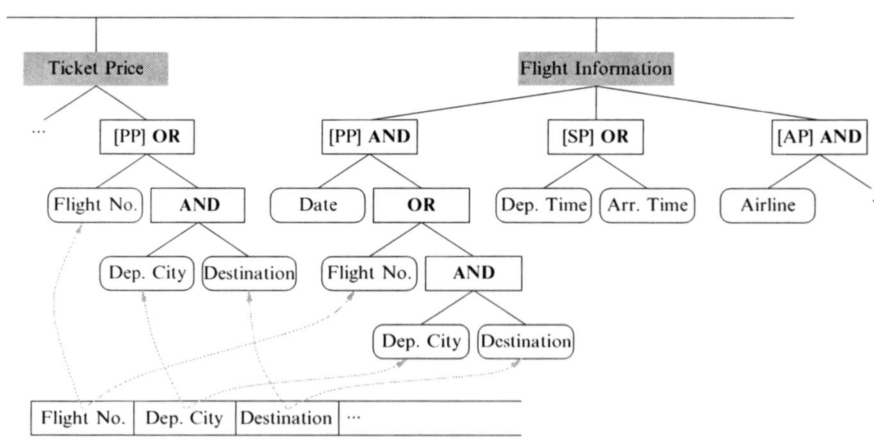

Fig. 3.3 Topic forest structure for plan-based dialogue management (Wu et al. 2001)

the respective information. Moreover, mid nodes may be labeled as either primary, secondary or additional property, depending on their importance within the topic. These nodes are traversed by the dialogue engine to fill the leaf nodes with semantic slots extracted from the user input where appending (storing multiple values for one field) and replacing operations are possible in a mixed-initiative dialogue.

A hybrid approach combining this topic forest with finite-state dialogue control is proposed in Wu et al. (2002). Each topic tree is appended a leaf node storing the current dialogue state. These dialogue states are organized in a state transition network the transitions in which are dependent on database operations or user input. The application of tree-based dialogue management is also described in Macherey and Ney (2003) where the dialogue manager integrates the user input into a knowledge tree based on the filling level of which the next dialogue action is determined. To accomplish that, the authors describe relevant features such as speech recognizer confidence measures, concept filling degrees, application/database feedback, etc.

Jokinen et al. (2002) define the adaptivity of a dialogue system by four basic system properties: an agent-based architecture supporting flexible component communication, natural language capabilities for linguistic analysis as well as for language generation, topic recognition and conversational abilities provided by the dialogue manager. Their dialogue manager consists of agents and evaluators. Each agent corresponds to possible system actions such as asking, informing or requesting confirmation and the evaluators select the agent which best suits the current (dialogue) situation (see also the Jaspis architecture presented by Turunen and Hakulinen 2001). Similarly, task management (interface to database and application) and presentation management (language generation) also consist of agents and evaluators. The whole architecture is illustrated in Fig. 3.4: the interaction manager constitutes the central component surrounded by and cooperating with input/output, dialogue, task and presentation management. On top of all, the information manager serves as a knowledge base to all other components.

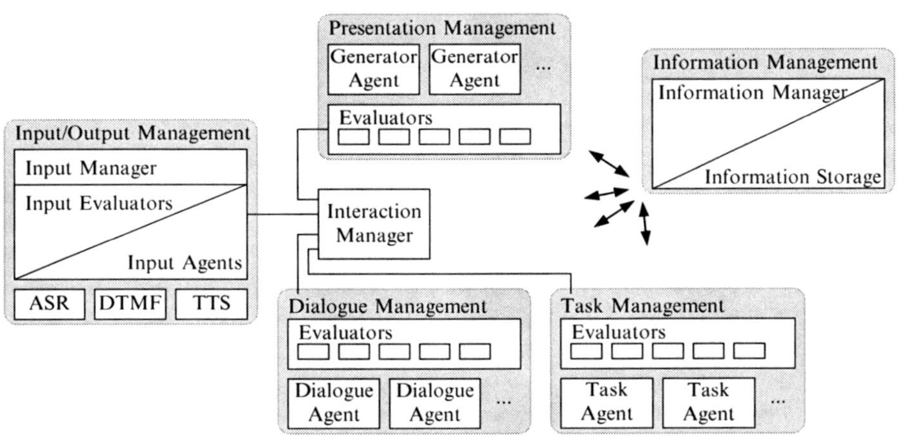

Fig. 3.4 Agent-based dialogue system architecture (Jokinen et al. 2002)

An expanded architecture of a natural multimodal SLDS for information retrieval is proposed by González-Bernal et al. (2002). As an alternative to the task management, they use an information and knowledge manager which handles the database queries by verifying the query's correctness, enriches the query based on domain knowledge or dialogue observations and selects the adequate database or search engine for information access.

An SLDS architecture integrating context tracking and pragmatic adaptation is presented in LuperFoy et al. (1998). Their architecture is star-shaped with the dialogue manager being the central component organizing the information flow from speech recognition via linguistic analysis, context tracking, pragmatic adaptation, application interface, etc., to speech synthesis. Their context tracking module keeps track of the dialogue context to support the interpretation of, e.g., *"this"* or *"the second one"*. The pragmatic adaptation module reasons over the user input and translates the input into application-conform commands. Both modules also operate on the output side translating application responses and bringing the output into a logical form. O'Neill et al. (2003) propose a Java-based dialogue manager which also distinguishes generic and domain-specific behavior in all components. In the center of the object-oriented Queen's Communicator architecture involving 17 classes is the dialogue manager which contains multiple expert classes for generic and domain knowledge and rules as well as a dialogue history class and a domain spotter maintaining the inquiry focus and selecting the appropriate expert class for the current dialogue state. A dialogue server provides an interface to the Galaxy hub which controls the interaction between dialogue system modules. As an add-on to the Queen's Communicator, a dialogue manager integrating dialogue strategy and problem-solving managers is presented in Chu et al. (2005). Their multi-strategy approach allows the selection among finite-state, frame-based and free-form strategies. A frame-based strategy is used to retrieve sets of data. If this strategy fails, or if the system needs to restrict the user's input the finite-state strategy is used. As long as no dialogue objective is established or when the user input is not relevant to the objective, the free-form strategy applies.

An information state approach to dialogue modeling is introduced by Larsson and Traum (2000) in conjunction with the description of the TRINDI (task oriented instructional dialogue) Dialogue Move Engine Toolkit (TrindiKit). They define an information state of a dialogue as the information which is necessary to distinguish this dialogue from other dialogues. Being described by previous actions, this information state motivates future action in the dialogue. Their dialogue model includes informational components and their formal representations as well as dialogue moves triggering updates of the information state on the basis of update rules selected by an update strategy. They propose an information state containing private information such as beliefs, an agenda of actions and long-term plans and shared information like shared beliefs, the latest dialogue move and questions under discussion. Such an information state is represented as a record structure where private and shared information form separate sub-records (Larsson et al. 2002). The fields in the sub-records contain values, sets or stacks together with the type of information like proposition, action, question or move. Dialogue moves are, e.g., "ask",

1 System: *Welcome to your virtual travel agency. How can I help you?*
2 User: *I'd like to book a flight to Munich.*
3 System: *Where would you like to depart from?*

Private =	Agenda	=	< >
	Plan	=	⟨ Raise(D^(date=D)), Raise(T^(time=T)), Respond(A^(availability=A)) ⟩
Shared =	Belief	=	{ (dest.=Munich),(how=plane)}
	QUD	=	< X^(departure=X) >
	Last_move	=	Ask(sys,Y^(departure=Y))

Fig. 3.5 Compact information state representation of a dialogue situation (Larsson and Traum 2000)

"answer", "repeat", "request_repeat", "greet", "goodbye", "quit", etc. The TrindiKit engine's main purpose is the updating of information states with respect to (previous) observed moves and the selection of appropriate moves to be performed. A dialogue example and the according information state are illustrated in Fig. 3.5. This compact representation captures the information state after turn 3, where D, T, A, X and Y represent local variables: according to the user's statement, the system knows (believes) that the destination (Munich) shall be reached by plane. The question under discussion (QUD) is about the departure city which coincides with the last move as this question has already been asked in turn 3. Furthermore, the plan foresees to ask for the travel date and time as well as to provide the user with availability information after all slots are filled.

The strengths and weaknesses of finite-state vs. form-filling (frame-based) vs. the augmented form-filling approach implemented in VoiceXML are described by Zinn (2004). For the use in an intelligent tutoring system, a three-layer structure for response generation in combination with an information state managed by TrindiKit is proposed. The information state avails itself of a knowledge base including domain knowledge, student model, curriculum and tutorial strategies and it controls the deliberative planning component, a sequencer for refinement of the plans, a controller for input (perception) and output (action) for the response generator.

A variety of aspects how and to which extent (spoken) dialogue systems can be adapted is covered by the above approaches. For the implementation of our dialogue manager as discussed in Section 5.3, we adopt the question under discussion idea as used in the TrindiKit architecture (Larsson et al. 2002). Our implementation envisages a hybrid framework adapting TrindiKit ideas while maintaining a simple and understandable dialogue description as provided by VoiceXML. In excess of these rule-based approaches, we focus on the integration of a stochastic component in our dialogue model to increase its flexibility to adapt to different applications or conditions.

3.1.2 Stochastic Approaches to Dialogue Modeling

Originally rather used in automatic speech recognition or linguistic analysis, statistical models meanwhile enjoy great popularity in dialogue management as well. As opposed to rule-based systems which need to be tailored manually to the application requirements, one major advantage of the stochastic approaches lies in their flexibility as the development of new or different applications only involves the collection of suitable training data. This, in turn, implies that the actual performance of these approaches, however, strongly depends on the quantity and quality of the collected training material.

Levin et al. (2000b) describe dialogue design as an optimization problem for achieving an application goal as efficiently as possible, i.e., minimizing the dialogue costs. They define the costs by the number of iterations (dialogue turns), the expected number of errors and the distance from the dialogue goal (i.e., how many fields are not filled in the end). Based on dialogue states s_t and actions a_t at time t, they describe the state change by a Markov Decision Process (MDP) the probability of which is given by $P(s_{t+1}|s_t, a_t)$ and analogously, respective cost c_t is distributed according to $P(c_t|s_t, a_t)$. The optimal strategy is then obtained by supervised learning (to create a user simulator) and reinforcement learning to estimate the optimal strategy on the basis of a sufficiently large number of dialogues between the dialogue system and the user simulator as illustrated in Fig. 3.6. An experiment conducted with the ATIS corpus (see also Glass et al. 1995) shows that the obtained strategies show a strong resemblance with manually created strategies for the same tasks.

Similarly, Litman et al. (2000) propose a method using reinforcement learning to build a dialogue system from a dialogue corpus without such a user simulator. Their approach includes the construction of MDPs from training data and an optimization of a reward function taking into account the state transition probabilities $P(s_{t+1}|s_t, a_t)$ and the associated reward $R(s_t, a_t)$ similar to the dialogue cost. The underlying NJFun dialogue system, implementing an information system for recreational activities in New Jersey, is used to collect dialogue data and is evaluated while implementing the newly learned dialogue strategies.

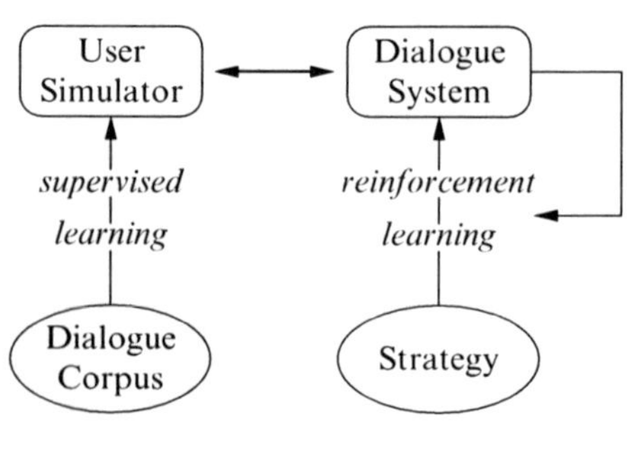

Fig. 3.6 Dialogue strategy learning paradigm as described in Levin et al. (2000b)

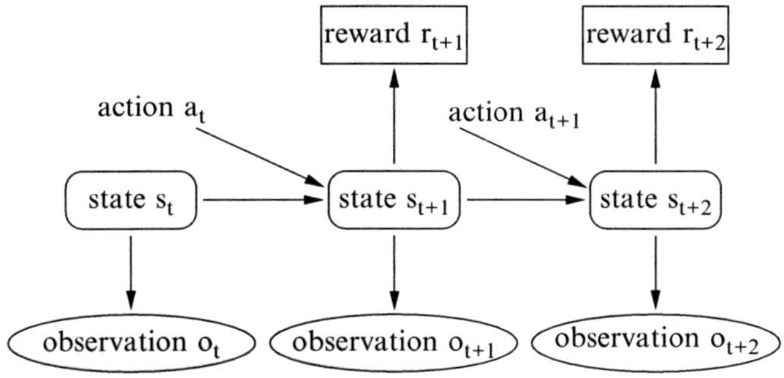

Fig. 3.7 Partially Observable Markov Decision Processes for adaptive dialogue management (Bui et al. 2006)

With respect to the fact that Markov Decision Processes have difficulties with noisy or ambiguous input, the use of Partially Observable Markov Decision Processes (POMDPs, see Fig. 3.7) is proposed in Roy et al. (2000). With this approach, the dialogue state is considered as the state of the user which is not known exactly but which can be inferred by the system on the basis of partial observations from the user input (keywords in utterances). The POMDP consists of an unobservable MDP characterized by states s_t, actions a_t, transition probabilities $P(s_{t+1}|s_t, a_t)$, a set of rewards/costs r_t/c_t and an initial state s_0. Furthermore, a set of observations o_t plus the respective observation probabilities $P(o_t|s_t, a_t)$ are added by what the rewards/costs also include the observations and the initial state is replaced by an initial belief $P(s_0)$. Beliefs include probability distributions over all states indicating the probability that the user is in each of these states. The planning occurs in the belief space where an optimal mapping of actions and belief is determined and where the selected action maximizes the expected reward. Instead of an optimal strategy, the search for which may be rather difficult, a near-optimal (in terms of performance) strategy can be computed faster. The POMDP problem can be simplified by the augmented MDP approach where domain-specific structures of belief states are exploited and a belief state's statistics are approximated by the belief state's entropy (approximated POMDP). Based on a prototype nursing home robot, the authors conduct experiments comparing conventional MDP, approximated POMDP and exact POMDP algorithms which show that for a restricted state space all three algorithms perform well whereat the POMDPs perform better than the MDP and the approximated POMDP is slightly better than the exact POMDP. Considering the full state space, no solution is found for the exact POMDP and the solution of the approximated POMDP outperforms the conventional MDP solution significantly.

A detailed description of POMDPs is given in Williams et al. (2005) and Williams (2006). Their POMDP approach consists of sets of states, actions, transition probabilities, reward observations and their probabilities differing from the above definition by discrete states but continuous observations. To integrate (continuous) speech recognizer confidence measures into the model, the observation is decomposed into a discrete component (words) and a continuous component (confidence measures). Experiments with a testbed dialogue management problem show that their proposed model shows a significant improvement compared to standard MDP approaches.

The reusability and adaptivity potential of strategies determined with reinforcement learning is challenged in Lecœuche (2001). Instead of reinforcement learning, the author proposes inductive logic programming to learn rule sets expressing and generalizing an optimal strategy. The resulting rules are more explicit and comprehensible in contrast to the decision tables generated by reinforcement learning and may be reused in different dialogue situations without having to learn a new strategy from scratch.

An alternative approach is described by Torres et al. (2003) who propose a semantic-driven approach including a bigram model using previous system and user dialogue acts to predict the following dialogue acts. Their system shows an average success rate of 80% despite the lack of sufficient training material.

The selection of the system answer can be also considered as a classification process based on the user input and the previous system turn. Hurtado et al. (2006) use neural networks which are trained on transcribed dialogue data. They consider a dialogue as a sequence of pairs s_t of system answer a_t and user turn u_t at the respective time t. To reduce the complexity, they introduce a what they refer to as dialogue register d_t which captures the current cumulated state (summarizing all previous a_ts and u_ts) regardless of how and in which chronological order this state has been reached. With this, they describe the selection of the following system answer a_{t+1} by maximizing the probability $P(a_i|d_t, s_t)$ over all possible system answers a_i. This search is accomplished with the aid of a multi-layer perceptron which classifies the previous user input and, thus, determines the associated system output.

Stochastic models typically involve a trade-off between complexity and accuracy. On the one hand, a high level of accuracy requires a higher complexity. On the other hand, a higher complexity requires more training data and involves a higher computational effort. For dialogue models like those discussed above, the complexity is not only dependent on the number of model parameters but also on the number of previous dialogue turns. Whereas (PO)MDPs only consider the previous dialogue turn, a more accurate model would include the whole dialogue history in the determination of the next dialogue state. One approach to solve this problem is the cumulative dialogue register by Hurtado et al. (2006) which, however, does not capture the exact order of turns and the cumulation process of which requires further effort. Our approach follows the observations in automatic speech recognition where bigrams and trigrams are commonly used for accurate stochastic language modeling. Accordingly, we use bi-turns and tri-turns involving the previous and the penultimate state to increase the model accuracy while maintaining a justifiably low model complexity.

Three key aspects of the dialogue system behavior adaptation with respect to the user's emotional state are summarized in Polzin and Waibel (2000), namely addressing prompting, feedback and dialogue flow adaptations. Their approach to emotion-sensitive dialogue management is able to change the prompting and feedback style (e.g., apologetic prompts for annoyed users, succinct prompts for users in a hurry, explicit feedback for frustrated users) as well as the assignment of different dialogue flows (specified beforehand) to cater for special needs in conjunction with the user's emotional state.

3.1.3 Emotions in Dialogue Systems

Holzapfel et al. (2002) propose the integration of emotions into multidimensional typed feature structures, which do not only contain semantic information but also additional information describing the speaker and situation. Accordingly, their dialogue state is characterized by seven variables including emotion type, speech act type, user's intention and confidence measures. They discretize the valence-arousal space into four categories plus a no-emotion residue class and bring together these types according to the OCC model. For the handling of emotions in, e.g., robot interaction, they propose a strategy operating in the seven-dimensional value space of the state variables. This strategy also decides how to interpret emotions, e.g., considering anger as a reaction to system failure.

Brown and Levinson (1987) discuss the influence of affect and politeness on linguistic style which is picked up by Walker et al. (1997a) to endow artificial agents with personality. They propose linguistic style improvisation to render these agents socially oriented and, thus, more credible. Their theory bases on speech acts for the abstract representation of utterances and plans for the improvisation. Variations are possible in the semantic content, the syntactic form and the acoustical realization. The strategy to realize a certain intention is chosen with respect to three parameters: social distance between user and system $D(S, U)$, power of the user over the system $P(U, S)$ and a ranking of imposition R (low for good news like acceptance, high for bad news like rejection) of the current speech act. These social variables are summed to determine the face threat θ to the user given by the speech act. Depending on the value of the threat, the agent chooses among four strategies: doing the speech act directly, orienting the act to the user's desire for approval, orienting the act to the user's desire for autonomy or pursuing an off-record strategy including hints or ambiguities. The option of not executing the speech act if the face threat is too high is overridden in conjunction with spoken language dialogue systems. This approach allows the inclusion of emotions as an orthogonal dimension to social variables.

Apart from the problem of retracing and verifying solutions proposed by the users, intelligent tutoring systems used for computer aided instruction of, e.g., students also need to be able to motivate their users to pursue taking part in and learning from the tutorial. An emotional model incorporating motivational cues for these systems is presented in Lopes Rodrigues and Carvalho (2004). Their emotional structure distinguishes primary, secondary and tertiary behaviors which all influence the temperament but which particularly influence the environment (primary behavior), teaching strategy (secondary) and the instant action

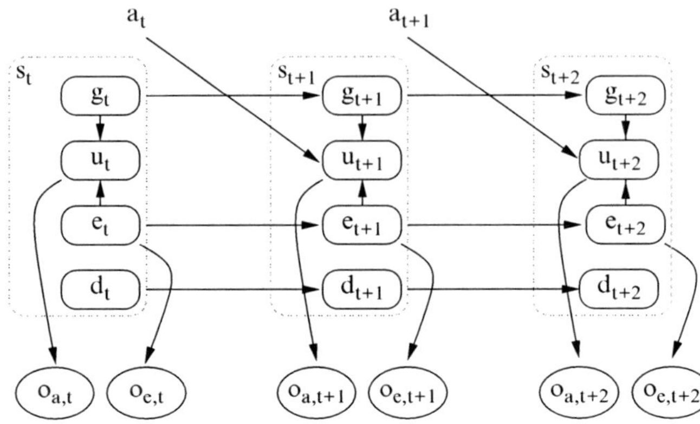

Fig. 3.8 Partially Observable Markov Decision Processes for affective dialogue modeling (Bui et al. 2006)

(tertiary). These also determine what to adapt to keep a user motivated whereas the temperament controls how this adaptation takes place. The actual strategies are defined by rule sets integrated into a tailored teaching strategy description language.

A stochastic dialogue model taking into account the user's emotional state is proposed by Bui et al. (2006). They extend the POMDP approach with an affective component. Thereby, a state s_t is broken down in the user's goal g_t, the user's emotional state e_t, the user's action u_t and the user's dialogue state d_t. An observation now consists of the observed user's action $o_{a,t}$ and the observed user's emotion $o_{e,t}$ and the reward r_t is omitted in their generalized model. These sub-state features and observations are integrated with the aid of a two time-slice Dynamic Bayesian Network as illustrated in Fig. 3.8. According to their model, the users' actions are influenced by their goals, their emotional states and the system's actions.

In our approach to adaptive dialogue management, we act on the ideas proposed by Polzin and Waibel (2000) which include prompting, feedback and dialogue flow adaptations. To accomplish that, we follow two ideas – a rule-based adaptation as described, e.g., in André et al. (2004) and a semi-stochastic dialogue model involving the MDP idea by Levin et al. (2000b) and, to some extent, the affective POMDP approach by Bui et al. (2006). However, we argue that although the whole affective POMDP idea seems quite suitable for emotion-based dialogue adaptation, it is too complex as measured by the actual purpose and does not seem to be scalable at first glance. Thus, in the remainder of this chapter we propose a semi-stochastic dialogue model consisting of states which represent fields, emotional characteristics and optionally further parameters in all possible combinations. The states in this predefined (rule-based) scaffolding are then related by transition probabilities which are determined on the basis of training dialogue data. Extending the transitions from plain (bi-turn) state-to-state transitions to what we refer to as tri-turn transitions, the model becomes more robust. E.g., assuming appropriate training material and applying plausibility checks on the chronological sequence of events, user input and detected emotional states in the dialogue history, the model is able to handle noisy (error-prone) emotion recognizer output (see also Sections 3.6–3.9).

3.2 User-State and Situation Management

In principle, we distinguish two basically different types of parameters based on which the dialogue manager adapts the interaction: user-state parameters which are closely connected with the current user on the one hand, and situation-based parameters which, independently of the user, influence the dialogue on the other hand.

An overview on user-state parameters is given in Fig. 3.9. Within the user-state parameters, we differentiate between three sub-categories:

- A user profile contains general information about the respective user. These details like name, address, etc., are typically stored in a central database. As soon as the user is identified in or before the dialogue, relevant information is retrieved from this database. I.e., if the user can not be identified, the appropriate data is queried from the user during the dialogue. Otherwise, the system uses the values from the database and does not query the user, except for an optional confirmation prompt like *"Would you like us to send the train tickets to 'Albert-Einstein-Allee 43, 89081 Ulm, Germany'?"*.

  During the dialogue, these parameters remain to greater or lesser extent untouched, e.g., it is not very likely that users change their names during the interaction. By that, the data in the user profile is mainly employed to accelerate the dialogue.

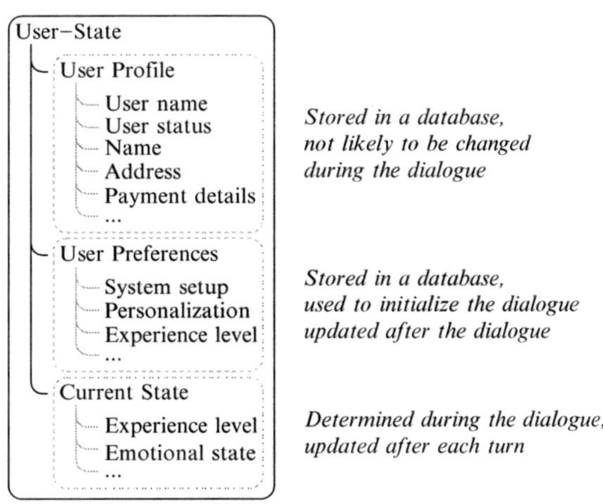

Fig. 3.9 Allocation of user-state parameters and overview on the handling of the different parameter types in human–computer interaction

- The user preferences are defined by the user and involve parameters how he/she wants to interact with the system. These parameters range from system settings like language or system voice (female/male) to personalization features and to the user's experience level.

  User preference parameters are gathered either by different means: either the user selects the preference in a setup dialogue or in an external (graphical) user interface, or the system requests the parameters explicitly, or the system determines the parameters by observation and automatic learning during the dialogue. The preferences may also be stored in a database, i.e., they are retrieved from the database to initialize the dialogue and the database is updated after the dialogue has finished.
- The user's current state is determined in each turn of the dialogue by means of signal analyses and dialogue measures. Among these parameters are the user's (current) experience level and emotional state. These parameters are not necessarily stored in a database but are rather contained in an extended dialogue history based on which parts of the user preferences are updated. As the parameters change within the dialogue, it is possible to adapt the dialogue flow to them.

The transitions between the sub-categories, however, are sometimes seamless as can be noticed for the user's experience level: On the one hand, the experience level may be regarded as a user preference parameter as it needs to be stored for further dialogues, e.g., a user who has used a dialogue system often enough would not be satisfied if he was treated as a novice at the beginning of each new dialogue. On the other hand, the experience level can be determined during the dialogue by means of turn measures such as number of no-input turns (pauses, i.e., the user does not reply to a prompt for a predefined time) or no-match turns (user turns the content of which does not match any rules/classes in the linguistic analysis). Thus, the prompt strategy is adapted to the (current) experience level after each turn and the (overall) experience level is stored in the user preferences after the dialogue so that the subsequent dialogue can be initialized with the suitable, i.e., the most recent, level.

As mentioned above, the user profile is mainly used to keep the number of dialogue turns, i.e., the overall time of interaction, which constitutes a "cost" measure in the evaluation of a dialogue system, as low as possible. The fields which would be normally queried by the system are filled from the database whereby also further misunderstandings are avoided which are especially likely to occur in the interrogation about names or address data.

The predominant purpose of user preference parameters is to make the user feel more comfortable with the dialogue system. This can be compared to the look and feel of a graphical user interface, where themes and skins are employed to provide a "familiar" interface. Here, it includes, e.g., the choice of the computer voice – depending on the application, some people like a female voice, other people prefer a low male voice. Furthermore, it can be defined how the computer addresses the user, e.g., *"Hello buddy, what's up?"* or *"Good morning, Professor, how may I help you today?"* and/or the level of the user's experience is included in the dialogue flow. A novice user who has not used the system before would be relatively helpless without any explanation what he is able to do and say in the interaction. In contrast, an experienced user would be rather annoyed when welcomed with a long explanation text like, e.g., *"Good morning and welcome to UTA, the automated air travel information and booking system. This system enables you to retrieve all kinds of information about flights on the dates which you specify, to manage your existing itineraries and to book flights. You may say, e.g., 'I want to book a flight from New York to Paris on Friday', 'Show me all flights to London on Thursday' or 'Open Itinerary number 2 4 2 5 6 7'. If you want to talk to an operator you can say 'Operator' at any time. Is there anything I can do for you?"*. Instead, a short *"Good morning, Sir, how can I help you?"* would be sufficient in this case. In addition to personal user preferences which are typically only stored for users who have already used the system, default values are defined for novices: E.g., this could be an "absolute novice" experience level, a female voice and an impersonal form of address like in the long prompt above.

These parameters can be considered as global parameters used as a rough estimate for the adaptation of the dialogue. The fine-tuning adaptation then takes place according to the user's current state which is determined on a turn-by-turn basis. E.g., assuming the user has already often used the system and is therefore considered as an advanced user, the system initiates the dialogue at the advanced level, i.e., with short prompts. However, if repeated no-inputs or no-matches occur during the interaction, the prompts are adapted immediately, and after the dialogue the experience level in the user preferences is downgraded for further use. A further aspect which we consider as an important factor within the user-state parameters is the user's emotional state as this has a considerable impact on the success of the dialogue and, before that, on the organization of the dialogue flow. In Chapter 4, we discuss different approaches to the recognition of emotions from the speech signal. In our considerations, these will serve as the predominant user-state parameters to be integrated into adaptive dialogue management.

Within the situation parameters, there exist also different categories as illustrated in Fig. 3.10. We address four of these, namely:

- Type of interaction and system, e.g., in a car, at home, information kiosk at an airport or train station, or telephony-based service (call center). The application type (information retrieval vs. booking system vs. customer tests or even job interview) also contributes to the situation, i.e., how important is the outcome of the interaction to the user.
- Surroundings – noisy or quiet environment, many people or few people around the user. This also includes the question whether there are other people waiting and somehow whether the user is under stress (time pressure).
- Sensor data: This is particularly relevant in automotive environments, where the data from the anti-lock braking system (ABS), electronic stability program (ESP) or distance sensors is evaluated to determine the driving conditions and thus the cognitive stress level for the user, i.e., the driver.
- Conflicts may arise from the interaction with multiple applications, e.g., in an integrated calendar, navigation and communication system in a car, such conflicts may occur when the user is not able to keep an appointment due to traffic

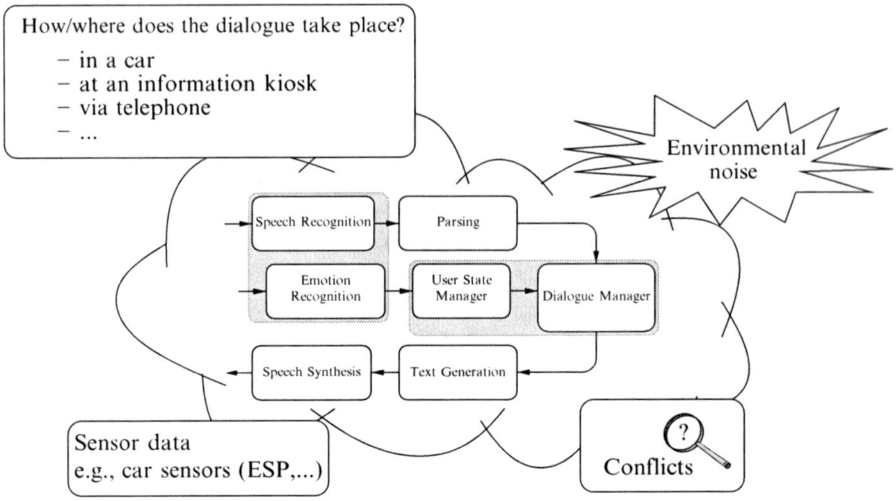

Fig. 3.10 Allocation of situation parameters in SLDSs

delays. The conflicts need to be resolved in the interaction between the user and the applications which calculate the new estimated time of arrival, integrate the new appointment in the calendar and communicate the changes to the other participants in the meeting.

Among these four sub-categories we distinguish between "given" situation parameters which are unchangeable and have to be accepted and "variable" situation parameters. The unchangeable parameters comprise the surroundings and the type of interaction which is typically defined by the application. Thus, the system and the dialogue need to be designed to account for these parameters during the entire dialogue. E.g., if users want to buy train tickets or retrieve timetable information at a train station where people are typically in a hurry, the dialogue system should be designed in such a manner that the interaction time is as short as possible. But, on the other hand, as places like train stations are likely to be quite noisy, the dialogue should also be constructed in an appropriate way that compensates robustness problems and misunderstandings of the automatic speech recognizer.

The variable parameters, determined by sensors or other modalities and issues from conflict management, in contrast, are very likely to change between or within the dialogue(s). A typical scenario can be seen in the automotive field. E.g., a user needs to schedule an appointment in a city which is 200 km away from his current location. With respect to the fact the user has another appointment later on the same day in another city, the navigation system determines the shortest route and a driving time of two hours so that the first appointment can be fixed accordingly after consultation with the user. If the user is caught up in a traffic jam on the way to the first appointment, a conflict arises if the user is not able to reach the first destination in time. This conflict is detected in the interplay of navigation system and calendar, and the reasoner or problem solving assistant is in charge of communicating the inferences to the user so that the conflict may be resolved in the interaction (Bühler and Riegler 2005):

1 System: *We won't be able to reach Munich in time due to this traffic jam. Would you like me to postpone the appointment or shall we go to Regensburg first?*
2 User: *How far is it to Regensburg from here?*
3 System: *It is about two hours driving time.*
4 User: *Then we go to Regensburg first.*
5 System: *OK, I will check with your partners...*

In the following sections we will discuss the integration of user-state and situation parameters into a framework of adaptive dialogue management. As a situation parameter it is imaginable to choose noise, i.e., the quality of the speech recognizer output. With respect to the focus of this work, we concentrate on emotions as the predominant user-state parameter.

3.3 Dialogue Strategies and Control Parameters

The variety of user-state, situation and further parameters suggest to summarize and to store these parameters for the use in adaptive dialogue management. Thus, we propose an approach in which we combine these parameters into the superset called *dialogue control parameters* (Pittermann et al. 2005). For dialogue systems which need to handle an arbitrary number N of dialogue control parameters p_i, $1 \le i \le N$, we suggest to combine these parameters into one single N-dimensional vector $\mathbf{P}_C(t)$ at discrete time t.

Together with the respective user input, all $\mathbf{P}_C(t)$ $t > 0$ form the comprehensive dialogue history $\underline{\mathbf{H}}_D$.

As opposed to the common definition of a dialogue turn, in the remainder of this chapter, we consider a turn as *one* semantic representation or its corresponding dialogue field and the respective control parameters. The function of the dialogue history and the attribution of the turns is illustrated with the following dialogue excerpt. Words the semantic representations of which correspond to dialogue fields are underlined. Further control parameters (regardless of their meaning) are given in square brackets.

1 System: *Good morning and welcome to UTA, your virtual travel agency. How can I help you?*
2 User: *I would like to travel <u>to London</u> please* [0.37 –20]

Analyzing this utterance, the content of this user's reply matches one semantic representation (i.e., the destination field), so that the dialogue history now is

$$\underline{\mathbf{H}}_D = \left[\text{destination} = \text{London } 0.37 \ -20 \right].$$

In the following part of the dialogue, the user's reply contains values for two dialogue fields:

3 System: *From where would you like to depart?*
4 User: *From Paris, if possible on Monday* [0.10 –17]

This reply is split up into two turns, both with the same control parameters, so that the dialogue history evolves to

$$\underline{\mathbf{H}}_D = \begin{bmatrix} \text{destination} = \text{London } 0.37 \; -20 \\ \text{departure\_city} = \text{Paris } 0.10 \; -17 \\ \text{date} = \text{Monday} \qquad 0.10 \; -17 \end{bmatrix}.$$

In this chapter, we address the formation of the dialogue flow and the system reaction according to the probability $P(o_S(T)|\underline{\mathbf{H}}_D)$ of the respective system output $o_S(T)$ given the history $\underline{\mathbf{H}}_D$. To accomplish that, we describe two approaches which, in analogy to the language model $P(W)$ of a speech recognizer, encompass rule-based methods and stochastic (corpus-driven) methods. These shall be discussed in the following sections. Limiting the history to semantic labels $p : p(1), \ldots, p(T)$ and emotions $e : e(1), \ldots, e(T)$ and assuming these parameters are independent, we can split up the holistic view of $P(o_S(T)|\underline{\mathbf{H}}_D)$ into

$$P(o_S(T)|\underline{\mathbf{H}}_D) \Rightarrow \begin{cases} P_p(o_S(T)|[p(1), \ldots, p(T)]) \\ P_e(o_S(T)|[e(1), \ldots, e(T)]) \end{cases}, \tag{3.1}$$

or suitable combinations of these parameters. In the following sections, we describe approaches to combine dialogue and emotions into a combined emotional dialogue model $P_pe(o_S(T)|[p(1), \ldots, p(T), e(1), \ldots, e(T)])$.

3.4 Integrating Speech Recognizer Confidence Measures into Adaptive Dialogue Management

Within the past years of research and development in the field of automatic speech recognition, a significant improvement could be achieved, not only with regard to vocabulary size and grammar flexibility but also concerning the recognition performance in terms of word and sentence accuracy (Lefèvre et al. 2001). Despite the improvements among ASR systems, there exist scenarios where the stochastic models perform poorly, e.g., in noisy environments like cars (Schmidt and Haulick 2006), outside or in crowded rooms/buildings. Further errors (dropouts) can occur with persons who have "non-standard" voices, who don't know how to pronounce certain words (Henton 2005) or who speak unclearly and with different dialects. Also, in telephony-based dialogue applications, one has to bargain for transmission errors, especially when the users call from their mobile phones.

In these cases, dialogue developers typically integrate confirmation prompts into the dialogue flow, urging the user to verify whether the input recognized by the system is correct. As described in Section 1.3, these verifications can be realized as explicit or implicit confirmations. Explicit confirmations constitute an independent system turn like

5 System: *I understood you wanted to travel on Sunday. Is that correct?*

limiting the user's reply to *"yes"*, *"no"* or variations like *"sure"*, *"nope"*, etc. Implicit confirmations are embedded in a regular system like

5 System: *At what time on Sunday would you like to depart from London?*

leaving it to the user whether to reply to the actual question (in case the user acknowledges *"Sunday"* and *"London"*), e.g., *"In the afternoon."* or to rectify the misrecognized items, e.g., *"On Monday."* or *"From Oslo."*.

With regards to dialogue cost and efficiency aspects, however, dialogue duration and thus the number of turns shall be kept as low as possible. I.e., it is not optimal if a user who is well understood by the system is excessively prompted for confirmation.

Bridging the gap between dialogue efficiency and robustness, (Litman and Pan 2002) propose an adaptive version of the TOOT train information system, which selects its confirmation strategy and dialogue initiative according to the confidence measures c provided by the speech recognizer, either for individual words or for the entire utterance.

A threshold θ which represents the confidence measure for which the predicted percentage of misrecognitions exceeds an arbitrary number is determined by machine learning. Each utterance, the confidence measure c of which is below that threshold is considered as "bad" and after a predefined number of "bad" turns, the system adapts to a more conservative mode. The system is initialized with user initiative and no confirmations. Then, with an increasing probability of errors, the system switches to mixed initiative or even system initiative and integrates implicit confirmations and explicit confirmations each time c or its weighted average $\bar{c}$ falls below the threshold θ.

The major advantage of adapting the dialogue initiative is the limitation of the input vocabulary which leads to a more robust speech recognition. In the worst case, the system could ask the user to spell the words, e.g.:

7 System: *Sorry, I could not understand you. Please spell your desired destination.*
8 User: *LONDON.*

Here, the vocabulary consists of 26 words ("A" to "Z") and the grammar is defined by the list of available cities, so that the recognition process performs significantly better. This, in turn, leads to the problem that the confidence parameters in the different initiative levels are not comparable. I.e., the confidence measures in the user initiative level are much lower than in the system initiative level although the situation itself (noise, unclearly speaking user, etc.) has not necessarily improved. By that, the adaptation can only occur as a one-way process, as there is no indicator on the actual situation.

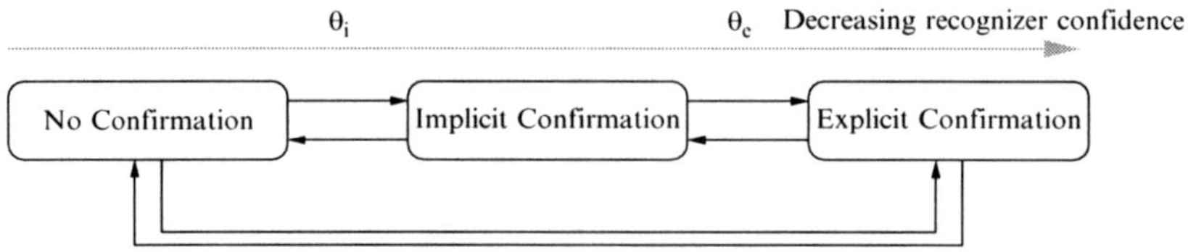

Fig. 3.11 "Two-way" adaptation of dialogue initiative and confirmation strategy (see Pittermann et al. 2007c)

Maintaining the initiative level as mixed initiative, this adaptation can also take place in two ways. I.e., once the recognition performance increases, e.g., due to lower background noise, the system returns to a less time-consuming and annoying confirmation strategy (see Fig. 3.11, Pittermann et al. 2005). Instead of one threshold θ, two thresholds θ_i and θ_e are used. With these thresholds, we distinguish three cases:

- $\bar{c} \geq \theta_i$: no confirmations, except for "standard" confirmations explicitly included in the dialogue.
- $\theta_e \leq \bar{c} < \theta_i$: implicit confirmations. These are included in all system turns (except for the standard confirmations), either by simply stating *"I understood you wanted to travel to London. When do you want to depart?"* or, more sophisticated, *"When would you like to depart to London?"*.
- $\bar{c} < \theta_e$: explicit confirmations.

The inclusion of explicit confirmations can be accomplished in different variations. These can range from the sporadical inclusion after an arbitrary number of user turns (relaxed strategy) to the (most conservative) case where the user is prompted for confirmation until each item that has been filled or changed is explicitly confirmed. For a successful confirmation, the user can acknowledge either by saying *"Yes."* or by repeating the previous item.

To include all types of explicit confirmations, e.g., in cases where the developer can not commit himself to one type, further thresholds $\theta_{e,1}$ to $\theta_{e,n-1}$ can be included, where n is the number of gradation levels, defining regions $\bar{c} < \theta_{e,n-1}$ (most conservative), $\theta_{e,n-1} \leq \bar{c} < \theta_{e,n-2}, \ldots, \theta_{e,2} \leq \bar{c} < \theta_{e,1}$ and $\theta_{e,1} \leq \bar{c} < \theta_e$ (least conservative, but explicit confirmations).

Alternatively, instead of a rule-based approach, which is rather inflexible in terms of adaptability to different situations, setups and speech recognizers, we will now describe our stochastic, corpus-based approach to integrate speech recognizer confidence measures into the confirmation strategy selection in adaptive dialogue management. A straightforward approach can be adopted from the above rule-based approach by calculating a weighted average $\bar{c}$ of the confidence measures and the confirmation strategy S is chosen among the model

$$P(S|\bar{c}) \Rightarrow \begin{cases} P(n|\bar{c}) \\ P(i|\bar{c}) \\ P(e|\bar{c}) \end{cases}, \tag{3.2}$$

where the probabilities $P(n|\bar{c})$, $P(i|\bar{c})$ and $P(e|\bar{c})$ are given for no confirmations, implicit confirmations and explicit confirmations. For the sake of overview, we do not consider the different types of explicit confirmations.

The probabilities are derived from training data consisting of the users' utterances, the recognized text (the relevant semantic representations are underlined) and the recognizer confidence measures.

Such a corpus may consist of recordings of existing human–computer dialogues, Wizard-of-Oz recordings or distilled human–human dialogues (see Jönsson and Dahlbäck 2000). Eventual confirmation prompts and the users' replies to these are omitted from the original dialogues in order to keep the annotators uninfluenced by these. Based on the objective speech recognizer performance, regardless of the actual values of the confidence parameters, the annotators decide whether a user utterance needs no confirmation (n), implicit confirmation (i) or explicit confirmation (e). E.g., in cases where the speech recognizer output perfectly matches the user's input, no confirmation is necessary and in cases where the important (underlined) information is not correctly recognized, typically explicit confirmation is required. The requirements for implicit confirmations are somewhere in between, i.e., the annotators can also decide according to the actual and previous utterances where applicable.

On the basis of the training data, probabilities $P(n|c)$, $P(i|c)$ and $P(e|c)$ are calculated for each value c of the confidence measure. Employing a large dialogue corpus, a higher resolution can be achieved so that probability curves can be interpolated. Such an exemplary probability distribution is shown in Fig. 3.12.

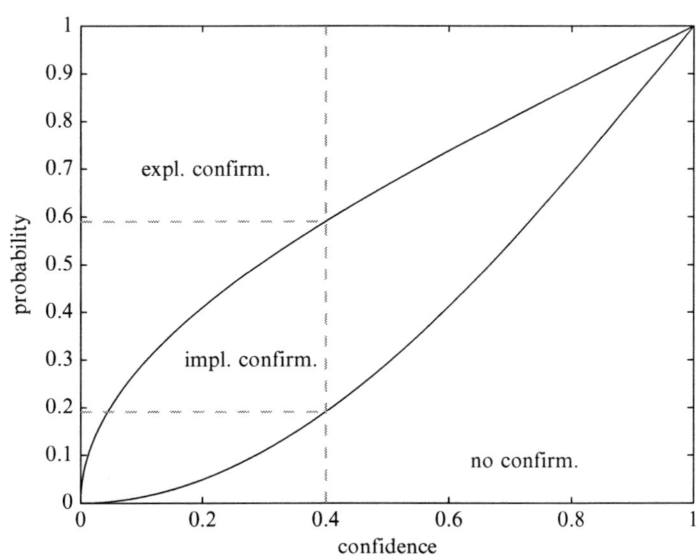

Fig. 3.12 Probability distribution based on processed dialogue data for explicit, implicit and no confirmations for one confidence measure parameter $c = 0.4$. The probabilities are determined for discrete confidence measure values and the curves are interpolated between these points

The curves are constructed such that there are three continuous zones. These three zones for no confirmations, implicit confirmations and explicit confirmations are separated by solid black lines The lower line separating no confirmations and implicit confirmations can be described by $P(n|c)$ and the upper line between implicit and explicit confirmations follows $1 - P(e|c)$. Confidence measures are plotted against the x-axis and probabilities are plotted against the y-axis. The probability of a certain confirmation strategy at a certain confidence measure is represented by the height of the respective zone at the respective confidence measure. E.g., for $c = 0.4$, the probabilities can be read off from the intersection points of the vertical dashed line with the solid lines. Following the horizontal dashed lines from the intersection point, the probability of explicit confirmations can be determined as $P(e|0.4) \approx 1 - 0.59 = 0.41$, $P(i|0.4)$ is approximately $0.59 - 0.19 = 0.40$ and the probability of no confirmations is $P(n|0.4) \approx 0.19 - 0.0 = 0.19$.

In order to determine which confirmation strategy S shall be applied, two approaches can apply. Either the maximum probability for a confidence measure $\bar{c}$ needs to be found as

$$S(\bar{c}) = \arg \max_{s \in \{\text{no,impl,expl}\}} P(S = s|\bar{c}), \tag{3.3}$$

which, however, has the disadvantage that for some curves, certain confirmation strategies might not be included as their probabilities are not high enough throughout the diagram. E.g., in curves like the one shown in Fig. 3.12, implicit confirmations would never occur. Alternatively, the strategy is selected randomly with accordance to the probability distribution of a random number r. This distribution can be directly obtained from a diagram as shown in Fig. 3.12.

A more sophisticated and more generalized approach taking into account the history of confidence measures considers the probabilities

$$P(S = s|\underline{\mathbf{H}}_D) = P(S = s|\mathbf{c}) = P(S = s|[c(1), \ldots, c(T)]). \tag{3.4}$$

This gives the advantage that outliers in the confidence values can be neglected if the surrounding values look reasonable. E.g., an utterance can be correctly recognized although its confidence measures are very low or, vice versa, there can be numerous misrecognitions despite a very high confidence. The one-dimensional case taking into account only one value for c has been described in the previous paragraphs. In accordance with the terminology used in statistical language modeling we refer to this case as "bi-turn" model as the current turn only depends on the previous turn. Analogously, we refer to the two-dimensional case as "tri-turn" model and to any other N-dimensional case as "$N+1$-turn" models. Basically, the probabilities can be trained with the same dialogue corpus as described above. The annotators decide on n, i and e depending on the comparison of speech recognizer output and actual utterance. Based on the labeled data, again, probabilities are calculated and the confirmation strategy is randomly selected with respect to the probabilities.

Being strongly dependent on random processes, this approach requires a more complex surveillance than rule-based approaches. This is particularly important to keep the number of confirmation prompts at a reasonable level with respect to the user-friendliness and the overall recognizer confidence measures. On the one hand, it is vital that the system ascertains whether it recognized the user correctly, on the other hand, however, the user shall not be annoyed by (eventually unneeded) numerous confirmation prompts. There exist several options to accomplish that, especially in cases where the speech recognizer performance is expected to be low (unclearly speaking user, noise, etc.):

- A predefined number (e.g., three) of consecutive items which require confirmation are pooled in one confirmation prompt, e.g.:

2 User:	*I want to travel to London tomorrow.* [0.189]
ASR:	*I want to go to Paris* (explicit confirmation required)
3 System:	*When do you want to depart?*
4 User:	*Tomorrow.* [0.553]
ASR:	*Tuesday* (explicit confirmation required)
5 System:	*At what time do you want to depart?*
6 User:	*At seven p.m.* [0.134]
ASR:	*Eleven a.m.* (explicit confirmation required)
7 System:	*I understood you wanted to travel to Paris on Tuesday at eleven a.m. Is that correct?*
	...

- The probability distributions are adapted to the number of recent confirmation prompts. I.e., if there has been no confirmation prompt within the past turns, the probabilities for implicit and explicit confirmations increase, if there have been more explicit confirmations than average within the past turns, the probability for explicit confirmations decreases.

Referring to the second option, Fig. 3.13 shows the adaptation of the probabilities depending on the number of confirmations in the past turns. Here, at the beginning of the dialogue, the probabilities are initialized with the trained distributions, e.g., as depicted in Fig. 3.12. If the number of explicit confirmation after a fixed number of turns is above average, the probability distribution curves tend to the extreme curves shown in Fig. 3.13 (a).

Here, the curves are shifted such that the probability of explicit confirmations decreases significantly. I.e., except for cases of very low speech recognizer confidence measures, the probability of explicit confirmations is very low. The probabilities of implicit confirmations remain more or less the same and the probabilities for no confirmations increase accordingly. If, contrariwise, the number of confirmation prompts is below average, the probability distribution curves tend towards the curves shown in Fig. 3.13 (b). Analogously, here, the curves are shifted such that the probability of explicit confirmations increases.

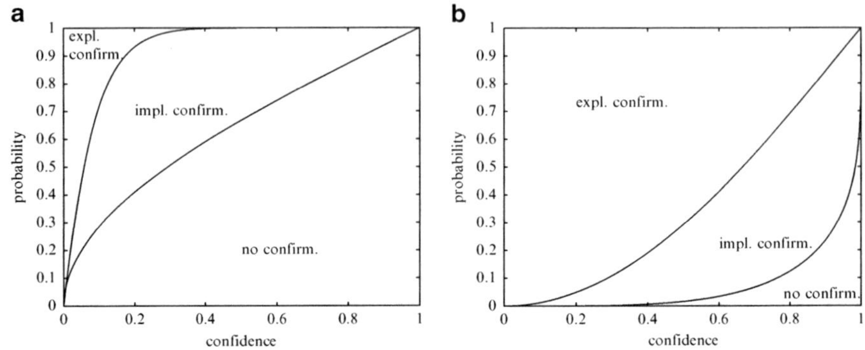

Fig. 3.13 Adaptation of probability distributions for the different confirmation strategies with respect to the number of previous confirmation prompts in situation where the speech recognizer performance is expected to be low. The initial probability distribution is the same like the one depicted in Fig. 3.12, with an increasing number of explicit confirmations in the previous turns, the initial curves tend to the curves depicted in the left figure (**a**), in cases where there haven't been any explicit confirmations in the past turns, the initial curves tend to the curves depicted in the right figure (**b**)

In state-of-the-art spoken language dialogue systems, which apply in varied call centers, a contingency plan stipulates that the user is connected to a human call center agent in cases where the (automated) dialogue is very likely not to finish as expected (see also Section 3.5). This can be the case, e.g., when the speech recognizer experiences severe problems due to unclearly speaking users or background noise (very low signal-to-noise ratio) and, thus, when even sophisticated confirmation strategies fail.

3.5 Integrating Emotions into Adaptive Dialogue Management

As discussed in Chapter 2, emotions are difficult to categorize, and therefore, difficult to handle in the user state and dialogue management. Firstly, the problems involve the choice of emotions which shall be recognized and considered. As shown in Section 2.2, there exist several emotion theories defining different sets of primary and secondary emotions some of which are overlapping in different theories and some of which are unique in other theories.

To make the variety of emotional states more utilizable, we pick up the idea of representing emotions by some numerical value as described in Section 2.2. One approach considers a two-dimensional vector containing numerical values for valence and arousal of an emotional state. The positions of selected emotions in the valence-arousal space are depicted in Fig. 3.14.

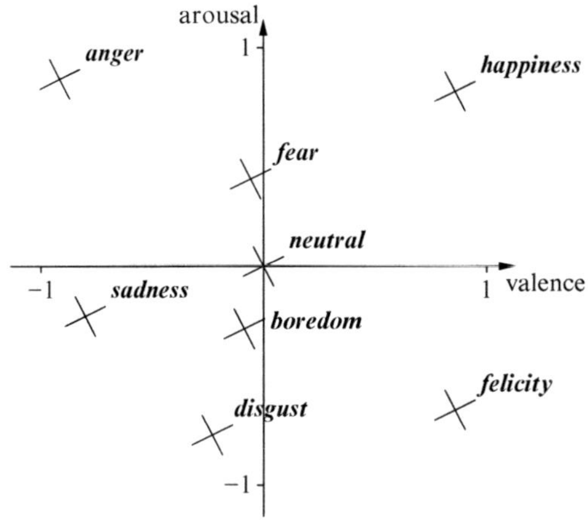

Fig. 3.14 Valence-arousal space representations of selected emotions. Due to the subjectivity in the different emotion theories and the uncertainty introduced by the emotion recognizer the cross markers do not represent the exact position of the emotions in the valence-arousal space, but give an indication of the region where the emotions are located

```
...
<field name="date">
  <prompt cond="valence==0.9 && arousal==0.8">
    Oh, that sounds fantastic. And when would you like to
    depart?
  </prompt>
  <prompt cond="valence==0 && arousal==0">
    When would you like to depart?
  </prompt>
  <prompt cond="valence==-0.8 && arousal==0.8">
    I am sorry to bother you again, but when did you say you
    wanted to depart?
  </prompt>
...
```

Fig. 3.15 Adapted excerpt of our VoiceXML dialogue description applying a simple approach to include valence and arousal values as dialogue control parameters

Limiting the range of values to $[-1, 1]$ for both valence and arousal, we find the neutral state at $(0, 0)$, happiness features a very positive valence and a high arousal and, thus, is situated, e.g., around $(0.9, 0.8)$. At a very negative valence but still high arousal, we find anger, and at low values for arousal, we find sadness (negative valence) or felicity (positive valence).

Using this two-dimensional vector as a dialogue control parameter, we are already able to construct simple mechanisms to adapt the dialogue flow by including conditions for the prompts of the dialogue fields as shown in Fig. 3.15. In this example, we can see that if valence equals 0.9 and arousal equals 0.8, the stylistic

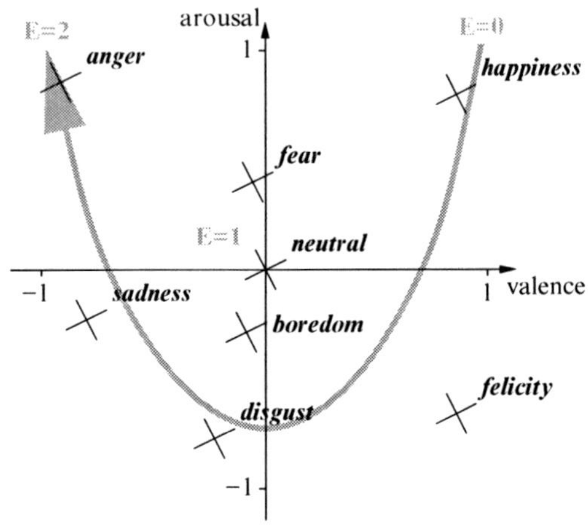

Fig. 3.16 Valence-arousal space representations of selected emotions and corresponding emotional values E. The value E increases along the grey arrow from 0 for happiness to 2 for anger (André et al. 2004; Pittermann and Pittermann 2007)

realization of the prompt is in a very happy way, if both valence and arousal equal 0, the prompt is formulated in a neutral way and if valence equals –0.8 and arousal equals 0.8, the prompt is designed to appease an angry user. Despite the fact that this approach features a very high flexibility in terms of adapting to various different emotional states, it also entails a very high complexity as the conditions need to cover a huge number of cases in the two-dimensional space, even if continuous areas are considered instead of discrete points (like in the example shown in Fig. 3.15), e.g., if $0.7 < \text{valence} < 1.0$ and $0.8 < \text{arousal} < 1.0$, the system reacts in a happy way.

In order to reduce the complexity, an approach described by André et al. (2004) can apply to reduce the emotional states to scalars instead of two-dimensional vectors. In their approach, an emotion is assigned a value E ranging from 0 to 2, where happiness is represented by 0, neutral is represented by 1 and anger is represented by 2 (see Fig. 3.16).

These values are chosen such that for a lower E it is easier for the system to interact with the user whereas a high E indicates that the user is difficult to handle, i.e., the system needs to find a suitable way to appease the user or to make the user feel more comfortable using the system. Such circumstances are modeled with the aid of the so-called threat θ resulting from an utterance, which is defined as

$$\theta = \Phi\left(\bar{\theta}\right), \tag{3.5}$$

with

$$\bar{\theta} = \frac{1}{3} \cdot E(U) \cdot (D(U,S) + P(U,S) + V), \tag{3.6}$$

where $E(U)$ is the user's current emotional state determined as shown in Fig. 3.16, $D(U,S)$ is the social distance between user and system, $P(U,S)$ is the power of the user over the system and V is the valence of the message to be communicated to the user (André et al. 2004). The values for $D(U, S)$ range from 0 (very low social distance – the user is familiar with the system and satisfied using the system) to 1 (high distance – the user is skeptical of using the system). Also, $P(U, S)$ ranges from 0 (the user has no power, i.e., can be easily influenced by the system) to 1 (the user has power over system, i.e., dominates the dialogue and is not influencable) and the valence V ranges from 0 to 1, where 0 represents a positive or pleasant message (e.g., there are flights available on the specified dates in an air travel information system scenario) and 1 represents a negative message (e.g., all flights are fully booked). The function $\Phi(\bar{\theta})$ is used to translate the range of its argument into the desired range. Here, the argument of $\bar{\theta}$ ranges from 0 to 2, and the desired range of θ is [0, 1]. Thus, to emphasize the effect of negative emotions, i.e., high values of $E(U)$, the function is defined as

$$\Phi[x] = \begin{cases} x \;\forall\, 0 \le x \le 1 \\ 1 \;\forall\, x > 1 \end{cases} \tag{3.7}$$

This approach formulating social behavior with the aid of mathematical terms also relates to the ideas of game theory described by von Neumann and Morgenstern (1944).

A further approach extending this idea includes a two-dimensional function $E = f(v, a)$ of valence v and arousal a. Applying such a function, we obtain a gradient of E as illustrated in Fig. 3.17. In order to replicate the curve shown in Fig. 3.16 and to determine surrounding values with a minimal complexity, in this example, we use the following function

$$E = f(v, a) = ((-v \cdot (a + 2)) + 3)/3 = 1 - v \cdot (a + 2)/3, \tag{3.8}$$

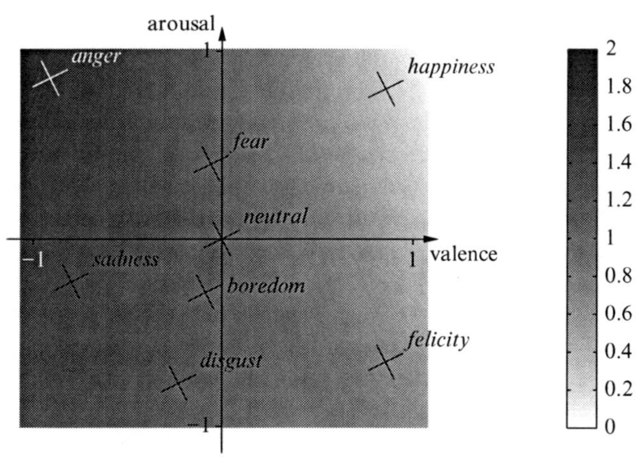

Fig. 3.17 Continuous gradient of E in the valence-arousal space. The value of E at a point in the space is represented by the respective color as indicated by the scale to the right

where $v \in [-1, 1]$ and $a \in [-1, 1]$ are the valence and arousal values of the emotion in the plane. Verifying the values of E, we obtain $E_{\text{anger}} = f(-1, 1) = 2$, $E_{\text{neutral}} = f(0, 0) = 1$ and $E_{\text{happiness}} = f(1, 1) = 0$ in the extreme cases. Furthermore, it can be observed that the difference between positive and negative valences at constant levels of arousal is higher for higher levels of arousal and lower for lower levels of arousal. E.g., at $a = 1$, the maximal difference is $|f(1, 1) - f(-1, 1)| = 2$ whereas at $a = -1$, the maximal difference is $|f(1, -1) - f(-1, -1)| = 2/3 \approx 0.67$.

In Fig. 3.1, the user-state manager cooperating with the dialogue manager receives the emotional control parameters directly from the emotion recognition component. This corresponds to the scenario where an independent (speech-signal-based) emotion recognition component is employed. When using a combined speech-emotion recognizer and/or a linguistic emotion recognizer, additional information is also obtained from the parsing (linguistic analysis) module. In this case, a fusion of the different recognizer outputs is required in or before the user-state management. As described in Section 5.2, we propose various approaches to combine the outputs, most of which resulting in soft emotion scores like, e.g., S_E ={ 0.554 ANGER, 0.188 NEUTRAL, 0.152 HAPPINESS, 0.106 DISGUST }. In order to apply a two-dimensional valence-arousal function like the one described in Eq. 3.8 and illustrated in Fig. 3.17, we need to define default (v, a) values for all emotions to be recognized by the system. If we predefine, e.g., anger at $(-0.8, 0.9)$, disgust at $(-0.2, -0.7)$, happiness at $(0.8, 0.8)$ and neutral at $(0, 0)$, we obtain an average (v, a) tuple $(\bar{v}, \bar{a}) = (-0.314, 0.546)$. With this, we can determine the user's overall emotional state as

$$E(U) = f(\bar{v}, \bar{a}) = ((-\bar{v} \cdot (\bar{a} + 2)) + 3)/3 = 1.266, \tag{3.9}$$

In analogy to the integration of recognizer confidence measures as described in the previous section, we initially describe our straightforward approach to include the user's emotional state $E(U)$ and/or exceeding parameters like the threat θ in dialogue management. These rules are integrated as condition statements in the prompt design as shown in Fig. 3.18. In the dialogue description excerpt, it can be observed that different stylistic realizations of the prompts apply for different ranges of $E(U)$ or θ. With respect to the large variety of user responses, it is more reasonable to generate (at least parts of) the prompts dynamically. In these examples, however, we use precompiled prompts used in these examples for the sake of a better overview. Considering only the user's emotional state as described in Fig. 3.18, these ranges cover a "very positive state" ($0 \leq E(U) < 0.4$), a "positive state" ($0.4 \leq E(U) < 0.8$), neutral ($0.8 \leq E(U) < 1.2$), a "negative state" ($1.2 \leq E(U) < 1.6$) and a "very negative state" ($1.6 \leq E(U) \leq 2$). Accordingly, the system reacts in a cheery manner to a happy user, responds rather "neutral", i.e., without any dispensable flowery phrases to a neutral user, tries to cheer up a sad ($1.2 \leq E(U) < 1.6$) user and affects conciliatorily on an angry or aggressive user.

As opposed to the user's emotional state $E(U)$ the whole range of which can apply in any dialogue context, the threat θ is also dependent on the message which needs to be communicated – if the valence of the message is negative, e.g., when

```
...
<field name="destination">
  <prompt cond="e_u<0.4">
    Wow, that sounds amazing. So, what is your destination?
  </prompt>
  <prompt cond="e_u>=0.4 && e_u<0.8">
    Oh, nice. And where would you like to travel to?
  </prompt>
  <prompt cond="e_u>=0.8 && e_u<1.2">
    Where would you like to go to?
  </prompt>
  <prompt cond="e_u>=1.2 && e_u<1.6">
    Hey, come on, where do you want to go to?
  </prompt>
  <prompt cond="e_u>=1.6">
    Please excuse me, I didn't get all of what you said.
    Where do you want to go to?
  </prompt>
...
```

Fig. 3.18 Excerpt of our VoiceXML dialogue description applying rules to include the user's emotional state $E(U)$ in dialogue prompts

a database query does not generate any pleasant results for the user (flights fully booked, hotels not available, etc.), the threat increases. Thus, following the idea of adapting the prompts according to $E(U)$ as described above, this would mean that if the user is already angry and needs to be appeased anyway, there needs to be even more appeasement in cases where the valence of the message is negative, challenging the purpose and practicability of stylistic prompt adaptation. In order to cope with these effects, André et al. (2004) propose a different concept realizing the prompt design based on four different strategies proposed by Walker et al. (1997a). Given a linear scale of θ, these strategies apply as follows:

- $\theta < 0.25$: *Direct* realization of the prompt or speech act: The output message or prompt is communicated in a more or less neutral way.
- $0.25 \leq \theta < 0.5$: *Approval oriented* realization (positive politeness) taking into account the user's demand for approval. This strategy can be subdivided into three more fine-grained substrategies which are:
  - $0.25 \leq \theta < 0.33$: Claiming common ground
  - $0.33 \leq \theta < 0.42$: Giving the user the impression that user and system are cooperating partners
  - $0.42 \leq \theta < 0.5$: Fulfilling the user's needs
- $0.5 \leq \theta < 0.75$: *Autonomy oriented* realization (negative politeness)
- $0.75 \leq \theta \leq 1$: *Off record* strategy

In this listing, the strategies are equally distributed along the scale of θ from 0 to 1, which is good and intuitive for the visualization of the process but not necessarily for the implementation in an actual dialogue system. However, the values may serve as a good starting point for (adaptive) thresholds which can be adjusted to the respective dialogue application with the aid of, e.g., machine learning algorithms. Especially, as stated in Walker et al. (1997a), off record strategies are difficult to realize in (spoken) human–computer interfaces. E.g., if we assume that there is no satisfactory solution for a user who wants to travel to London on the specified dates and who is already quite angry due to misunderstandings during the previous turns of the dialogue, it is up to the system to suggest to travel to another destination – either to the nearest alternative (e.g., London Gatwick instead of London Heathrow) or to somewhere completely different (*"Lhasa has been voted 'Most beautiful city' by the readers of the 'Tibetan Travel' magazine and it features perfect traveling conditions during that time. Wouldn't you like to travel there instead?"*).

In order to make this approach more practical, a two-level differentiation is presented in André et al. (2004), manually selecting the four different strategies described in the list above and using adaptive threat thresholds to choose among "claiming common ground", "cooperating partners" and "fulfilling user's needs" substrategies. To accomplish that, diverse types of threat are introduced.

As with many rule-based approaches, like for language modeling, dialogue flow design or adaptation of the confirmation strategy, the rule-based adaptation of prompts and dialogue behavior features a very high degree of flexibility in terms of accurateness covering any imaginable aspect and situation for any value of $E(U)$ or θ. This, however, entails a very high effort in setting up these rules and, moreover, also requires a profound knowledge and experience to judge in which way a message is communicated appropriately to the user.

3.6 A Semi-Stochastic Dialogue Model

In order to release the dialogue developer from having to evaluate the proper assignment of emotional or affective control parameters and to setup an enormous number of rules, in the remainder of this section, we propose a "semi-stochastic" approach for the integration of these parameters in the design of the dialogue. We call this approach semi-stochastic as it reconciles a set of dialogue states and rules which are predefined beforehand and a stochastic model describing the transitions between these states (Pittermann and Pittermann 2007). In terms of model complexity, comprehensibility and design effort, our approach constitutes a trade-off between typical rule-based systems like the ones employing VoiceXML as description language and stochastic systems which are, e.g., employing Partially Observable Markov Decision Processes (see Williams et al. 2005; Williams and Young 2007).

The structure of a simple semi-stochastic dialogue model disregarding emotional parameters is illustrated in Fig. 3.19. This dialogue is represented by a network formed by a set $\mathcal{S}$ consisting of four states $\mathcal{S}_1$, $\mathcal{S}_2$, $\mathcal{S}_3$ and $\mathcal{S}_4$, i.e., nodes which represent the dialogue fields P_i, $i = 1, 2, 3, 4$, to be filled during the dialogue and appropriate transitions between the states. Considering only dialogue fields without any further control parameters, we have a direct mapping

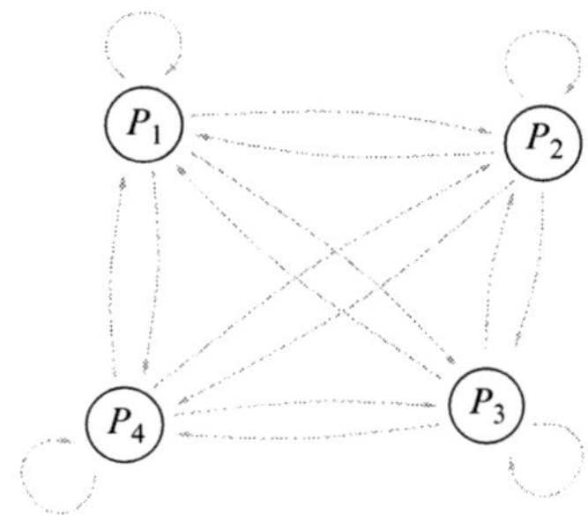

Fig. 3.19 Semi-stochastic dialogue model represented by a network consisting of four dialogue states representing dialogue fields

```
...
<field name="date">
  <prompt>
    On what date would you like to travel?
  </prompt></field>
<field name="departure_city">
  <prompt>
    From where would you like to depart?
  </prompt></field>
<field name="destination">
  <prompt>
    Where would you like to travel to?
  </prompt></field>
<field name="time">
  <prompt>
    At what time do you want to depart?
  </prompt></field>
...
```

Fig. 3.20 Excerpt of our standard VoiceXML dialogue description predefining dialogue fields and prompts for four dialogue states

$$\mathcal{S}_i \rightarrow (P_i), \quad i = 1, \ldots, M, \tag{3.10}$$

where $M = |\mathcal{S}|$ is the cardinality of $\mathcal{S}$, i.e., number of dialogue fields to be filled during the dialogue. If we choose a travel agency scenario, we can define, e.g., P_1 as the departure city, P_2 as the destination, P_3 as the travel date and P_4 as the time of departure.

Pursuing this travel agency dialogue application, the states are defined beforehand by the fields and the corresponding prompts in the dialogue description as shown in Fig. 3.20.

Having set up the scaffolding of the dialogue consisting of the dialogue field states, the transitions between these states are summarized in the set of all edges $\mathcal{E}$. The probabilities of the respective transitions are obtained on the basis of existing dialogue data. In analogy to the adaptation of the confirmation strategy described in the previous section, we consider the probability of a state (field) $P\left(\mathcal{S}_i | \underline{\mathbf{H}}_D\right)$, $1 \leq i \leq M$, given the dialogue history $\underline{\mathbf{H}}_D$. Recapitulating Eq. 3.1 at the beginning of this section, for the moment, we limit our considerations to dialogue fields, i.e., taking into account the probability of a state and the corresponding dialogue field P_i, $i = 1, \ldots, M$ for the system action p_S at time T described by $P_p(p_S(T) = P_i|[p(1), \ldots, p(T)])$ given the aggregate user input, i.e., previously filled dialogue fields $p(1), \ldots, p(T)$.

The complexity of the stochastic share of such a dialogue model increases drastically with the size of the history to be considered as the number of turns increases continuously during the course of the dialogue. For the first system turn, we need a bi-turn model $P(p_S(1)|p(1))$ when the user's utterance contains one utilizable item (turn) $p(1)$. With a total of M states in the network, and assuming that the system will not ask *"Where do you want to travel to?"* after the user states *"I want to go to Oslo."*, the bi-turn model contains $M \cdot (M - 1)$ transition probabilities. Thus, in the ideal case, the dialogue would end after a maximum of M turns and the maximum complexity of the model would include $M \cdot (M - 1) \cdot \ldots \cdot 1 = M!$ possible transition probabilities. In reality, however, the number of turns can easily exceed M due to repetitions, misunderstandings and/or confirmation prompts and it can be assumed that certain fields are addressed multiple times, e.g., when the user answers under pressure and, thus, becomes uncertain.

Hence, we can presume that the number of possible n-turn probabilities is approximately $|\mathcal{S}|^n = M^n$ constituting a high model complexity which is hardly feasible in terms of computational efforts and the required training material. On this account, we reduce the stochastic component of the dialogue model to bi-turn and tri-turn combinations, i.e., $P(p_S(T)|p(T))$ and $P(p_S(T)|p(T-1), p(T))$, incorporating the previous or the previous and the penultimate user turn, respectively.

The bi-turn and tri-turn probabilities are determined by training on the basis of preprocessed dialogue corpora. These corpora normally contain human–computer dialogues, recorded in real-world situations (people who call service centers are typically informed that the phone call may be recorded and analyzed to improve the service) or in WOZ scenarios where a human supervisor (the wizard) simulates the behavior of the system making the users believe that they are interacting with a computer. In order to render automated SLDSs more natural, i.e., not making the users feel that they are talking to a "stupid" computer, we suggest to employ human–human dialogues for the training. As humans tend to exhibit a different behavior in terms of language, style and communication of information when talking to humans or computers, these dialogues need to be prepared for their use in a computerized environment. Jönsson and Dahlbäck (2000) describe the process of distilling human–human dialogues in order to adapt these for further use in SLDSs. This becomes especially important when more complex dialogue models, requiring more than just an analysis of semantic labels, are applied.

For the travel agency example, by definition, dialogue data obtained in the domain of travel agencies, airline call centers, airport information desks, etc., no matter whether distilled or not, are required. In the following paragraphs, we describe the process of preparing the data and the training of the dialogue model on the basis of excerpts of three example dialogues in the travel agency domain collected in a corpus to verify the feasibility of dialogue concepts.

Agent: *Good morning, Sir. How are you?*
Customer: *Good, thanks. Can I book a flight to Frankfurt here?*
Agent: *Absolutely. From where would you like to depart?*
Customer: *From New York.*
Agent: *When would you like to leave?*
Customer: *On Monday.*
Agent: *OK, what time would you prefer?*
...

Customer: *Hi, Can you tell me whether there are any flights to Oslo today?*
Agent: *Departing from here?*
Customer: *Yes, from Copenhagen.*
Agent: *Let me see... Yes, there are seats available on eight flights. At what time do you want to leave?*

Agent: *Hi, how can I help you?*
Customer: *Hello, we would like to book a flight.*
Agent: *Good, do you know where you want to travel to?*
Customer: *Hm, we thought of just leaving this place for quite a while as soon as possible.*
Agent: *I see. Would you prefer a destination in the Southern hemisphere?*
Customer: *Oh, Buenos Aires would be great.*
Agent: *Let me check... There are seats available on flight leaving on Monday at 6 p.m.. Would you like to book that flight?*

In these dialogues, the determining parts of the customer (corresponding to the user) utterances are underlined and the resulting agent (system) reactions are wavily underlined. Extracting these semantic labels, the dialogues can be "compressed" in to a sequence of the labels contained in both customer and agent utterances. E.g., the first dialogue is rewritten as

Agent: *[—]*
Customer: *[destination]*
Agent: *[departure_city]*
Customer: *[departure_city]*
Agent: *[date]*
Customer: *[date]*
Agent: *[time]*
...

From these compressed dialogues, we can determine bi-turn and tri-turn combinations of customer (user) turns leading to a specific agent (system) reaction. The first dialogue starts with a customer utterance containing the destination and the agent reacts by asking about the departure city, i.e., the first bi-turn combination is

```
destination                  -> departure_city.
```

The customer's reply to the agent's question contains the departure city and the agent asks for the travel date. Now we obtain the second bi-turn combination

```
departure_city               -> date,
```

and (including the previous turn) the first tri-turn combination

```
destination, departure_city -> date.
```

Analogously, the following bi-turn and tri-turn combinations are

```
date                         -> time
departure_city, date         -> time.
```

Following this procedure for each turn in the three example excerpts, we obtain the processed training data as shown in Fig. 3.21. It should be noted that the first customer utterance in the second dialogue contains two turns, destination and date. I.e., in combination with the agent's reaction (departure city?), we obtain two bi-turn (destination $\rightarrow$ departure_city, and date $\rightarrow$ departure_city) combinations and one tri-turn (destination, date $\rightarrow$ departure_city) combination. The third dialogue starts with an empty customer utterance, so that the agent needs to take the initiative by asking about the desired destination. Here, we obtain a uni-turn, i.e., a possible system reaction in cases where users of an SLDS do not know what to say. In the training data, the uni-turn is described as a bi-turn containing a zero state ($*$).

```
destination                  -> departure_city
departure_city               -> date
destination, departure_city -> date
date                         -> time
departure_city, date         -> time
destination                  -> departure_city
date                         -> departure_city
destination, date            -> departure_city
departure_city               -> time
destination, departure_city -> time
date, departure_city         -> time
*                            -> destination
departure_city               -> destination
date                         -> destination
date, departure_city         -> destination
destination                  -> time
destination, departure_city -> time
destination, date            -> time
```

Fig. 3.21 Dialogue model training data extracted from three excerpts of human–human dialogues in the travel agency domain

In the case where the dialogue model consists of $M = |\mathcal{S}|$ states, there are M uni-turns, $M \cdot (M-1)$ bi-turns and $M \cdot (M-1) \cdot (M-2)/2$ tri-turn combinations, i.e., in this example model, there are 4 uni-turns, 12 bi-turns and 12 tri-turns. This implies that we do not include the order of occurrence in the left part of the tri-turn combinations. E.g., (departure_city, destination → date) and (destination, departure_city → date) are treated as one combination. Some of these combinations occur in the training data, other combinations do not occur. Moreover, the dialogue developer can exclude certain combinations from the model manually by prefixing the respective lines with an exclamation mark (!). E.g., if we did not want to allow the system to prompt for the departure city after the user has mentioned date and time, we could add

```
!date, time                     -> departure_city
```

to the training data.

Based on the frequency of occurrence of these bi-turn and tri-turn combinations, the respective (transition) probabilities are determined in the training process. To accomplish that, at first, for all possible bi-turn and tri-turn combinations the frequencies of occurrence of which are initialized with zeros as shown in Fig. 3.22. This "skeleton" model corresponds to the dash-dotted grey transitions shown in Fig. 3.19. For the sake of clarity, the illustration only describes the bi-turn transitions, i.e., transitions between single states. A complete representation of the model is illustrated in Fig. 3.23. Here, the left network represents uni-turn originating from the extra zero state ($*$) and bi-turn transitions between all states. The right network represents the tri-turn transitions from two connected states to the remaining (single) states. The connection between two states is presented by a solid black line and a grey dot forming the origin of the respective transitions. Still not knowing the actual transition probabilities, in this figure, the transitions are represented by grey dash-dotted lines.

In this scenario, there are no self-transitions from one state to the same state as we assume a simple dialogue model without any repetitions or confirmations.

```
*                               -> date              0
...
*                               -> time              0
date                            -> departure_city    0
date                            -> destination       0
...
time                            -> destination       0
date, departure_city            -> destination       0
date, departure_city            -> time              0
date, destination               -> departure_city    0
...
destination, time               -> date              0
destination, time               -> departure_city    0
```

Fig. 3.22 Untrained skeleton dialogue model

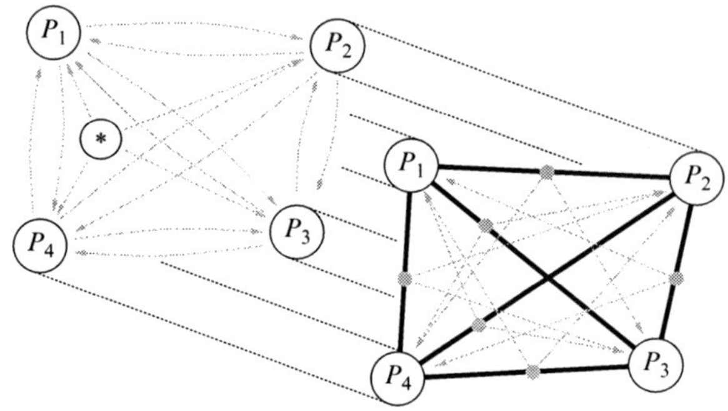

Fig. 3.23 Representation of uni-turn, bi-turn and tri-turn transitions in a semi-stochastic dialogue model network. Uni-turn transitions are represented by arrows originating from the ⋆ node and bi-turn transitions are represented by the arrows between the states in the left part of the figure. Tri-turn transitions are represented by the arrows originating from two states combined by thick black lines (*origin indicated by grey circles*) in the right part. The unity of both parts is indicated by dotted lines

Moreover, for the same reason, there are also no tri-turn transitions from a state pair to one of the connected states, e.g., (departure_city, destination → destination). In the general case, these transitions can also be included in the model, even tri-turn combinations with transitions from one state connected to itself to the same state, e.g., (date, date → date). With this, assuming M states, the number of uni-turns is still M, but the number of bi-turns is M^2 and the number of tri-turns is M^3 or $M^2 + M^2 \cdot (M - 1)/2 = M^2 \cdot (M + 1)/2$ depending on whether the chronological order within the tri-turns is taken into account or not. The possible application of these can be flexibly predefined by the dialogue developer in the training process.

Having set up the skeleton model in any implementation, for each line of the training data, the counter of the respective combination in the model increases by 1. I.e., after the first training step, each uni-turn, bi-turn and tri-turn combination is assigned its frequency of occurrence in the training data.

Depending on the corpus size, especially for smaller corpora, it is likely that some of the combinations are assigned 0 as frequency of occurrence. On the one hand, this can emerge due to the low significance of the respective combinations, on the other hand, this may lead to deadlock situations in the dialogue later on. To avoid possible divisions by zero in the field selection process, we add an infinitesimal number ε to each value in the trained data. Having accomplished the training with the data shown in Fig. 3.21, we obtain a dialogue model as shown in Fig. 3.24.

In this example, there is only one transition originating from the $*$ state. I.e., independent of ε, in most of the cases, the dialogue manager would ask for the destination when the user does not know what to say. However, if the question about the destination was already answered, there would be a deadlock situation as the system would not be able to determine what to ask next. Adding the same ε to all

```
*                            -> date               0.001
*                            -> departure_city     0.001
*                            -> destination        1.001
*                            -> time               0.001
date                         -> departure_city     1.001
date                         -> destination        1.001
date                         -> time               1.001
departure_city               -> date               1.001
departure_city               -> destination        1.001
departure_city               -> time               1.001
destination                  -> date               0.001
destination                  -> departure_city     2.001
destination                  -> time               1.001
time                         -> date               0.001
time                         -> departure_city     0.001
time                         -> destination        0.001
date, departure_city         -> destination        1.001
date, departure_city         -> time               2.001
date, destination            -> departure_city     1.001
date, destination            -> time               1.001
date, time                   -> departure_city     0.001
date, time                   -> destination        0.001
departure_city, destination -> date               1.001
departure_city, destination -> time               2.001
departure_city, time         -> date               0.001
departure_city, time         -> destination        0.001
destination, time            -> date               0.001
destination, time            -> departure_city     0.001
```

Fig. 3.24 Dialogue model trained on the data shown in Fig. 3.21

numbers, here, ensures that all of the other transitions are considered as equally probable, increasing the robustness of the dialogue manager.

Based on the skeleton depicted in Fig. 3.23 and the model data shown in Fig. 3.24, the dialogue model is structured as illustrated in Fig. 3.25. In this figure, transitions are represented by dashed grey lines when their frequency of occurrence is zero or ε, respectively. Thin solid black lines represent transitions with an average frequency of occurrence (here: once) and thick solid black lines represent transitions that occur more often (here: more than once). In our following considerations we summarize all transitions/combinations regardless of their probabilities into sets: $\mathcal{U}$ contains all uni-turn combinations ($* \rightarrow p_S$), $\mathcal{B}$ contains all bi-turn combinations ($p(T) \rightarrow p_S$) and $\mathcal{T}$ contains all tri-turn combinations ($p(T-1), p(T) \rightarrow p_S$), $p_S = P_1, \ldots, P_n$ of the emotional model. Furthermore, we define bi-turn subsets $\mathcal{B}(y) \subseteq \mathcal{B}$ and tri-turn subsets $\mathcal{T}(x, y) \subseteq \mathcal{T}$ for arbitrary previously uttered user turns x and y, where

$$\begin{aligned}\mathcal{B}(y) &= \{(p(T) \rightarrow p^\star) | p(T) = y\},\\ \mathcal{T}(x, y) &= \{(p(T-1), p(T) \rightarrow p^\star) | (p(T-1) = x \wedge p(T) = y) \vee \\ &\quad (p(T-1) = y \wedge p(T) = x)\},\end{aligned} \tag{3.11}$$

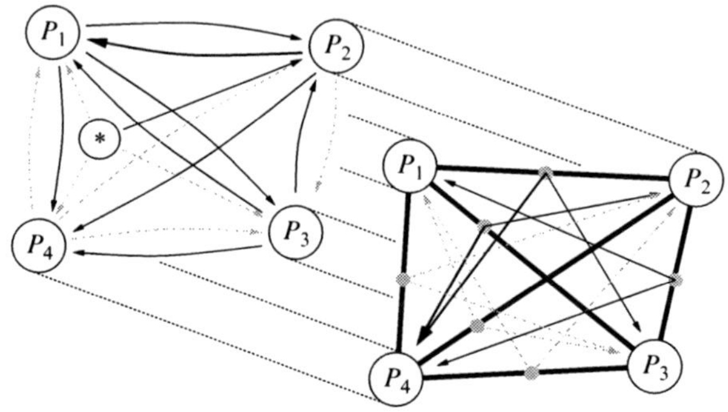

Fig. 3.25 Illustration of the dialogue model described by Fig. 3.24. The structure of the model representation is the same like in Fig. 3.23: Uni-turn and bi-turn transitions are shown in the left part, tri-turn transitions are shown in the right part of the figure. The probability of the transitions is indicated by the respective line thickness and style. In this example, state P_1 represents the departure city, P_2 is the destination, P_3 is the travel date and P_4 is the time of departure

where $p^\star \in \mathcal{S}$ (or $p^\star \in \mathcal{S} \setminus \{p(T-1), p(T)\}$) can be any (other) state. Here, it should be noted that $p^\star$ does not take states the fields of which have already been filled during the dialogue. This rule or restriction, among others, is one reason why we call the model semi-stochastic and not stochastic.

For the sake of clarity, in our considerations, we avoid using the uni-turn set $\mathcal{U}$ and we describe the set as special cases of bi-turns, i.e., $\mathcal{U} \equiv \mathcal{B}(*)$. By that, we obtain the set of all transitions $\mathcal{E}$ as $\mathcal{E} = \mathcal{B} \cup \mathcal{T}$. Furthermore, for a bi-turn combination $(p(T) \rightarrow p_S = p)$ or a tri-turn combination $(p(T-1), p(T) \rightarrow p_S = p)$, we define its frequency of occurrence in the dialogue model as $N(p(T) \rightarrow p_S = p)$ or $N(p(T-1), p(T) \rightarrow p_S = p)$, respectively. With this, the sum of the frequencies of occurrence of all combinations in a set $\mathcal{B}(y)$ is calculated as

$$N(\mathcal{B}(y)) = \sum_{(p(T) \rightarrow p_S = p) \in \mathcal{B}(y)} N(p(T) \rightarrow p_S = p), \tag{3.12}$$

and for $\mathcal{T}(x, y)$ as

$$N(\mathcal{T}(x, y)) = \sum_{(p(T-1), p(T) \rightarrow p_S = p) \in \mathcal{T}(x,y)} N(p(T-1), p(T) \rightarrow p_S = p). \tag{3.13}$$

Using such a dialogue model, the current field is determined among possible combinations that include the previous and, optionally, the penultimate user turn in the history. I.e., for each system turn, a set $\mathcal{V} = \mathcal{B}(p(T)) \cup \mathcal{T}(p(T-1), p(T))$ is formed and for each combination $(p(T-1), p(T) \rightarrow p_S = p)$ or $(p(T) \rightarrow p_S = p)$ in $\mathcal{V}$, the respective probability $P(p(T-1), p(T) \rightarrow p_S = p)$ or $P(p(T) \rightarrow p_S = p)$ is calculated with respect to its frequency of occurrence $N(p(T-1), p(T) \rightarrow p_S = p)$ or $N(p(T) \rightarrow p_S = p)$. For implementation reasons, in cases where the user has not provided enough information to apply the tri-turn model, $\mathcal{V}$ only contains bi-turn combinations. This technique coincides with the back-off technique which is used in stochastic language models in speech recognition.

In statistical language modeling it has been shown that tri-gram models provide a significantly better representation of the structure of a language than bi-gram or uni-gram models as the tri-gram models involve more information. Thus, in automatic speech recognition, tri-gram models also contribute to a more robust recognizer performance. Similarly, tri-turn models represent a more detailed structure of the dialogue flow. Thus, we argue that if there are tri-turns among $\mathcal{V}$, their influence on the field selection process should be increased. This can be accomplished by multiplying their frequency of occurrence with an arbitrary factor $\alpha \geq 1$. Depending on the size of the model and the dimensions of the training data, α ranges from 1 (few training data for a large model) to 10 (huge amount of training data compared to the model complexity). With this, we calculate the probability of a bi-turn combination as

$$P(p(T) \rightarrow p_S = p) = \frac{N(p(T) \rightarrow p_S = p)}{N(\mathcal{B}(p(T))) + \alpha \cdot N(\mathcal{T}(p(T-1), p(T)))}, \tag{3.14}$$

and of a tri-turn combination as

$$P(p(T-1), p(T) \rightarrow p_S = p) = \frac{\alpha \cdot N(p(T-1), p(T) \rightarrow p_S = p)}{N(\mathcal{B}(p(T))) + \alpha \cdot N(\mathcal{T}(p(T-1), p(T)))}. \tag{3.15}$$

To avoid confusion, we assign each bi-turn and tri-turn combination either a unique label or an arbitrary number i, $1 \leq i \leq |\mathcal{V}|$ and we label the respective probability $P(i)$. With this, there exist two approaches to select the current field under discussion:

A simple approach would be to find the combination i with the highest probability among all combinations in the set $\mathcal{V}$

$$i = \arg \max_{1 \leq i \leq |\mathcal{V}|} P(i), \tag{3.16}$$

and choose the respective field from the ith combination. This procedure, however, neglects finer details in the dialogue model, e.g., if the model contains

```
departure_city, destination -> date             11.001
departure_city, destination -> time             12.001,
```

there is a 100% chance, that the system will ask for the departure time, although the dialogue model also allows for a high probability of asking about the travel date. Moreover, this selection process lacks robustness, especially in cases when several combinations are equally probable, e.g.,

```
date                        -> departure_city   1.001
date                        -> destination      1.001
date                        -> time             1.001.
```

Such a case typically requires a random selection process among the remaining combinations with the highest probability. Having to do so anyway in the given cases, we argue that we obtain a higher robustness and, moreover, a higher flexibility if we accomplish the entire selection procedure on the basis of a random process.

Thus, with respect to the fact that the probabilities $P(i)$ of all bi-turn and tri-turn combinations add up to $\sum_i P(i) = 1$, we assign each of these combinations i a range $[t_i, t_i + P(i)]$, with $0 \leq t_i < 1$ and $t_i + P(i) \leq 1$. Then, a random process generates a number $r \in [0, 1]$ which is uniformly distributed in the interval between 0 and 1. Depending on r, the respective combination i is selected according to

$$i = \arg\{t_i | t_i \leq r \leq t_i + P(i)\}, \tag{3.17}$$

and the current field is chosen from the ith combination. The random selection process for a general set $\mathcal{V}$ is illustrated in Fig. 3.26. The diagram shows the uniform distribution along the different combinations i.

Coming back to our travel agency scenario, we can exemplify the selection process on the basis of the following dialogue taken from our dialogue collection:

	...
2 User:	*I want to book a flight to Copenhagen.*
3 System:	*From where would you like to depart?*
4 User:	*From Paris.*

At this point the dialogue history contains $p(T-1) =$ destination and $p(T) =$ departure_city. Based on these previous turns, the initial set $\mathcal{V}$, as defined above, consists of the following combinations, taken from the model shown in Fig. 3.24:

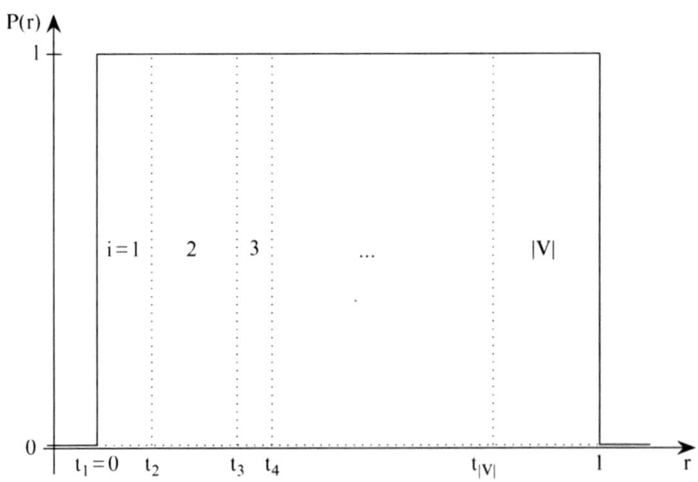

Fig. 3.26 Illustration of the field selection process using the dialogue model described by Fig. 3.25

```
1 departure_city                   -> date          1.001
2 departure_city                   -> time          1.001
3 departure_city, destination -> date          1.001
4 departure_city, destination -> time          2.001
```

Here, the bi-turn combination (departure_city → destination (1.001)) is not included in $\mathcal{V}$ as the destination field has already been filled. Each of the four combinations is assigned a unique number (1–4). To emphasize the importance of tri-turn combinations, we presume an amplification factor $\alpha = 2$. With this, we can calculate the individual probabilities $P(i), i = 1, \ldots, 4$ as follows:

$$
\begin{aligned}
P(1) = P(2) &= \frac{1.001}{1.001 + 1.001 + 2 \cdot (1.001 + 2.001)} \approx \frac{1}{8} = 0.125 \\
P(3) &= \frac{2 \cdot 1.001}{1.001 + 1.001 + 2 \cdot (1.001 + 2.001)} \approx \frac{1}{4} = 0.25 \\
P(4) &= \frac{2 \cdot 2.001}{1.001 + 1.001 + 2 \cdot (1.001 + 2.001)} \approx \frac{1}{2} = 0.5,
\end{aligned} \tag{3.18}
$$

and we obtain

$$
t_1 = 0, \quad t_2 = 0.125, \quad t_3 = 0.25, \quad t_4 = 0.5. \tag{3.19}
$$

Adding up the probabilities, we can see that the overall probability for a combination leading to time as the next field is 0.625 and the probability for date is 0.375. The probability distribution for the four combinations is illustrated in Fig. 3.27.

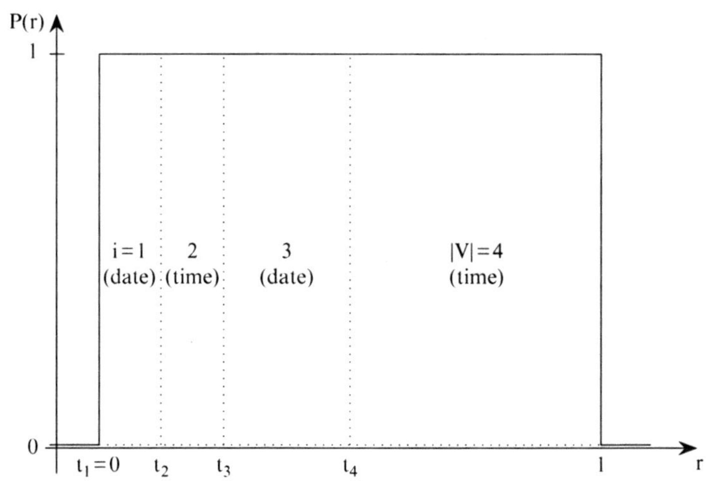

Fig. 3.27 Illustration of the field selection process in the travel agency scenario consisting of four dialogue fields. Combinations 1 and 2 are bi-turn combinations, 3 and 4 are tri-turn combinations with correspondingly higher probabilities assuming an amplification of $\alpha = 2$

The order of the combinations in this diagram seems quite arbitrary as it corresponds to the order of appearance in the model. Each combination i is also labeled with the name of next dialogue field which would be prompted if the respective combination was randomly selected. E.g., if we obtain $r = 0.3973$ (date), the next system turn in the dialogue is

5 System: *On what date would you like to travel?*

3.7 A Semi-Stochastic Emotional Model

In analogy to the setup of the dialogue model described in the previous section, the user's emotional state can also be conceptualized in a semi-stochastic model (Pittermann and Pittermann 2006c). The aim of such a model is to enable the system to react appropriately to the emotional state and, if applicable, to previous states determined by an automatic emotion recognizer. As shown for the plain dialogue model, the emotional model consists of a predefined number N of states $\mathcal{S} = \{\mathcal{S}_1, \mathcal{S}_2, \ldots, \mathcal{S}_N\}$ with $N = |\mathcal{S}|$. Given one arbitrary field P_i from the above dialogue model, the states in $\mathcal{S}$ represent the respective emotional states of the system. The structure of the emotional model disregarding the dialogue states is illustrated in Fig. 3.28.

This example model is represented by a network formed by a set $\mathcal{S}$ consisting of three states, i.e., nodes which represent the emotional states E_i, $i = 1, 2, 3$, in which the system reacts to the user's emotional states. Considering only emotional states without any information about dialogue states and further control parameters, we have a direct mapping

$$\mathcal{S}_i \rightarrow (E_i), \quad i = 1, \ldots, N, \tag{3.20}$$

where $N = |\mathcal{S}|$ is the cardinality of $\mathcal{S}$, i.e., number of emotions to be considered in the model. Depending on the application, the values of E_i can be strings like "anger", "happiness", etc., or floating point values, e.g., $0.0 \leq E(U) \leq 2.0$ as described above. Featuring a higher flexibility, for our example, we choose the use of numerical values $E(U)$ and we assign $E_1 = 0.3$ (strong and positive emotion

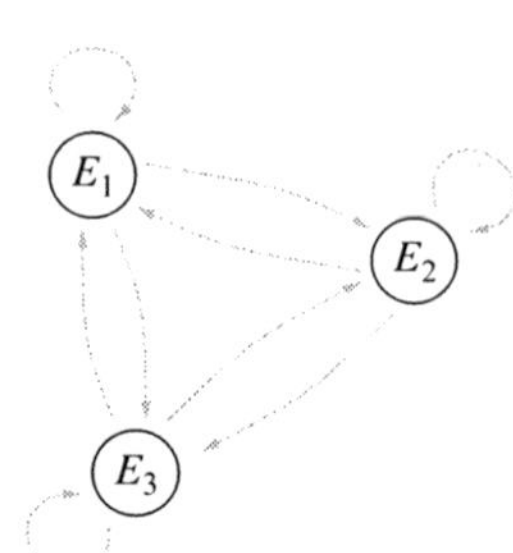

Fig. 3.28 Semi-stochastic emotional model represented by a network consisting of three emotional states representing the respective system output

```
...
<field name="destination">
  <prompt cond="e1">
    Excellent, and what's your destination?
  </prompt>
  <prompt cond="e2">
    Where would you like to travel to?
  </prompt>
  <prompt cond="e3">
    I'm sorry to bother you again. Where did you say you
    wanted to travel to?
  </prompt></field>
...
```

Fig. 3.29 Excerpt of our VoiceXML dialogue description predefining prompts for the respective (three) emotional states of one dialogue field (here: destination)

like happiness), $E_2 = 1.0$ (neutral) and $E_3 = 1.7$ (strong and negative emotion like anger). Pursuing the travel agency example from the previous section, the emotional states are defined beforehand by the prompts as shown in Fig. 3.29 for the destination field of the dialogue description.

The transitions between the states are, again, summarized in the set of all edges $\mathcal{E}$. For each transition, its probability is determined by the probability of the target state $P\left(\mathcal{S}|\underline{\mathbf{H}}_D\right)$ given the dialogue history $\underline{\mathbf{H}}_D$. As opposed to Section 3.6, now, we are not interested in the previously filled dialogue fields $p(t)$ but in the previously recognized emotional states $e(1), \ldots, e(T)$. Consequently, see also Eq. 3.1 (page 66), we describe the probability of a transition by the probability $P_e(e_S(T) = E_i|[e(1), \ldots, e(T)])$ of a system output in emotional state E_i given the "emotional user turns" $e(1), \ldots, e(T)$. It should be noted, that when we say that the system is in a certain state, e.g., an angry state, this does not mean that it behaves angrily and yells at the user, but that it reacts (or tries to react) appropriately to the user's anger. Thus, system and user state are not identical, i.e., if the system is in state $e_S = E_i$, this means the system reacts as if the user were in this state E_i.

To reduce the model complexity, the number of emotional turns in the dialogue history is limited to one or two, so that we only consider bi-turn and tri-turn probabilities, i.e., $P(e_S(T)|e(T))$ and $P(e_S(T)|e(T-1), e(T))$, incorporating the previous or the previous and the penultimate user turn, respectively. As opposed to the plain dialogue model, however, we do not include uni-turns, as these are only important for the first system turn which is typically uttered in a neutral style, e.g., *"How can I help you?"* and for further turns, the system can always rely on data from the emotion recognizer.

These probabilities are also determined by training, either on dialogue data obtained from processed human–human dialogues or on recorded human–computer dialogues, typically already employing an automatic emotion recognizer. As mentioned before, the exact stylistic realization of the user and system turns is not important for the model, also the domain of the dialogue data, e.g., travel domain, pizza ordering system, etc., is irrelevant. In the following excerpts from the training data, the emotional states $e(t)$ of the user and the emotional states $e_S(t)$ of the system are represented by numbers $0 \leq e(t), e_S(t) \leq 2$. In general, $e(t)$ and $e_S(t)$ can take any value in the given range, however, in order to adapt to the predefined model structure, the values need to be quantized according the values of E_i, $i = 1, \ldots, N$. E.g., given a set $\mathcal{S}$ of states mapped to emotions E_i where $E_1 = 0.1$, $E_2 = 0.2$, $E_3 = 0.3$, ..., $E_{20} = 2.0$, a user's emotion $e(T) = 0.6345$ would be quantized to $\bar{e}(T) = 0.6 = E_6$, whereas given the three-state model as described above ($E_1 = 0.3$, $E_2 = 1$, $E_3 = 1.7$), $e(T) = 0.6345$ would be quantized to $\bar{e}(T) = 0.3 = E_1$.

Due to the fact that in principle any transition from any emotion to any other or the same emotion is possible, the emotional model is more complex than the simple four-state dialogue model described in the previous section. Tapping the full range of possible transitions, we obtain N^2 emotional bi-turn combinations and N^3 emotional tri-turn combinations adding up to a total of $N^2{\cdot}(N+1)$ combinations, i.e., the three-state model includes $9+27 = 36$ transitions. In this case, we do not only allow combinations, where one state transits to itself, e.g., $P(e(T) = E_1 \rightarrow e_S = E_1)$, but also multiple occurrences of the same state within a tri-turn combination, e.g., $P(e(T-1) = E_1, e(T) = E_2 \rightarrow e_S = E_1)$, $P(e(T-1) = E_1, e(T) = E_1 \rightarrow e_S = E_2)$ or even $P(e(T-1) = E_1, e(T) = E_1 \rightarrow e_S = E_1)$. Moreover, we take into account the chronological order of user's emotions in the tri-turn combinations, i.e., $P(e(T-1) = E_1, e(T) = E_2 \rightarrow e_S = E_3)$ and $P(e(T-1) = E_2, e(T) = E_1 \rightarrow e_S = E_3)$ are considered as different combinations. This is especially important when the temporal aspect of how the user's emotional state changes shall be included – e.g., if the user's state changes from happiness to anger, the system should not react in the same way as if the user's state changes from anger to happiness.

The complete interrelationship in such a model is illustrated in Fig. 3.30. Here, the upper network represents bi-turn transitions between all states and the couple of lower right network represents the tri-turn transitions from two connected states to any other state. Moreover, the tri-turn combinations in which the previous and the penultimate emotional state of the user are the same are represented by the lower left network. In the lower networks, the connections between the two user states are presented by a solid black line and a grey dot forming the origin of the respective transitions. The chronological order and the fact that the upper and the lower networks on the right are not identical are emphasized with the aid of triangular lines. In this illustration, the transitions which occur in the training data (in this case processed dialogue data) are represented by solid lines – those which occur more often than average are represented by thicker lines. All remaining transitions are represented by grey dashed lines. Apart from the remarkably high number of grey dashed transitions, it is striking that the majority of transitions are self-transitions, i.e., from a single state to itself, e.g., $(0.3 \rightarrow 0.3)$, or from a couple to one of the states belonging to the same couple, e.g., $(1.0, 0.3 \rightarrow 0.3)$. This is a rather typical behavior, as humans also tend to respond to their dialogue partner's emotional state.

From this point of view, looking at the simple example, it seems rather unpractical to use such a complex model instead of just creating a set of rules which,

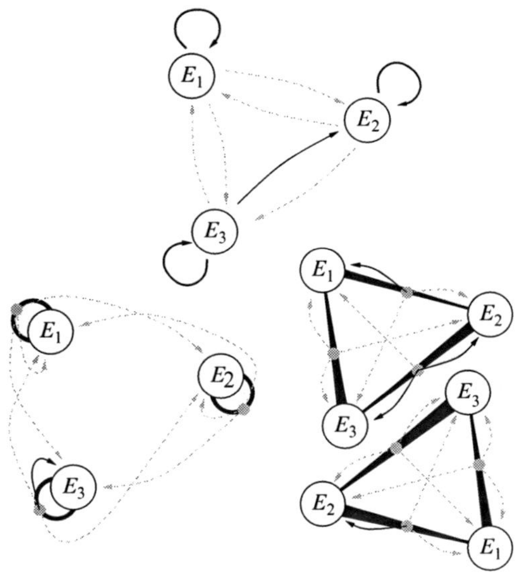

Fig. 3.30 Illustration of a trained emotional model. The probability of transitions is indicated by the respective line thickness and style. State $E_1 = 0.3$ represents a happy emotional state, $E_2 = 1.0$ corresponds to neutral and $E_3 = 1.7$ represents anger

e.g., contain if the user is angry, apply some appeasement strategy. The actual advantages of this semi-stochastic model become apparent when a finer quantization structure is applied, i.e., if we distinguish more different levels of positive, neutral or negative states. Even here, given a sufficient amount of training data, it can be observed how this corpus-based approach provides a more flexible and less complex model generation, as opposed to a complex rule set which needs to be defined manually.

The major advantage of such an emotional model, however, can be found in the interplay with any type of automatic emotion recognizers like these described in Chapter 4, which can not provide 100% error-free recognition results. Especially due to the fact that emotions actually cannot be judged on an objective basis, the automatic emotion recognizers can not please everybody. These discrepancies and the inaccurateness, however, can be integrated into the model in such a way that the system reacts appropriately even if the recognizer output does not quite match the user's actual emotional state.

To accomplish this, the system needs to include previous user states, assuming these are more or less correctly recognized. In practice, on the one hand, this means that in addition to bi-turns and tri-turns, also n-turns including a higher number n of previous states need to be considered. By that, it is possible to detect outliers in the recognizer output assuming that it is quite unlikely that the user's state abruptly changes between two different states.

On the other hand, the dialogue model needs to be trained on "noisy" data, i.e., in the training material, the customers' real emotional states need to be replaced by what the automatic emotion recognizer determines from the respective speech signal (or video, biosignals, etc.). This, however, entails that the higher the emotion recognizer's error rate the more training data is required to obtain a robust emotional model.

The selection process during the dialogue is similar to the process described in Section 3.6. Given the previous one or two recognized emotional state(s) $e(T)$ and optionally $e(T-1)$, we can define a set $\mathcal{V} = \mathcal{B}(e(T)) \cup \mathcal{T}(e(T-1), e(T))$ which includes all possible bi-turn and tri-turn combinations, based on which we can calculate probability distributions for all emotional states. Then, a random selection process determines the following emotional state.

Picking up our travel agency scenario again, we can exemplify the emotional state selection process on the basis of an invented dialogue, in which the system is now about to ask for the user's destination. This dialogue is now extended by one emotional control parameter:

	...
2 User:	*I want to book a flight tomorrow.* [0.8954]
3 System:	*From where would you like to depart?*
4 User:	*From Paris.* [1.5446]

This additional number in each user turn is the user's emotional state $E(U)$ ranging from 0 to 2, determined with the aid of an automatic emotion recognizer. To simplify matters, we presume that the emotion recognizer has correctly estimated the user's state. After the second user turn, the dialogue history contains $e(T-1) = 0.8954$ which is quantized to $e(T-1) = 1.0 = E_2$ and $e(T) = 1.5446$ which corresponds to $e(T) = 1.7 = E_3$. Based on these previous turns, the initial set $\mathcal{V}$, as defined before, now consists of the following combinations, taken from the model described by Fig. 3.30:

```
1.7 (E3)                  -> 0.3  (E1)      0.001
1.7 (E3)                  -> 1.0  (E2)      1.001
1.7 (E3)                  -> 1.7  (E3)      2.001
1.0 (E2), 1.7 (E3)        -> 0.3  (E1)      0.001
1.0 (E2), 1.7 (E3)        -> 1.0  (E2)      1.001
1.0 (E2), 1.7 (E3)        -> 1.7  (E3)      1.001
```

Here, again, we argue that tri-turn combinations contribute to a more robust decision in the model and, thus, we presume an amplification factor $\alpha = 2$. With this, we can calculate the individual probabilities $P(E_i), i = 1, \ldots, N = 3$ as follows:

$$P(E_1) = \frac{0.001 + 2 \cdot 0.001}{(0.001 + 1.001 + 2.001) + 2 \cdot (0.001 + 1.001 + 1.001)} = \frac{0.003}{7.009} \approx 0, \tag{3.21}$$

$$P(E_2) = \frac{1.001 + 2 \cdot 1.001}{7.009} \approx 0.43, \tag{3.22}$$

$$P(E_3) = \frac{2.001 + 2 \cdot 1.001}{7.009} \approx 0.57. \tag{3.23}$$

Looking at the three probabilities, we can see that, e.g., the probability for a combination leading to an appeasing reaction to an angry user is relatively high

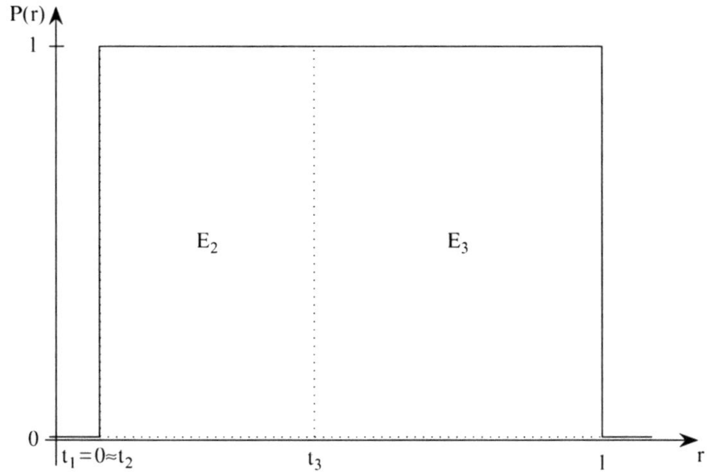

Fig. 3.31 Illustration of the emotional reaction selection process in the travel agency scenario allowing three emotional states

(approx. 0.57) whereas the probability for a cheerful reaction which would be more suitable for a happy user is negligible. The overall probability distribution for these combinations is illustrated in Fig. 3.31. If we obtain, e.g., $r = 0.7645$, the following system reaction would be realized according to E_3 as follows:

5 System: *I'm sorry to bother you again. Where did you say you wanted to travel to?*

3.8 A Semi-Stochastic Combined Emotional Dialogue Model

Having introduced both dialogue and emotional models separately, we will now describe our combined approach to an emotional dialogue model taking into account both dialogue fields and the appropriate stylistic realization of the respective prompts (Pittermann et al. 2007b). Such an emotional dialogue model consists of a predefined number O of states $\mathcal{S} = \{\mathcal{S}_1, \mathcal{S}_2, \ldots, \mathcal{S}_O\}$ with $O = |\mathcal{S}| \leq M \cdot N$, where M is the number of dialogue fields and N is the maximum number of emotional states which are "attached" to each dialogue field. I.e., basically each state $\mathcal{S}_i$ is represented by a certain dialogue field the prompt of which is realized according to a certain emotional state of the user. The structure of the emotional dialogue model is illustrated in Fig. 3.32.

For the sake of clarity, this illustration only contains an extract of a network consisting of $O = 12$ states in total. I.e., combining the example models in Sections 3.6 and 3.7, there are $M = 4$ dialogue fields P_i and $N = 3$ emotional states E_j which now coincide in all possible field-emotion combinations $P_i : E_j, i = 1, \ldots, M$,

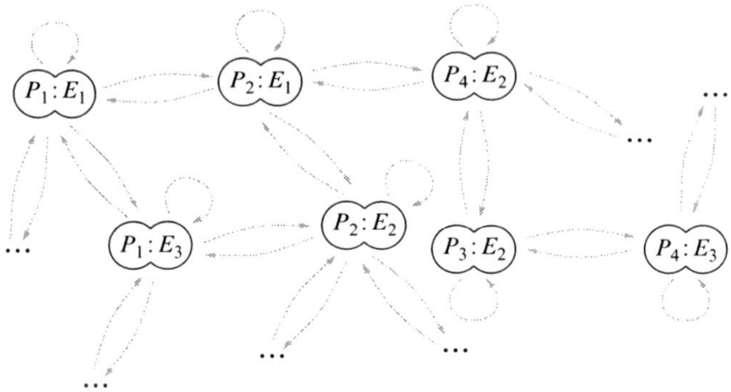

Fig. 3.32 Extract of a network representing a semi-stochastic emotional dialogue model. Each state represents a combination of dialogue plus emotional state

$j = 1, \ldots, N$. Following the notations in the above sections, we can now define a mapping of states

$$\mathcal{S}_k \to \left(P_i : E_j\right), \tag{3.24}$$

where $k = (i-1) \cdot N + j$. As described in the previous section, the emotional attachment can either be a (quantized) numerical value or an arbitrary string like "anger", "neutral", etc. To avoid confusion with the field labels which are typically names like "departure_city" or "date", we limit our considerations to numerical values so that a state in the model can be, e.g., "destination:0.7". It should be noted that there are a predefined number M of dialogue fields, whereas the number of emotional states per field can definitely differ from field to field. E.g., for whatsoever reason there can be "date:0.3", "date:1.0" and "date:1.7" but also "time:0.7" and "time:1.3". The correct mapping of the emotional states happens by appropriate quantization of the training data which are typically not quantized yet. Using different numbers of emotional states per field and/or different quantization steps, however, increase the model complexity significantly. Thus, to simplify matters, we describe our approach by means of a model in which there are the same emotional states E_j, $j = 1, \ldots, N$ and the respective values attached to each field. The fields and states are then defined beforehand in the dialogue description as shown in the VoiceXML excerpt in Fig. 3.33.

Based on this scaffolding, we describe the transitions between the states by the probabilities $P\left(\mathcal{S}|\underline{\mathbf{H}}_D\right)$ of a target system state given the dialogue history $\underline{\mathbf{H}}_D$. Here, the history includes the previously filled dialogue fields as well as the previously recognized emotional states. I.e., expanding Eq. 3.1, we consider now the probability $P_{pe}(p_S(T) = P_i, e_S(T) = E_j|[p(1), \ldots, p(T), e(1), \ldots, e(T)])$ of a system reaction for field P_i in emotional state E_j given the previous user turns referring to fields $p(1), \ldots, p(T)$ in the respective emotional states $e(1), \ldots, e(T)$. Here, it should be noted that $p_S(T)$ is not only dependent on $p(1), \ldots, p(T)$ but also on $e(1), \ldots, e(T)$, and $e_S(T)$ also depends on both fields and emotional states in the

```
...
<field name="date">
  <prompt cond="e1">
    Great, when would you like to depart?
  </prompt>
  <prompt cond="e2">
    On what date would you like to travel?
  </prompt>
  <prompt cond="e3">
    Excuse me, I need to know on what date you would like to
    travel?
  </prompt></field>
<field name="departure_city">
  <prompt cond="e1">
    OK, from where will you leave?
  </prompt>
  <prompt cond="e2">
    From where would you like to depart?
  </prompt>
  <prompt cond="e3">
    Could you please tell me from where you would like to
    depart?
  </prompt></field>
<field name="destination">
  <prompt cond="e1">
    Excellent, and what's your destination?
  </prompt>
  <prompt cond="e2">
    Where would you like to travel to?
  </prompt>
  <prompt cond="e3">
    I'm sorry to bother you again. Where did you say you wanted
    to travel to?
  </prompt></field>
<field name="time">
  <prompt cond="e1">
    Lovely, at what time could you leave?
  </prompt>
  <prompt cond="e2">
    At what time do you want to depart?
  </prompt>
  <prompt cond="e3">
    Could you please also tell me, when you would like to
    depart?
  </prompt></field>
...
```

Fig. 3.33 Excerpt of a VoiceXML representation of our dialogue description predefining (four) dialogue fields and prompts for the respective (three) emotional states of the dialogue fields

past user turns. Again, our terminology is such that if we say that the system is in emotional state E_j, this means that the system reacts as if the user was in this emotional state E_j.

As discussed for the dialogue and the emotional models, the complexity of the stochastic share of the emotional dialogue model increases exponentially with the size of the history to be considered, as the number of turns increases continuously during the course of the dialogue. Thus, we limit our considerations to bi-turn and tri-turn combinations. Given the number of combinations in the individual models from the previous sections, we calculate the overall number as $M \cdot (M - 1) \cdot N^2$ bi-turns and $M \cdot (M - 1) \cdot (M - 2)/2 \cdot N^3$ tri-turns in the emotional dialogue model. In order to account more accurately for interdependencies between fields and emotional states, we propose a threepart model structure including a plain dialogue model, a plain emotional model and the emotional dialogue model as illustrated in Fig. 3.34. Within each of the three sub-models, there exist bi-turn and tri-turn combinations as described for the single models in Figs. 3.25 and 3.30. The same types of transitions also occur in the combined model, however, the illustration of which would be too complex and confusing.

Consequently, the transition probabilities are determined on the basis of training data. As discussed before, for the plain dialogue model, we do not include any temporal aspects in the tri-turn combinations, whereas the chronological sequence is very important in emotional combinations to keep track of the user's state. Thus, the plain dialogue and emotional sub-models concur with the ones illustrated in Fig. 3.25 and Fig. 3.30. As described there, the two previous user states are connected by thick solid lines and the transition occurs from the grey point between these states to the possible following system state. In the plain emotional and combined sub-models, the temporal aspect is illustrated with the aid of triangular connection lines.

With respect to the structure of the model, the field and emotional state selection process during the dialogue is subdivided into two sub-processes:

1. The current field under discussion P_i is determined on the basis of the plain dialogue sub-model and the combined emotional dialogue sub-model.
2. the corresponding emotional state is selected among the combinations occurring in the plain emotional sub-model and also the combined emotional dialogue sub-model.

Although both selection sub-processes are based on the same dialogue history and (partly) the same dialogue and emotional sub-models, the respective selection processes are independent of each other. In principle, both selection sub-processes bear a strong resemblance with the processes described in the previous sections.

Given the previous one or two dialogue and recognized emotional state/s $p(T)$, $e(T)$ and optionally $p(T - 1), e(T - 1)$, we can form two sets $\mathcal{V}_p$ and $\mathcal{V}_e$ which contain all possible bi-turn and tri-turn combinations separated into dialogue state (p) and emotional state (e) in all sub-models.

For each bi-turn or tri-turn combination in each model, we define the frequency of occurrence, and, as described for the plain dialogue and emotional models, we employ an amplification factor $\alpha \geq 1$ to emphasize the significance of tri-turn combinations as opposed to bi-turn combinations within one model. Moreover, we argue that the combined emotional dialogue sub-model plays a more important role in the overall model than the other two plain submodels and that the plain dia-

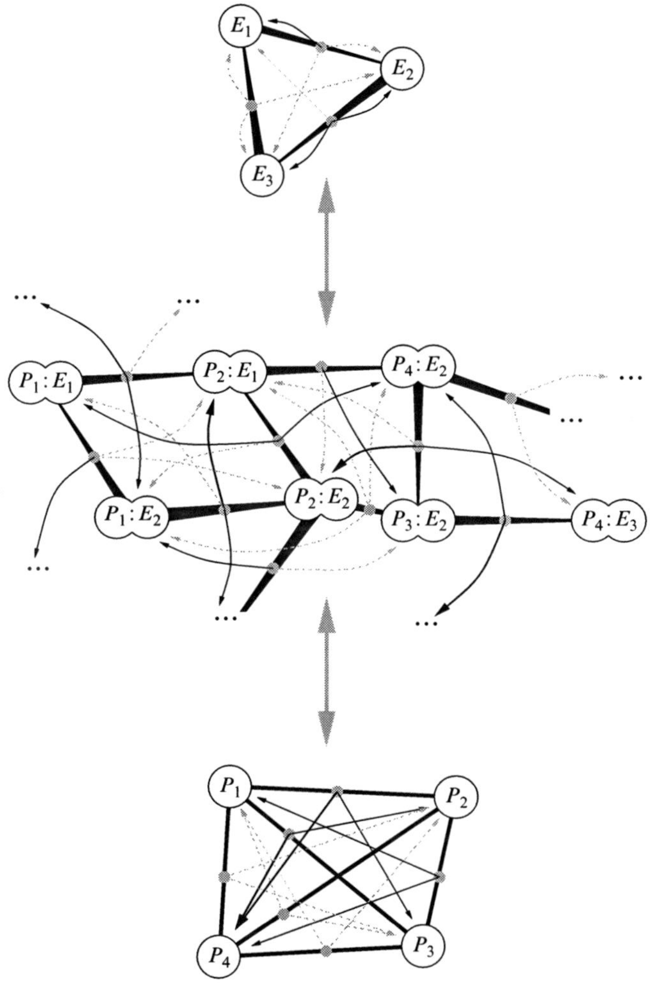

Fig. 3.34 Partial illustration of the trained emotional dialogue model. Instead of the full model representation, some selected tri-gram combinations are shown. P_1 represents the departure city, P_2 is the destination, P_3 is the travel date and P_4 is the time of departure. E_1 represents a happy state, E_2 is rather neutral and E_3 is an angry state

logue and emotional sub-models should only contribute to a more robust adaptation process in cases where the combined model is not trained well enough. Thus, we introduce a second amplification factor $\beta \geq 1$ which emphasizes the significance of the combinations in the combined emotional dialogue sub-model. Taking into account these variations, we can calculate the probabilities of each field P_i and each emotion E_j.

Based on the calculated probability distributions $P(P_1), \ldots, P(P_M)$ and $P(E_1), \ldots, P(E_N)$, the system's new dialogue and emotional states are determined by random selection.

In order to illustrate both selection sub-processes, we extend the example from Section 3.6 by the emotional control parameters $E(U)$ $(0 \leq E(U) \leq 2)$ as already done in the example in the previous section:

	...
2 User:	*I want to book a flight to Copenhagen.* [0.8954]
3 System:	*From where would you like to depart?*
4 User:	*From Paris.* [1.5446]

Here, after the second user turn, the dialogue history contains $p(T-1)$ = destination, $p(T)$ = departure_city, $e(T-1) = 0.8954 = E_2$ and $e(T) = 1.5446 = E_3$.

Then, e.g., if the random processes return the combination $P_4 : E_2$, the respective system turn would be

5 System: *At what time do you want to depart?*

3.9 Extending the Semi-Stochastic Combined Emotional Dialogue Model

In the previous section, we have presented our approach to integrate one dialogue control parameter (in this case the user's emotional state) into a semi-stochastic dialogue model. Having described the summarization of an arbitrary number of dialogue control parameters in the beginning of this section, we will now give a short overview on how the semi-stochastic model approach could theoretically be extended to include $N \geq 1$ extra dialogue control parameters $p_1, \ldots, p_N$. Assuming M_0 plain dialogue states and M_i states per parameter $p_i, 1 \leq i \leq N$, we obtain a total of $M_0 \cdot M_1 \cdot \ldots \cdot M_N$ states in the combined model. In addition to this combined sub-model integrating all parameters, there also exist $\binom{N+1}{1} = N+1$ plain dialogue and parameter sub-models, $\binom{N+1}{2}$ sub-models combining two of the parameters out of $p_0, \ldots, p_N$, $\binom{N+1}{3}$ sub-models combining three of the parameters, ..., and $\binom{N+1}{N} = N+1$ sub-models combining N parameters. I.e., the entire model consists of a total of $2^{N+1}-1$ sub-models. The excerpt of such a model is illustrated in Fig. 3.35. Here, the states of a parameter p_i are labeled as $P_i^{(j)}$, where $0 \leq i \leq N$ and $1 \leq j \leq M_i$. The model which we consider in this example includes $N = 3$ parameters with $M_0 = 4$, $M_1 = 3$, $M_2 = 2$ and $M_3 = 3$ states per parameter. For the sake of a better overview, only 4 of the 15 sub-models are depicted in the illustration.

Within the individual sub-models there exist a variable number of state transitions. Assuming the same limitations for the plain dialogue sub-model like in the above sections and assuming full flexibility among the other parameters, the numbers of bi-turns, tri-turns and states are as follows:

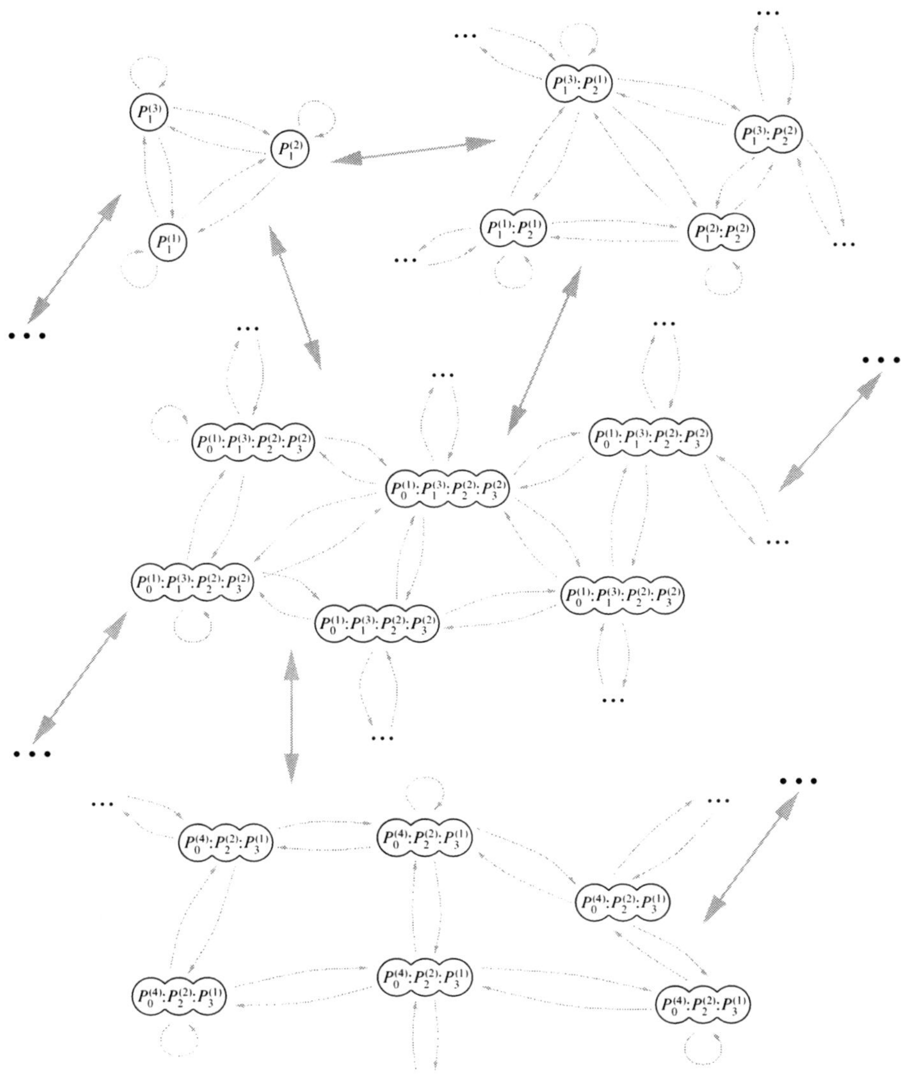

Fig. 3.35 Illustration of the model approach including a higher number N of additional dialogue parameters by means of $N = 3$ parameters with $M_0 = 4$, $M_1 = 3$, $M_2 = 2$ and $M_3 = 3$

- $M_0 \cdot (M_0 - 1)$ bi-turns, $M_0 \cdot (M_0 - 1) \cdot (M_0 - 2)/2$ tri-turns in the plain dialogue sub-model consisting of M_0 states.
- M_k^2 bi-turns, M_k^3 tri-turns in each of the other plain parameter sub-models, $1 \leq k \leq N$, each consisting of M_k states.
- $M_0 \cdot (M_0 - 1) \cdot M_k^2$ bi-turns, $M_0 \cdot (M_0 - 1) \cdot (M_0 - 2) \cdot M_k^3/2$ tri-turns in a sub-model combining dialogue fields and one additional parameter, $1 \leq k \leq N$, consisting of $M_0 \cdot M_k$ states.

- $M_k^2 \cdot M_l^2$ bi-turns, $M_k^3 \cdot M_l^3$ tri-turns in a sub-model combining two additional parameters, $1 \leq k, l \leq N$, $k \neq l$, consisting of $M_k \cdot M_l$ states.
- $M_0 \cdot (M_0 - 1) \cdot M_j^2 \cdot M_k^2$ bi-turns, $M_0 \cdot (M_0 - 1) \cdot (M_0 - 2) \cdot M_k^3 \cdot M_l^3/2$ tri-turns in a sub-model combining dialogue fields and two additional parameters, $1 \leq k, l \leq N$, $k \neq l$, consisting of $M_0 \cdot M_k \cdot M_l$ states.
- ...
- $M_0 \cdot (M_0 - 1) \cdot M_1^2 \cdot \ldots \cdot M_N^2$ bi-turns, $M_0 \cdot (M_0 - 1) \cdot (M_0 - 2) \cdot M_1^3 \cdot \ldots \cdot M_N^3/2$ tri-turns in a sub-model combining dialogue fields and all additional parameters, consisting of a total of $M_0 \cdot M_1 \cdot \ldots \cdot M_N$ states.

Summarizing all terms appropriately, we can calculate the total number of transitions in the over-all model as

$$\left[(M_0 \cdot (M_0 - 1) + 1) \cdot \prod_{i=1}^{N} \left(M_i^2 + 1 \right) \right] - 1 \text{ bi-turns,} \tag{3.25}$$

and

$$\left[(M_0 \cdot (M_0 - 1) \cdot (M_0 - 2)/2 + 1) \cdot \prod_{i=1}^{N} \left(M_i^3 + 1 \right) \right] - 1 \text{ tri-turns.} \tag{3.26}$$

The total number of states is calculated as $\prod_{i=0}^{N} (M_i + 1) - 1$.

Considering the number of parameters of our example, we obtain 12 bi-turns and 12 tri-turns in the plain dialogue sub-model, 9 bi-turns and 27 tri-turns in the first plain parameter sub-model, 4 bi-turns and 8 tri-turns in the second plain parameter sub-model and 9 bi-turns and 27 tri-turns in the third plain parameter sub-model. Among the sub-models including dialogue fields and one of the additional parameters, there are, e.g., 108 bi-turns and 324 tri-turns in the sub-model combining parameters 0 and 1.... The sub-model combining all three additional parameters (1, 2 and 3) contains 324 bi-turns and 5832 tri-turns, and finally, the sub-model including all parameters contains 3888 bi-turns and 69984 tri-turns. In total, this integrated dialogue model consists of 15 sub-models, a total of 239 states, 6499 bi-turn and 91727 tri-turn state transitions, requiring a sufficiently large dialogue training database to obtain a robust dialogue behavior. In cases where the training data is not that extensive, the ε parameter, also used in the previous sections, helps to avoid dead-lock situations and can, in any case, contribute to a successful dialogue.

The training of the model is accomplished by counting the frequency of occurrence of each possible bi-turn or tri-turn combination of the respective sub-model in the training data. Labeling the states of a parameter i as $P_i^{(j)}$, $0 \leq i \leq N$, $1 \leq j \leq M_i$, the model can be described as shown in Fig. 3.36. The figure describes the representation of bi-turns and tri-turns in the plain sub-models as well as in differently combined sub-models up to a tri-turn in the sub-model combining all parameters.

```
P0(1):::                          -> P0(2):::                    83.01
P0(1):::                          -> P0(3):::                    32.01
...
P0(1):::, P0(2)                   -> P0(3):::                    39.01
...
P0(3):::, P0(4)                   -> P0(2):::                    13.01
P0(1):P1(1)::                     -> P0(2):P1(1)::               27.01
...
P0(4):P1(3)::                     -> P0(3):P1(3)::               54.01
P0(1):P1(1)::, P0(2):P1(1)::      -> P0(3):P1(1)::               28.01
P0(1):P1(1)::, P0(2):P1(1)::      -> P0(3):P1(2)::                5.01
...
P1(3):P2(2)::, P1(2):P2(1)::      -> P1(1):P2(2)::                2.01
...
P2(2):P3(2)::, P2(2):P3(1)::      -> P2(1):P3(2)::               26.01
...
P0(2):P2(2):P3(2):                -> P0(1):P2(1):P3(2):          14.01
...
P1(2):P2(2):P3(2):, P1(3):P2(2):P3(2):
                                  -> P1(1):P2(1):P3(1):           3.01
...
P0(4):P1(2):P2(2):P3(2)           -> P0(3):P1(1):P2(1):P3(1)  12.01
...
P0(4):P1(3):P2(2):P3(3), P0(4):P1(3):P2(2):P3(2)
                                  -> P0(4):P1(3):P2(2):P3(3)  17.01
```

Fig. 3.36 Excerpt of a trained extended dialogue model including three additional parameters ($\varepsilon = 0.01$)

During the dialogue, the selection of the current states / parameters is subdivided into $N + 1$ random processes in which $p_{i\,S}$ is determined on the basis of the probabilities of a state $P_i^{(j)}$ given the previous $(p_0(T) : p_1(T) : \cdots : p_N(T))$ and optionally the penultimate $(p_0(T-1) : p_1(T-1) : \cdots : p_N(T-1))$ user state / turn. In the calculation of these probabilities, an amplification factor $\alpha \geq 1$ is used to emphasize the contribution of tri-turn combinations as opposed to bi-turn combinations. Moreover, a further amplification factor $\beta \geq 1$ is integrated to emphasize the significance of sub-models containing a higher number of parameters. Actually, such a weighting can be fitted individually for each sub-model, however, to keep things simple, we suggest to use β^{n-1} as an amplification factor for a sub-model containing n parameters. I.e., the contributions from the plain sub-models would be multiplied by 1 and the probabilities from the sub-model including all parameters would be multiplied by β^{N-1}.

Having calculated all $M_0 + M_1 + \cdots + M_N$ possible probabilities, $N + 1$ probability distributions are given for a random process generating a number $r_i \in [0, 1]$, $0 \leq i \leq N$ which determines the respective system state. Four exemplary distributions for $N = 3$ additional dialogue control parameters are illustrated in Fig. 3.37.

Each region in the distributions is labeled with the respective $P_i^{(j)}$ which will be selected if the value of r_i is in that region. The width of each region equals the probability $P(P_i^{(j)})$ and the values of the region boundaries t_j are rounded to two

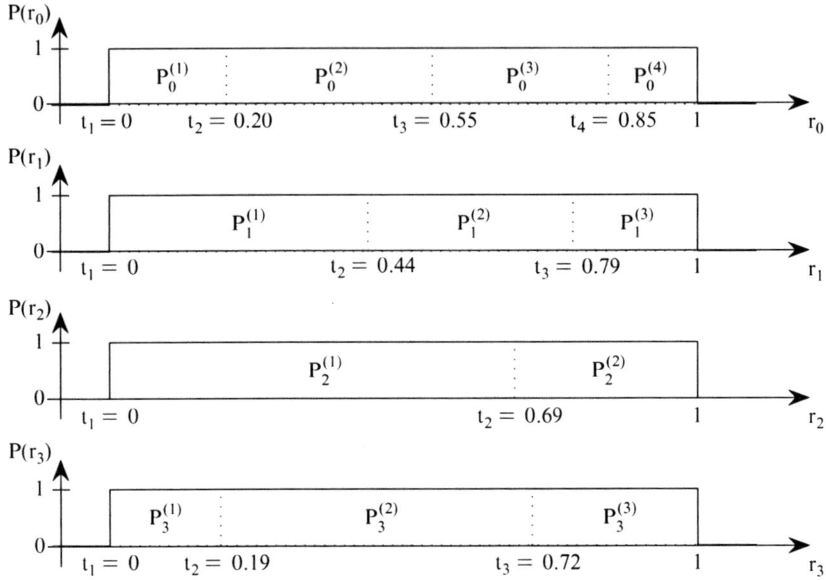

Fig. 3.37 Illustration of the $N + 1$ selection sub-processes in an extended dialogue model with $M_0 = 4$, $M_1 = 3$, $M_2 = 2$ and $M_3 = 3$

digits after the decimal point. If, e.g., the random processes generate $r_0 = 7834$, $r_1 = 0.5397$, $r_2 = 1265$ and $r_3 = 0.8943$, the respective system output would be according to $P_0^{(3)} : P_1^{(2)} : P_2^{(1)} : P_3^{(3)}$.

3.10 Discussion

There exist a large variety of approaches to integrate dialogue-influencing parameters in spoken dialogue. In this chapter we have picked up existing approaches to adaptive dialogue management and we have proposed stochastic dialogue models exceeding the flexibility of rule-based dialogue strategies. The stochastic model behind this idea constitutes a basis for our emotional dialogue model described in Sections 3.8 which we refer to as semi-stochastic model due the fact that all states (fields and emotions) including their properties are predefined whereas the transitions between the states are defined by probabilities. Instead of proposing the ultimate fashion how to integrate emotions in the dialogue, we propose this model as a basis for dialogue adaptation.

All of proposed models have in common the probability $P(O_t|I_{t-1}, I_{t-2}, \ldots)$ of a system reaction O_t at time t given previous input I_{t-1}, I_{t-2}, ... In the first case, O_t is the confirmation strategy which depends on the previous confidence measures. In the latter case, O_t is a dialogue field with an attached emotional state the probability of which depends on the fields previously addressed by the user and the previously recognized emotional states. In both scenarios, we face the problem that the output of upstream modules such as speech and emotion recognizers is error-prone which leads to problems in rule-based approaches whereas stochastic models can be trained (using appropriate material) to handle recognition errors as well. Nevertheless, the quality of these models stands or falls with the complexity and quality as well as the availability of the training material.

Chapter 4
Hybrid Approach to Speech–Emotion Recognition

In order to be able to be responsive to the user, it is vital for human–computer interfaces to correctly determine a user's current emotional state. To perform this emotion recognition, a large variety of ideas and approaches utilizing different modalities can be implemented. Among these, typically, audio and video data or signals from biosensors apply in state-of-the-art systems. With the aid of cameras, the user's physical behavior, including facial expressions or gestures, is captured and correlated to the respective emotional state (Ioannou et al. 2005; Busso and Narayanan 2006). Similarly, physiological signals like skin conductance, heart rate, blood pressure, finger temperature or an electromyogram are also involved in the recognition of the user's emotional state (Nasoz et al. 2003; Peter and Herbon 2006).

Depending on the application, different modalities are utilized. Emotion recognition based on gestures or facial expression, e.g., applies in tele-learning or in techniques enhancing drivers' or pilots' safety where users typically do not permanently talk. Measuring physiological signals is only possible in scenarios where users explicitly agree to be monitored as there are sensors attached to their wrists, upper arms or faces. In many applications, however, the use of biosensors or cameras is rather impractical or even impossible, e.g., in telephony-based SLDSs. Thus, in this chapter we discuss the recognition of emotions based on features extracted from the speech signal. At first, we describe the signal processing and the extraction of emotion-related features. Then we give details on classification methods and the actual recognition process.

Due to the fact that our emotion recognizer is closely related to speech recognizers, especially in terms of feature extraction and classification, we will also go into detail about algorithms and approaches, which have been originally developed for speech recognition and are now typically used in this area, before describing their application in our emotion recognizer. We propose a system architecture as illustrated in Fig. 4.1. In addition to stand-alone speech and/or emotion recognizers, described in Sections 4.4 and 4.5, we describe our integrated approach combining both speech and emotion recognition, and we show, how both speech and emotion recognizers can benefit from such a cooperation. This proposed combination is also motivated by the similarity of speech and emotion recognition regarding signal processing and classification. Thus, we also describe selected aspects of plain speech recognition before going into detail about our emotion recognizers.

J. Pittermann et al., *Handling Emotions in Human-Computer Dialogues*,
DOI 10.1007/978-90-481-3129-7_4,

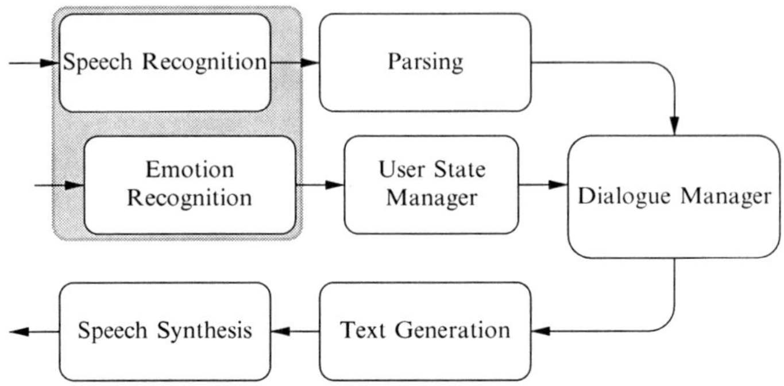

Fig. 4.1 SLDS applying a combined speech–emotion recognizer (Pittermann et al. 2007b)

Independent of the different speech signal analysis and classification methods, emotional states, moods or at least tendencies may also be determined from the linguistic content of texts. In Section 4.7, we outline an approach to recognize emotions with the aid of grammars. This includes the integration of such a recognizer with our speech–emotion recognizers as described in Sections 4.5 and 4.6.

4.1 Signal Processing

Before the actual recognition of speech or emotions is performed, the speech signal needs to be processed, i.e., relevant features need to be extracted.

In regular speech recognition the commonly used features are either coefficients calculated with the aid of a linear prediction analysis (Makhoul 1975) or cepstral features determined from the output of a filterbank analysis or perceptual linear prediction coefficients (Hermansky 1990) in different variations (Hermansky and Morgan 1994). Furthermore, energy measures and time derivatives of the features are included. Profound explanations of the feature extraction techniques are given in Fallside and Woods (1985), Deller et al. (1993), Robinson (1997a) and Young et al. (2006).

For the recognition of emotions, further types of features such as prosodic ones are used. Thus, having outlined the speech signal preprocessing and the characteristics of the speech signal leading to linear prediction, we concentrate on the description and calculation of cepstral coefficients and on the extraction of prosodic features as these are used in our system.

Considering speech signal processing from the information theoretical point of view, the speech signal may be regarded as a message consisting of information, redundancy and irrelevance. Here, we assume that the message originates from a source which produces events $e_1, e_2, \ldots$ with different probabilities $p_1, p_2, \ldots$.

Constituting the relevant proportion of such a message, the information $I(e_i)$ of an event e_i is described by

$$I(e_i) = -\log_2 p_i, \tag{4.1}$$

and the expectation value of the information of all events is defined as the entropy $H(X)$ of the source where

$$H(X) = \sum_i p_i \cdot I(e_i) = -\sum_i p_i \cdot \log_2 p_i. \tag{4.2}$$

The entropy is maximal ($H(X) = H_{\max}$) when all probabilities are equal. Redundancy is then defined as the difference between $H_{\max}$ and $H(X)$. Irrelevance can not be described mathematically as its definition actually depends on the perception of a human viewer or listener. This plays a role in so-called lossy compression techniques which apply in image, video or audio container formats like JPEG, MPEG or MP3 (cf. Schroeder et al. 1979; Wallace 1991) where irrelevance is removed to a certain extent. For pure data transmission it is vital that all information is transmitted without corruption, i.e., there, lossless compression techniques only reducing redundancy apply. The use of such source coding methods for information transmission is circumstantially described in Johannesson (1988) or Reza (1961).

While the above mentioned coding techniques could theoretically be applied for speech or emotion recognition as well, their effect on the recognition performance would be mainly negative. This is not due to the low compression provided by such methods but mainly due to characteristics of speech signals which differ strongly from those of a regular random signal as typically presumed from an information source. Thus, contrarily, different coding techniques are applied to extract relevant characteristics (features) from the speech signal. The most prominent representatives for speech recognition are Mel-frequency cepstral coefficients (MFCCs). These are calculated every 10 ms and the feature vector typically contains 39 elements (including energy, first and second order regression coefficients) which seems to be suitable to describe phoneme state transitions in Hidden Markov Models (HMMs) (Young et al. 2006).

Features such as MFCCs are typically used in automatic speech recognition, as these cater for a robust and reliable recognition performance independent of the speakers and the accompanying different characteristics of voices which also change according of the speakers' emotional states. Intuitively, one would now argue that such features are rather not useful for the recognition of emotions and, thus, instead, rather prosodic or acoustic characteristics like pitch, intensity (energy, loudness), hamonicity, jitter, etc., could be used. Depending on the classifier different time intervals for the calculation of the features may apply. Whereas for HMMs which are used in speech recognition and which allow for temporal aspects the features can be calculated at theoretically any rate from once per sample to once per utterance, classifiers like feedforward neural networks settle for one feature vector per utterance.

Thus, various approaches to optimizing the feature set with respect to computational complexity and recognizer performance have been presented in literature. A qualitative investigation on how prosody and its attitudinal effects relate to certain emotions has been presented in Wichmann (2000). Among these intonationally relevant cues are, e.g., pitch (a wide range is commonly associated with strong emotions, a narrow pitch range plus positive orientation might indicate boredom) or nasalized sounds and pitch contour (high fall vs. low fall). Similar relations between tones and emotional or attitudinal labels have been discussed by Cauldwell (2000) based on observations presented in, e.g., Crystal (1969). Among these findings are that, e.g., falling tones relate to anger, impatience, irritation or satisfaction, rise-falling tones relate to excitement, pleasure or amusement whereas a leveled constant tone indicates boredom. The same line is taken by Potapova and Potapov (2005), who describe different emotional states by the average pitch level of an utterance, pitch level of stressed syllables and the pitch range as well by tendencies of the speech rate. For the distinction of emotional / non-emotional utterances, the pitch contour on word and turn levels is considered in Rotaru and Litman (2005). Their results show that compared to turn-level features, the use of word-level pitch features leads to a better emotion recognition performance. Fotinea et al. (2003) suggest to determine the characteristics in four different sub-bands of the speech signal and to include vocal cord openings in the estimation of the pitch contour. Even in tonal languages such as Chinese, the shape of pitch curves correlates with emotional states as described in Yang and Campbell (2001).

As for the selection of features used in our experiments we stick with the majority of the approaches described above limiting our considerations on acoustic, prosodic, spectral and linguistic characteristics. Particular attention is paid to the robustness as well as to an efficient computation of these features. Our choice of features for plain emotion recognition includes prosodic and acoustic features such as pitch, intensity and formants plus their computational statistics such as mean, minimum, maximum, variance, etc. For our hybrid approach to speech–emotion recognition, we also include Mel-frequency cepstral coefficients as these benefit a robust word recognition performance of our system.

4.1.1 Preprocessing

Before the actual features are calculated, the speech signal undergoes certain preprocessing steps. Among these are pre-emphasis and windowing (Nuttall 1981).

When people speak, the speech signal experiences a certain spectral roll-off due to the radiation effects of the sound from the mouth. In the spectrum the signal energy of speech generally decreases as the frequency increases. Especially in the calculation of linear prediction coefficients this leads to the problem that important information about specific sounds are lost, as the analysis wrongly focuses on the predominant low frequencies. To avoid that effect, the signal is flattened with the aid of a low-pass ("pre-emphasis") filter described by

$$y(n) = x(n) - \alpha \cdot x(n-1), \tag{4.3}$$

where the factor α typically, also in our system, defaults to 0.97.

In the calculation of the features, on the one hand, the temporal aspects of the speech signal shall be accounted for. But also, on the other hand, some operations require a stationary signal. Thus, the quasi-stationarity of speech is exploited and the speech signal is subdivided into small blocks in which the signal is assumed to be stationary. The signal is multiplied by a window function, like the Hamming window, which is a raised cosine defined as

$$w(n) = 0.54 - 0.46 \cdot \cos\left(\frac{2\pi n}{N-1}\right), \qquad 0 \le n \le N-1, \tag{4.4}$$

where N is the window size. Outside the defined range, $w(n)$ is set to 0. Considering the dependency between adjacent samples, overlapping windows are used as shown in Fig. 4.2. Within these frames, defined by the respective windows, feature parameters are calculated. For speech signals, the quasi-stationarity is fulfilled within time spans of approximately 20 to 30 ms. Thus, in speech applications, the window size (window duration) is typically 25 ms and the frame period, i.e., the time between successive frames, is 10 ms.

Further preprocessing steps include the analog-to-digital conversion (quantization) as well as noise reduction procedures like spectral subtraction, where an estimated spectrum of the noise is subtracted from the spectrum of the noisy speech signal (Linhard and Haulick 1999).

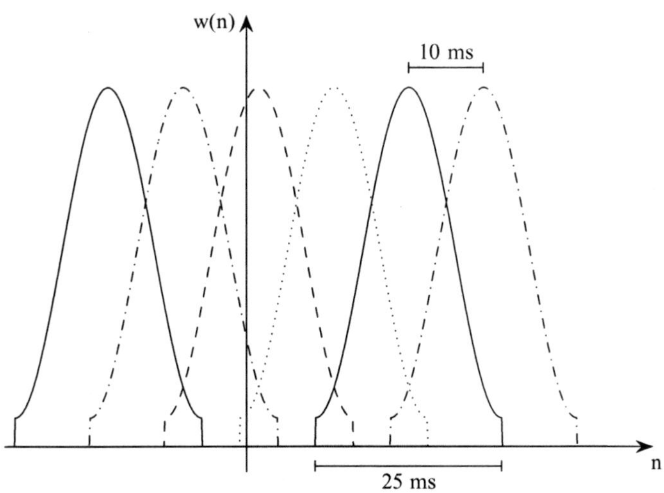

Fig. 4.2 Overlapping window function for speech signal preprocessing

4.1.2 Linear Prediction

The linear prediction (LP) analysis is employed to estimate future values $s_p(n)$ of the discrete-time speech signal as a linear function of previous samples' values. This linear function is typically implemented as a finite impulse response filter with impulse response $h(n)$, so that $s_p(n)$ is calculated as follows:

$$s_p(n) = \sum_{i=1}^{m} h(i) \cdot s(n-i). \tag{4.5}$$

The optimal filter coefficients are calculated every 10 or 20 ms based on a minimum mean square algorithm, demanding that the power of the prediction error $e(n) = s(n) - s_p(n)$ is minimal. This demand can be met by employing the autocorrelation of the signal as shown in Yule (1927) or Hermansky (1990).

4.1.3 Mel-Frequency Cepstral Coefficients

The human ear perceives frequencies non-linearly across the spectrum. In a perceptual experiment conducted by Stevens et al. (1937) listeners were asked to define an equidistant scale of pitches according to their subjective impressions. Based on the results, the mel scale has been defined as

$$m = 2595\text{mel} \cdot \log_{10}\left(1 + \frac{f}{700}\right), \tag{4.6}$$

where f is the frequency in Hz.

This scale applies in the filterbank analysis which is used to obtain a non-linear frequency resolution of a speech signal. In this filterbank, p filters, typically triangular filters, are distributed according to the mel scale of equidistant frequencies as shown in Fig. 4.3 (see also Young et al. 2006).

In the analysis, the windowed speech signal is Fourier-transformed and the filterbank coefficients m_i, $1 \leq i \leq p$ are calculated by correlating the magnitudes of transformed signal $|S(N)|$ with the respective triangular filter. I.e., if one of the triangular filters is described by $T_i(N)$ in the frequency domain, m_i is calculated as follows:

$$m_i = \sum_{N=-\infty}^{\infty} |S(N)| \cdot T_i(N). \tag{4.7}$$

Alternatively, instead of the signal magnitude, also the signal power $|S(N)|^2$ or the logarithm of the signal magnitude $\log |S(N)|$ (leading to "log filterbank coefficients") can be included in the formula.

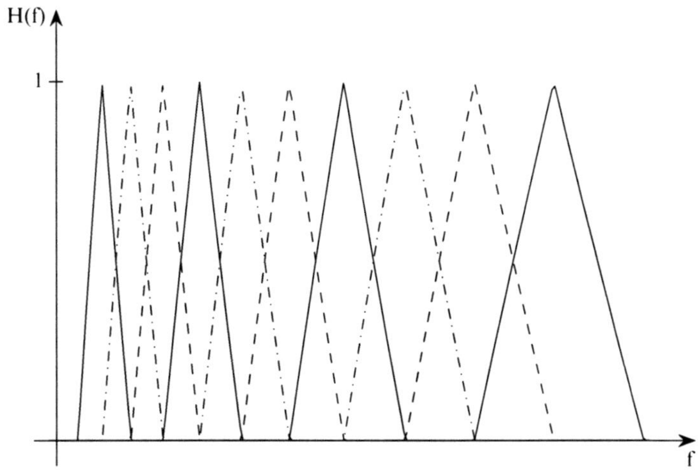

Fig. 4.3 Mel-scale filterbank

In most speech (recognition) applications, however, cepstral features are employed. The cepstrum analysis (see Bogert et al. 1963; Oppenheim and Schafer 2004) typically applies when the effects of the signal source and of a (eventually time-variant) system's transfer function need to be distinguished. The cepstrum is defined as the inverse Fourier transform of the logarithm of the Fourier transform of a time signal.

Mel-frequency cepstral coefficients combining mel-filterbank analysis and cepstrum analysis are obtained by applying a discrete cosine transform to the log filterbank coefficients m_i as

$$c_j = \sqrt{2/p} \sum_{i=1}^{p} m_i \cdot \cos\left(\frac{\pi j}{p}(i - 0.5)\right), \tag{4.8}$$

where p is the number of filterbank coefficients (see Young et al. 2006).

4.1.4 *Prosodic and Acoustic Features*

Conventional speech recognition exclusively aims at the text regardless of the manner and the speaker. In order to recognize emotional states, however, it is particularly necessary to determine how the respective text is expressed. Most of the relevant acoustic properties are described by prosodic parameters like intonation and stress (see Bolinger 1989; 't Hart et al. 1990; Burzio 1993; Kochanski et al. 2005).

In tonal languages like Chinese (cf. Sun et al. 2006) or Thai, pitch (tone) is used to distinguish different meanings of a word or a syllable. E.g., in Mandarin, the syllable "ma" has four different meanings including "mother" and "horse" depending on the applied pitch contour (high pitch, rise, fall & rise, fall) and it may also denote a question when pronounced in a neutral tone (low pitch). In contrast, European languages like English or German are referred to as intonation languages as in these language pitch is primarily used in a syntactical context, e.g., to distinguish questions from statements (Ladd 1996; Gussenhoven 2004; Jun 2005).

Moreover, in intonation languages, prosody is also employed, among syntax distinctions including irony (see Tepperman et al. 2006) and surprise, to express the speaker's emotional state. In SLDSs this is, on the one hand, utilized in speech synthesis to make the utterance sound more "natural" (Rank and Pirker 1998; Iida et al. 2003), but also, on the other hand, to detect the user's current emotional state (McGilloway et al. 1995; Koike et al. 1998; Fujisawa and Cook 2004).

Without having to extract the (textual) content or the meaning of a speech signal, prosodic and acoustic features can be extracted from the signal. Among these features are pitch (fundamental frequency) and intensity (volume), constituting the most relevant ones, as well as formants, jitter, shimmer, harmonicity, duration and speech rate. The formants represent the harmonics, i.e., integer multiples of the fundamental frequency, created by the resonance properties of the human vocal tract. Jitter, shimmer and harmonicity describe the voice "quality", where jitter considers the variation of the fundamental frequency and shimmer considers the peak amplitudes (cf., e.g., Baken and Orlikoff 2000). Harmonicity is the harmonics-to-noise ratio defined by energies of the periodic part and the noise in a speech signal. In general, for periodic signals, harmonicity is equivalent to the signal-to-noise ratio. Further processing is required, e.g., to determine the speech rate and the duration. To obtain these parameters, it is essential to know the text content of the utterance to compare the length of the contained words or the number of words per time unit to standard values listed in tables or averaged over language- and domain-specific corpora.

The pitch distribution functions of different emotions are displayed in Figs. 4.4 for female speakers and 4.5 for male speakers. The diagrams in both figures are obtained by calculating the probability density functions of the pitch parameters of all utterances in the respective data subset (male/female, anger/.../sadness) taken from the Berlin Database of Emotional Speech (Burkhardt et al. 2005), also described in Section 6.3, provided by the Technical University of Berlin. Comparing the pitch distributions, it can be observed that, for both female and male speakers, there exist differences but also similarities between certain emotion pairs: anger and happiness as well as boredom and neutral show strong similarities, whereas, e.g., sadness and happiness show strong differences. The pitch distributes in a larger range for anger, fear and happiness and the most frequent frequencies are around 200 Hz – 300 Hz. For boredom, neutral and sadness, the pitch distributes in a smaller range and the most frequent frequencies are around 100 Hz to 200 Hz. Comparing female and male speakers, it can be observed that the respective distributions have similar shapes but are shifted due to the fact that men typically speak with a lower-pitched voice than women.

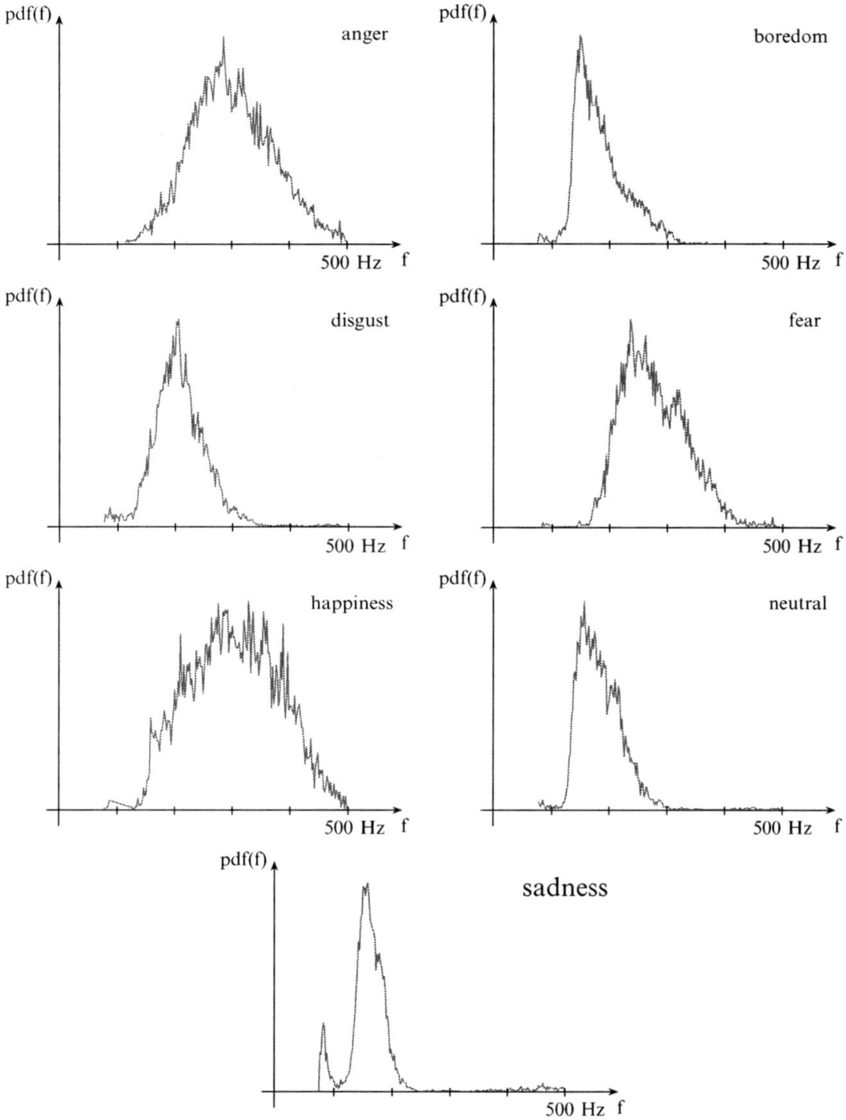

Fig. 4.4 Pitch distributions of different emotions (female speakers)

The application of speech rate and duration in emotional speech is illustrated in Fig. 4.6. Both diagram show the waveform, also taken from the Berlin Database, of a woman pronouncing the sentence *"Die wird auf dem Platz sein, wo wir sie immer hinlegen"* (*"It will be on the place where we always put it."*) in an angry and a sad fashion. At a first glance, it is striking that the angry utterance (2.52 s) is of considerably shorter length than the sad utterance (3.93 s). This is, on the one

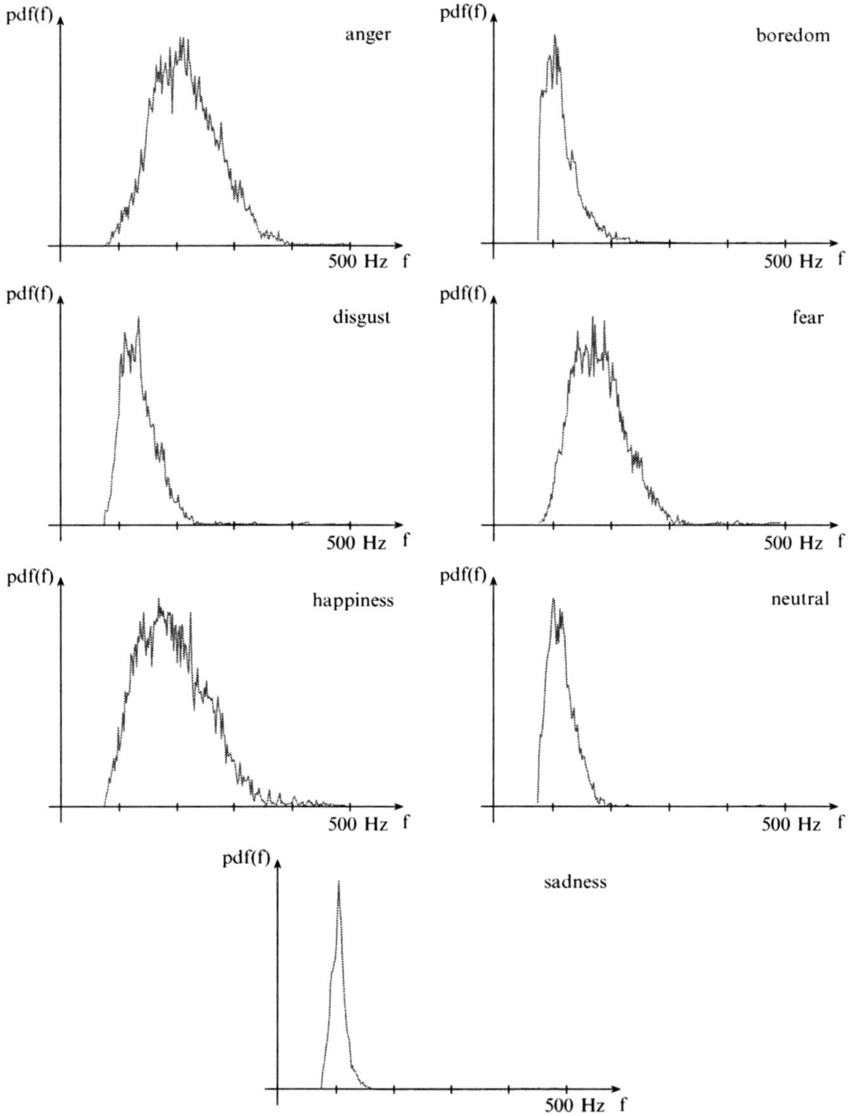

Fig. 4.5 Pitch distributions of different emotions (male speakers)

hand due, to the long pauses between certain words (*"Die wird auf dem* PAUSE *Platz sein,* PAUSE *wo ..."*) leading to a lower speech rate, but also, on the other hand, due to longer duration of single words. In this specific sentence with eleven words, the speech rate is 4.4 words per second for anger and 2.8 words per second for sadness (Fig. 4.6).

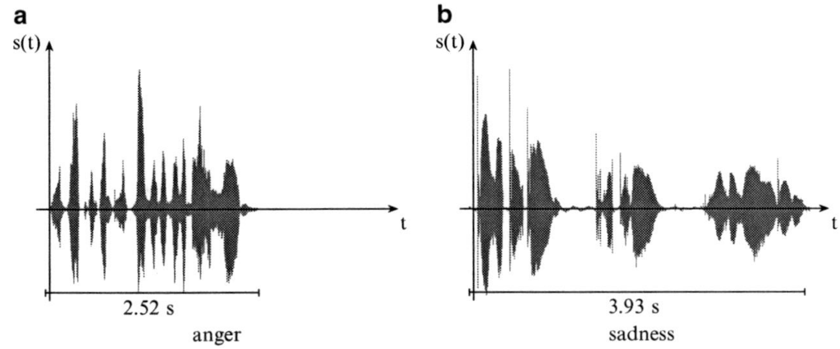

Fig. 4.6 Duration comparison of a sentence uttered in an angry and a sad fashion

In the majority of current speech recognition systems, MFCCs, perceptual linear prediction (PLP) coefficients or optimized parameters based on these principles are employed. To enhance these systems' performance, time derivatives are appended to the feature parameters (Veeravalli et al. 2005). The "regular" speech recognition feature vectors typically consist of 12 MFC coefficients plus an energy measure. First and second order regression (delta and acceleration) coefficients are calculated as difference quotients, typically including 2 previous and 2 succeeding frames (see Young et al. 2006).

In addition to the acoustic and prosodic features described in the previous section, also their computational statistics are of interest in the recognition of emotions. These parameters include minimum, mean and maximum values as well as standard deviation and range of the respective features. A comparison of minimum, mean and maximum values of pitch and intensity for different emotions is shown in Fig. 4.7. In both diagrams the overall range minimum, mean and maximum values of all female or male speakers are represented by black bars (female speakers) and white bars (male speakers). For the sake of clarity, the values of two representative speakers are indicated with 'x' markers. Looking at the pitch diagram in Fig. 4.7 (a), it can again be observed that anger and happiness are among the emotions which contrast most with neutral, whereas boredom is particularly similar to neutral. Moreover, it can be seen that female and male speakers do not only differ in pitch in general but also in the way how an emotion is expressed. E.g., looking at the pitch range, i.e., the difference between minimum pitch and maximum pitch – for male speakers the pitch range of sad utterances is smaller than the range of neutral utterances, whereas for female speakers the pitch range of sad utterances is larger than that of neutral utterances.

Emotions are also distinguishable regarding intensity as shown in Diagram 4.7 (b). As expected it can be observed that angry or happy persons tend to speak louder than bored or sad persons. However, in many speech corpora, the recordings are normalized which means that the intensity is more or less equal for all emotions. Also when the speakers do not keep to a predefined distance to the microphone, the intensity varies significantly. Consequently, in the respective

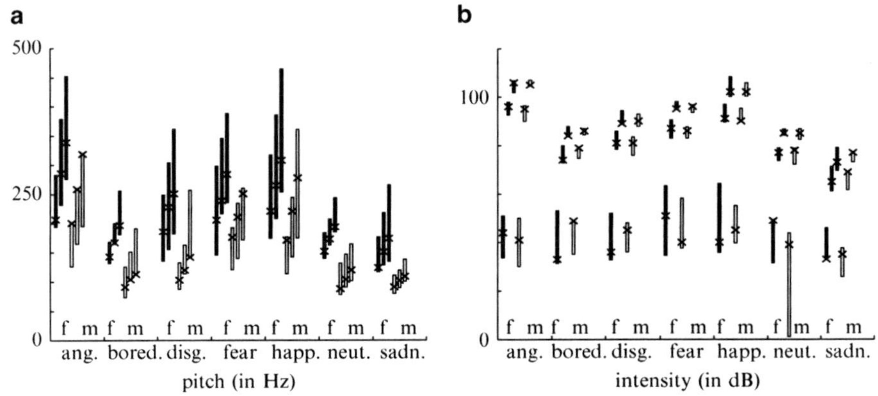

Fig. 4.7 Computational statistics of pitch and intensity for different emotions

experiments, the intensity feature will not have a strong impact on the recognizer performance, however, it can be observed that, e.g., the intensity range of an angry or happy speaker is larger than the range of a neutral utterance. Similarly, for the other features, also not only minimum and maximum values but range is partially included in the feature sets.

The parameters described above and their "derivatives" sum up to a total of 63, $3 \cdot 13$ MFC coefficients plus 24 acoustic/prosodic parameters. Thus, theoretically, there exist $2^{63} - 1 \approx 10^{19}$ combinations that may form feature vectors for speech and/or emotion recognition. As it is obviously impractical to carry out experiments with all these feature combinations, we restrict our considerations to a limited number of combinations (feature vectors) as described in Fig. 4.8.

There exists a large variety of software tools to extract certain features from the speech signal. For our experiments, we use the HCopy tool included in the Hidden Markov Model Toolkit (HTK, Woodland and Young 1993; Young et al. 2006), which allows single file and batch conversions from standard RIFF waveforms or HTK-parameter waveforms to MFCCs, LPCs, PLP coefficients and further feature types which are commonly used in automatic speech recognition. The prosodic and acoustic features are extracted with the aid of Praat (see Boersma 2001, 2002), a software for phonetical purposes. Praat allows the calculation of multiple features like the ones described above, as well as their statistical computations. The calculations can either be performed and controlled from a graphical user interface or in batch mode controlled by Praat scripts containing menu and action commands. Apart from speech analysis and feature calculations, the software can also be used for speech synthesis, labeling and to illustrate the calculated features in a variety of diagram types.

The periodic change of pitch, intensity and the MFCC features of an angry and a sad utterance (both containing the same text), taken from the Berlin Database of Emotional Speech (Burkhardt et al. 2005), are illustrated in Fig. 4.9. At first glance, obvious differences between both emotionally uttered sentences in terms of the

MFCC-13: 13 MFC coefficients

$mfcc_0$	$mfcc_1 \dots mfcc_{12}$

MFCCDA-39: 13 MFC coefficients plus their first (delta, d) and second time derivatives (acceleration, a)

mfcc-13	$mfccd_0$	$mfccd_1 \dots mfccd_{12}$	$mfcca_0$	$mfcca_1 \dots mfcca_{12}$

PAC-24: 24 prosodic and acoustic coefficients: pitch (p), formants (f), intensity (i), jitter (j), harmonicity (h), pitch in voiced parts (vp) and the respective computational statistics including minimum, maximum, range, mean and standard deviation (cs)

p	f_1	f_2	f_3	i	j	h	vp	p_{cs}	i_{cs}	vp_{cs}

MFCPAC-40: 39 MFC-DA coefficients and pitch

mfcc-39	p

MFCPAC-41: 39 MFC-DA coefficients, pitch and intensity

mfcc-39	p	i

MFCPAC-44: 39 MFC-DA coefficients, pitch, intensity and formants

mfcc-39	p	i	f_1	f_2	f_3

MFCPAC-46: 39 MFC-DA coefficients, pitch, intensity, formants, jitter and harmonicity

mfcc-39	p	i	f_1	f_2	f_3	j	h

MFCPAC-48: 39 MFC-DA coefficients, pitch, intensity, formants, minimum pitch, maximum pitch, mean pitch and pitch deviation

mfcc-39	p	i	f_1	f_2	f_3	p_{min}	p_{max}	p_{mean}	p_{dev}

MFCPAC-52: 39 MFC-DA coefficients, pitch, intensity, formants, minimum pitch, maximum pitch, mean pitch, pitch deviation, minimum intensity, maximum intensity, mean intensity and intensity deviation

mfcc-39	p	i	f_1	f_2	f_3	p_{min}	p_{max}	p_{mean}	p_{dev}	i_{min}	i_{max}	i_{mean}	i_{dev}

MFCPAC-56: 39 MFC-DA coefficients, pitch, intensity, formants, minimum pitch, maximum pitch, mean pitch, pitch deviation, minimum intensity, maximum intensity, mean intensity, intensity deviation, minimum voiced pitch, maximum voiced pitch, mean voiced pitch and voiced pitch deviation

mfcc-39	p	i	f_1	f_2	f_3	p_{min}	p_{max}	p_{mean}	p_{dev}	i_{min}	i_{max}	i_{mean}	i_{dev}

vp_{min}	vp_{max}	vp_{mean}	vp_{dev}

MFCPAC-63: 39 MFC-DA coefficients and all 24 PA coefficients

mfcc-39	pac-24

Fig. 4.8 List of feature combinations considered in our experiments (Pittermann and Pittermann 2006d; Pittermann et al. 2007b)

respective features can be observed as previously discussed: The duration of the sad utterance is approx. 55% longer than the duration of the angry utterance, the pitch is significantly lower and spans a very small range. Also, differences in the MFCC parameter values can be noticed. Concerning intensity, no particular difference can be observed which is due to the fact that the utterances have been normalized to a

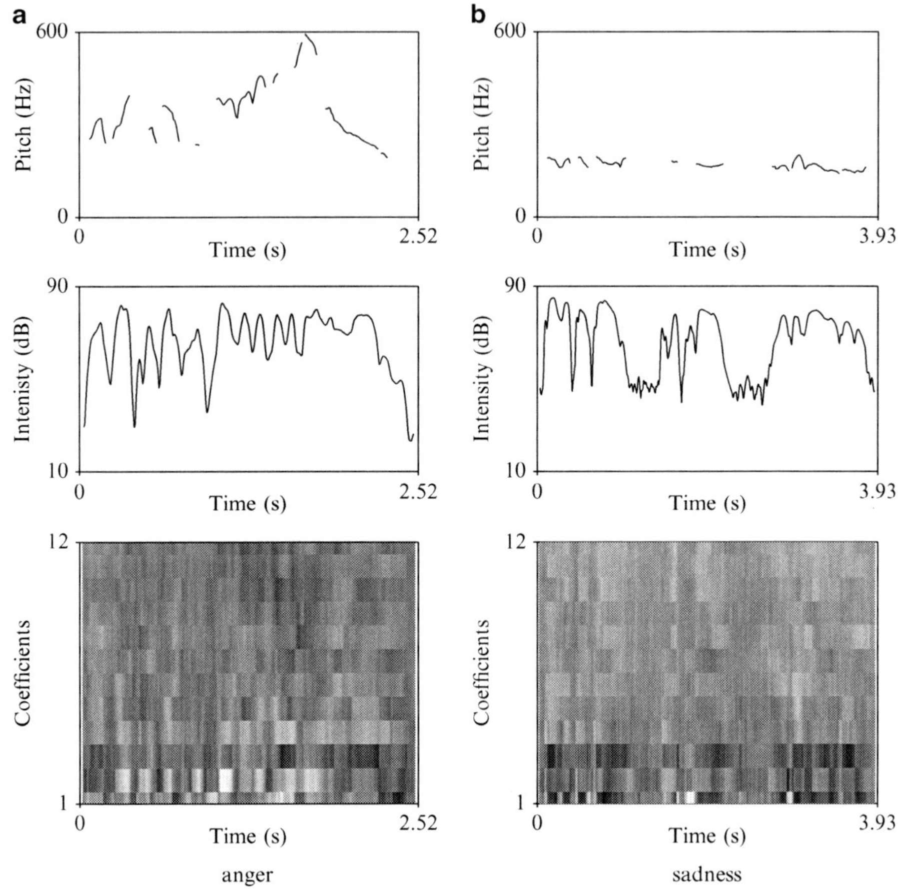

Fig. 4.9 Illustration of pitch, intensity and MFCC features of an angry and a sad utterance

common power density (Pittermann and Pittermann 2006a). The visible discontinuities and the omission of values in the pitch curves occur in unvoiced parts of the utterance where noise is the predominant share in the speech signal. To avoid unwanted side-effects in the further processing steps, the pitch curves are commonly smoothed over a short period.

4.2 Classifiers for Emotion Recognition

Having extracted relevant and representative features X, these need to be assigned to an emotion E such that the probability $P(E|X)$ is maximized. Here, a large variety of classifier types seem suitable, a few of which are actually implemented in speech and emotion recognizers: Hidden Markov Models, artificial neural networks, support vector machines (SVMs, cf. Cristiani and Shawe-Taylor 2000), k-nearest neighbor (KNN, cf. Dasarathy 1991) classifiers, etc. In this section, we give a short overview on HMMs and ANNs as these constitute the most prominent classifiers in current emotion recognizers and as we also use HMMs in our speech–emotion recognizers the remainder of this chapter.

4.2.1 Hidden Markov Models

Hidden Markov Models are currently applying in speech recognition and understanding (see Rabiner 1989), image classification (see Li et al. 2000), bioinformatics (see Durbin et al. 1999) and various other disciplines. With respect to our speech–emotion recognizer which also employs HMMs, we shortly describe the most prominent characteristics of this classifier.

A time discrete stochastic process $X(n), n \geq 0$ taking non-negative integer values $i_n, n \geq 0$, is called a *Markov process* if it holds that

$$\begin{aligned} &P\left[X\ (n_0+1) = i_{n_0+1} | X(n_0) = i_{n_0}, X(n_0-1) = i_{n_0-1}, \ldots, X(0) = i_0\right] \\ &\quad = P\left[X(n_0+1) = i_{n_0+1} | X(n_0) = i_{n_0}\right], \end{aligned} \tag{4.9}$$

for all $n_0 \geq 0$ (Markov 1907; Howard 1971). Furthermore if the above described probability satisfies

$$P\left[X(n_0+1) = i | X(n_0) = j\right] = P\left[X(1) = i | X(0) = j\right] = a_{ij}, \tag{4.10}$$

for all $n_0, i, j \geq 0$, this Markov process is called *homogeneous*.

Assuming that the values i_n of a Markov process $X(n)$ form a countable set $\mathcal{S}$, a sequence $X(1), X(2), \ldots$ is called a *Markov chain*. Such a Markov chain is typically defined by its state space $\mathcal{S} = \{s_1, s_2, \ldots\}$. If $\mathcal{S}$ is a finite set, the transition probabilities may be combined to a matrix $\underline{\mathbf{A}}$.

Markov chains or models are extensively used in various disciplines of research, not only for pattern recognition but also to describe dependencies in languages, e.g., used for data compression (see Shannon 1948), or to model complex dynamical systems. One of the most active and proliferous implementations of a Markov chain is Google's PageRank (cf. Page et al. 1998) which considers the whole world wide web as a large state space and calculates scores for all websites (i.e., states) depending on how popular these are among other websites.

In most applications like pattern recognition, the dimension of Markov chains is considerably smaller. Therefore we assume that, unless otherwise indicated, the Markov chains are (time-)homogeneous and that their state space $\mathcal{S}$ is finite in the following considerations. Then the chain can be described by $\mathcal{S} = \{s_1, \ldots, s_n\}$ and $\underline{\mathbf{A}} = \left[a_{ij}\right]$ where $i, j \in \mathcal{S}$.

Fig. 4.10 Directed graph of a three-state Markov chain

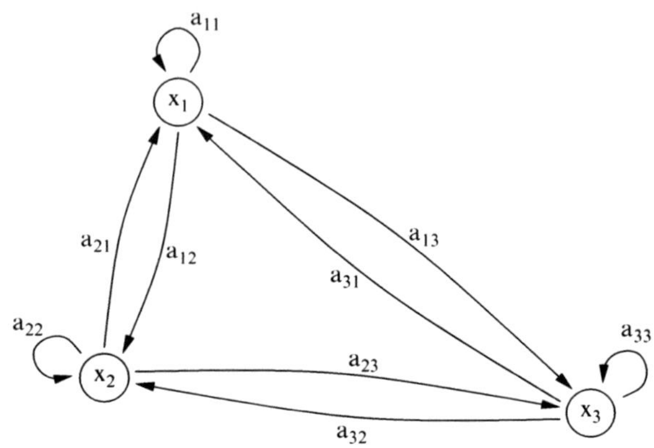

As shown for a simple example in Fig. 4.10 a Markov chain may be represented by a directed graph. Here $\mathcal{S}$ is formed by three states $\{1, 2, 3\}$ and $\underline{\mathbf{A}}$ is a 3×3 matrix

$$\underline{\mathbf{A}} = \begin{bmatrix} a_{11} & a_{12} & a_{13} \\ a_{21} & a_{22} & a_{23} \\ a_{31} & a_{32} & a_{33} \end{bmatrix} . \qquad (4.11)$$

An HMM is a statistical model involving two stochastic processes: a Markov chain characterized by $\mathcal{S}$ and $\underline{\mathbf{A}}$ and a stochastic process generating output symbols according to a state-dependent probability density function (pdf). The parameters of this model, i.e., $\underline{\mathbf{A}}$ and the pdfs, however, are unknown (hidden) and the state is not visible implicating the task to determine these parameters on the basis of observed output.

An HMM may also be represented by a directed graph (see Fig. 4.11) and it is formally defined by a five-tuple

$$M = (\mathcal{S}, \underline{\mathbf{A}}, \pi, \mathcal{O}, \mathcal{B}) , \qquad (4.12)$$

where $\mathcal{S}$ and $\underline{\mathbf{A}}$ are state space and transition probabilities of the Markov chain, π is a pdf which includes probabilities π_i of $i \in \mathcal{S}$ being the entry state. $\mathcal{O}$ is the feature space, i.e., a finite set of possible observations $o(t)$ at instant $t \geq 0$ and $\mathcal{B} = \{b_1, \ldots, b_n\}$ is the set of output pdfs of the states.

Depending on the application, different types of HMMs can be employed. An HMM is called ergodic, when every state j of the model can be reached from every other state $i \neq j$ in a finite number of steps. E.g., the model shown in Fig. 4.11 is a special case of an ergodic HMM as all states are adjacent to each other and can therefore be reached within one step. In this case all elements a_{ij} of the transition matrix are non-zero, $a_{ij} > 0$.

Ergodic models may be used, e.g., to model stationary signals the properties of which do not change over time. However, in case of time-variant signals like speech, so-called left-to-right (LR) models (see Fig. 4.12) have been found more suitable

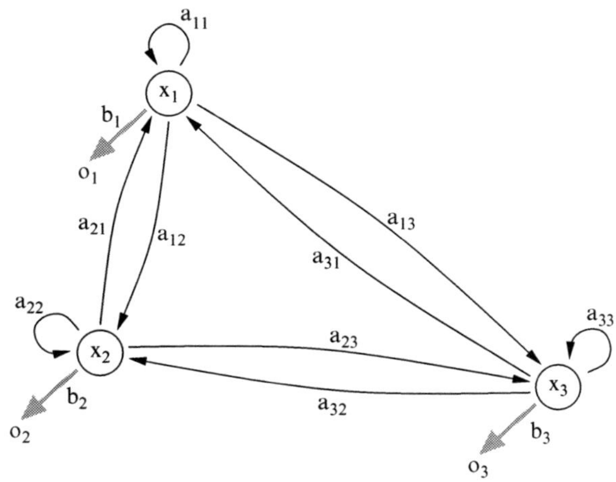

Fig. 4.11 Directed graph of a three-state Hidden Markov Model

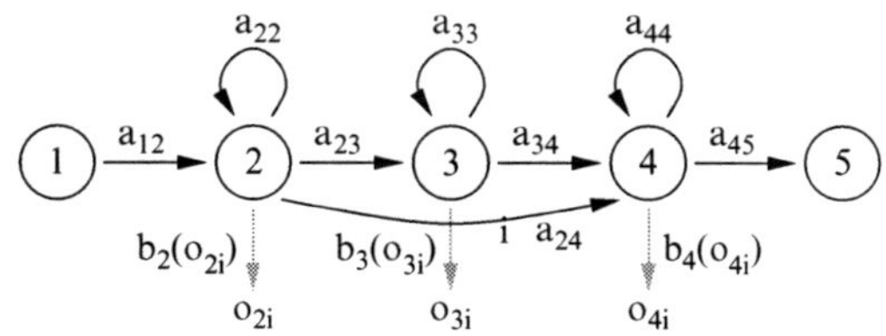

Fig. 4.12 A left-to-right Hidden Markov Model

(Bakis 1976). In the state sequence of such a model, the states proceed from left to right, i.e., the property of the observation changes, as time increases (Rabiner 1989).

When employing HMMs in classification applications such as speech (or emotion) recognition, there arise three fundamental problems (see Baum et al. 1970; Rabiner 1989) that need to be solved:

1. *Evaluation/Decoding:* How can the probability $P(\mathbf{O}|M)$ of an observation sequence $\mathbf{O} = o_1, o_2, \ldots, o_T$ given a model M be calculated?
2. *Unveiling hidden characteristics:* How can the optimal corresponding state sequence $\mathbf{S} = s_1, s_2, \ldots$ be determined given an observation sequence $\mathbf{O} = o_1, o_2, \ldots, o_T$?
3. *Training:* How can the model parameters $(\underline{\mathbf{A}}, \pi, \mathcal{B})$ be adjusted to optimize $P(\mathbf{O}|M)$?

The evaluation or decoding problem poses the question how well an observation sequence matches a given model. I.e., the solution to this problem facilitates the selection of the best-matching model among a set of competing models and can be described by the "best" path through a graph formed by one or multiple HMMs. As a brute-force approach, i.e., testing all possible state and observation combinations, is more or less unfeasible, the forward part of the *Forward-Backward* algorithm is applied (Baum and Eagon 1967; Baum and Sell 1968). The probability of a partial observation sequence $o_1, \ldots, o_t$ and state i at time t

$$a_t(i) = P\left[o_1, \ldots, o_t, s_t = i\right], \tag{4.13}$$

is called forward probability. Starting with $a_1(i)$, which is the product of the initial probability π_i and the probability of observation o_1 for state i, all $a_t(i)$ are iteratively estimated traversing the trellis of the model's possible state transitions until $a_T(i)$ is determined. Then the solution is

$$P(\mathbf{O}|M) = \sum_{i=1}^{n} a_T(i). \tag{4.14}$$

Such a trellis representing all possible state transitions in a Hidden Markov Model for an observation sequence length T is shown in Fig. 4.13.

In order to solve the second problem, the term "optimal" needs to be defined. One approach includes the choice of the most likely state for each time t. This, however, may lead to improbable or even impossible state sequences, especially when certain transitions probabilities a_{ij} are zero. Thus usually the Viterbi algorithm is used to find the single best state sequence, i.e., to maximize $P(\mathbf{S}|\mathbf{O}, M)$ (Viterbi 1967). Also being implemented by a trellis of the model's state transitions, the procedure is similar to the forward part of the Forward-Backward algorithm, where the summation in Equation (4.14) is replaced by a maximization of path metrics.

The training problem can not be solved analytically, however, iterative procedures may be employed to locally maximize $P(\mathbf{O}|M)$. For the training of HMMs for automatic speech recognition (see Young et al. 2006), typically, the Baum-Welch algorithm (Welch 2003) is used to re-estimate the model parameters once they are initialized. For the initialization it is assumed that the observations are equally distributed among the states, e.g., if there are three states and nine observations, the first three observations are assumed to relate to the first state, and so on, and observation probability functions (pdfs) are calculated for each state. Once a first estimate is determined, it can be refined by realigning the observations using the Viterbi algorithm or the Baum-Welch algorithm and recalculating the pdfs.

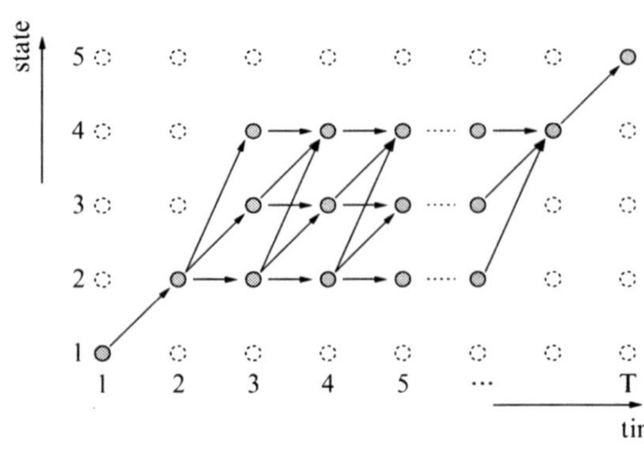

Fig. 4.13 Trellis of possible state transitions in an HMM

For the observation probability density function (pdf) there exist several models ranging from a single Gaussian distribution to highly complex shapes involving a huge number of parameters. In HMMs used in signal processing, speech recognition or econometrics typically include mixture models

$$f^{\star}(x) = \sum_{i=1}^{M} w_i \cdot f_i(x), \tag{4.15}$$

where a complex pdf is approximated by $f^{\star}(x)$ which is a combination of M "simple" pdfs $f_i(x)$, called *mixtures*. These pdfs are differently weighted with factors w_i, where $\sum_i w_i = 1$. When $f_i(x)$ are Gaussian distributions, which are in the one-dimensional case described by

$$f_i(x) = g(x, \mu_i, \sigma_i) = \frac{1}{\sqrt{2\pi} \cdot \sigma_i} \cdot \exp\left(\frac{(x - \mu_i)^2}{\sigma_i^2}\right), \tag{4.16}$$

the model is called *Gaussian Mixture Model* (GMM).

Figure 4.14 shows three mixtures and the resulting pdf of an exemplary GMM. A detailed description of mixtures of distributions including model parameter estimation algorithms is given in Marin et al. (2005).

4.2.2 *Artificial Neural Networks*

Inspired by the network structure of the human brain, an ANN is commonly described as a network of processing elements (artificial neurons) and is typically used

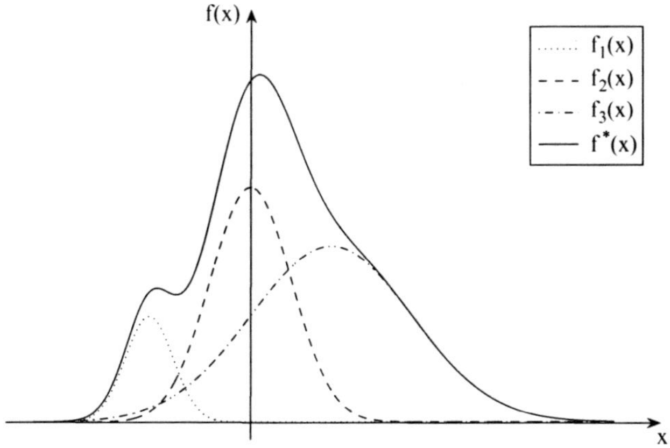

Fig. 4.14 Probability density functions of a GMM

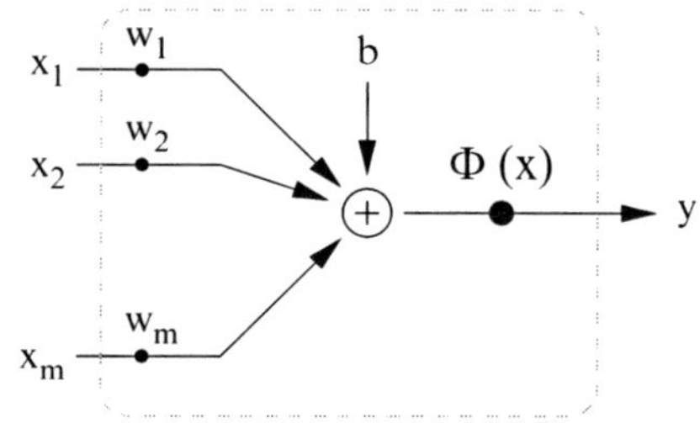

Fig. 4.15 Structure of an artificial neuron

to recognize patterns or to describe complex relationships between (observed) M-dimensional input and N-dimensional output.

An artificial neuron (see Fig. 4.15) constitutes the basic unit of an ANN receiving an m-dimensional vector of input signals $x_1, \ldots, x_m$ and producing an output signal y according to

$$y = \Phi\left(b + \sum_{i=0}^{m} w_i \cdot x_i\right), \tag{4.17}$$

where b is a bias value and $w_1, \ldots, w_m$ are the input weights. $\Phi(x)$ is a typically non-linear function called activation function. Depending on the application, different activation functions can be applied. The simplest function as used by McCulloch and Pitts (1943) is a threshold function

$$\Phi(x) = \begin{cases} 0 & x < \theta \\ 1 & x \geq \theta \end{cases}, \tag{4.18}$$

where θ is a specified threshold. Other activation functions include the sigmoid function or the hyperbolic tangent.

Formed by multiple artificial neurons which are interconnected and which are able to operate in parallel, an ANN can be described by a function $y = f(x)$ mapping input x to output y. This function may be composed of other functions, these again can be composed of functions, etc., and these functions and subfunctions can be represented by one or more artificial neurons.

The structure of an exemplary artificial neural network with four input nodes and three output nodes is depicted in Fig. 4.16. There can be a different number of layers between input and output nodes, these layers are commonly called hidden layers. ANNs that can be described as directed graph without cycles are called feedforward ANNs as they propagate the data linearly from the input nodes to the output nodes. Feedforward networks are typically arranged in layers putting together neurons that operate in parallel. Popular models among feedforward networks are single- and multi-layer perceptrons. ANNs that show cycles and thus, unlike feedforward networks, include certain temporal dependencies are called recurrent ANNs.

One characteristic of ANNs which is especially taken advantage of in pattern or sequence recognition is their learning capability. I.e., using an appropriate ANN

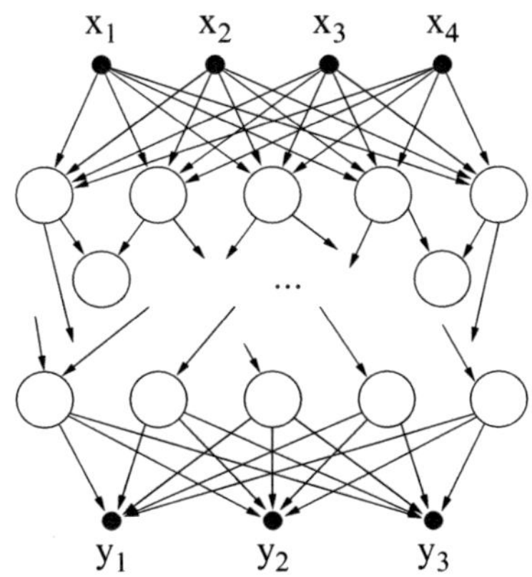

Fig. 4.16 Structure of an Artificial Neural Network

and given a set of observations x and the desired output y, a function $\hat{f}(x)$ can be deduced which approximates an unknown and/or complex function $y = f_0(x)$ best. To accomplish the learning process, first a cost function $c(f)$ which is minimal for $\hat{f}(x)$ needs to be defined. E.g., given a corpus of N labeled observations $x_1, \ldots, x_N$ and the respective output $y_1, \ldots, y_N$, a cost function can be defined as the mean-squared error, and $\hat{f}(x)$ can be determined as

$$\hat{f}(x) = \arg\min_{f \in \mathcal{F}} \left(\frac{1}{N} \sum_{i=0}^{N} |f(x_i) - y_i|^2 \right), \tag{4.19}$$

where $\mathcal{F}$ is a class of functions to be considered. This type of learning process is commonly applied in classification applications and is called supervised learning. Further learning methods used in coding or control engineering are unsupervised learning where y is not given and reinforcement learning where x is not explicitly given but generated by a Markov process. Concise descriptions of artificial neural networks and their use in signal processing and classification applications are given in Abdi et al. (1999); Masters (1994), and Ripley (1996).

4.3 Existing Approaches to Emotion Recognition

In current research there exist several approaches to classify and recognize emotions, ranging from gestures and facial expressions interpretation in multimodal systems (cf., e.g., Martin et al. 2006) to physiological measurements (see Picard 2000b; Bosma and André 2004; Kim et al. 2004a), semantic analyses or combinations of these modalities. With respect to the integration of emotions into SLDSs, we focus our considerations on the recognition of emotions from spoken user input. This includes recognition on the basis of speech signals as well as analyses of the textual content determined by a preceding speech recognizer. The bar is raised by Schröder (2000) who gives a concise overview on the human emotion recognition performance on affect bursts where an average emotion recognition rate of 81% is achieved for ten different emotions.

A good overview on the performance of speech-based emotion recognizers is given in Kwon et al. (2003). The comparison includes support vector machines, linear discriminant analysis, quadratic discriminant analysis and Hidden Markov Models for a neutral-stressed discrimination and a classification of five emotions concluding that support vector machines and Hidden Markov Models yield better recognition results (of approx. 43%). An approach to directly distinguishing only positive and negative emotions on the basis of prosodic values like duration (longer for negative), pitch (higher for negative), however, facing problems with significant variances, is described in Swerts and Krahmer (2000). Decision trees allow the classification of emotions on the basis of decisions and their sub-decisions which are represented by a tree, in each node of which the system decides between two or more emotions or emotion groups.

In order to improve the performance of weak classifiers, so-called boosting algorithms are commonly used. The idea behind boosting is to iteratively learn multiple weak classifiers and to combine these to a stronger classifier. Freund and Schapire (1997) describe the AdaBoost (adaptive boost) algorithm which repeatedly calls the same weak classifier while increasingly concentrating on the classifier's misclassifications. For the classification of four different states of (un)certainty, Ai et al. (2006) use AdaBoost with a decision tree with the aid of which an accuracy of 59% is achieved.

A commonly used tool which implements a large variety of machine learning algorithms is the WEKA software (see also http://www.cs.waikato.ac.nz/~ml/weka/, Witten and Frank 2000). Apart from the boosted decision tree, it also includes, e.g., the C4.5 decision tree learner, k-nearest neighbor classifiers, decision rules, perceptrons, support vector machines, Bayes classifier, etc. An overview on selected algorithms provided by the tool is given by Oudeyer (2003).

Nogueiras et al. (2001) use Hidden Markov Models for the recognition of emotions. On the basis of seven emotions to be distinguished, they show that even with single state HMMs, a recognition rate significantly above chance level can be achieved and that for speaker-dependent models accuracies of 80% are obtained with 64-state HMMs and eleven vector quantized features. In Luengo et al. (2005) experiments with HMMs and support vector machines are conducted on a single-speaker emotional database containing seven emotions. Their experiments lead to the conclusion that HMMs perform significantly better despite the fact that support vector machines with a lower complexity in terms of feature space dimensions also lead to an overall accuracy of 92%. Similar observations are found in Pao et al. (2004) where HMMs perform better than linear discriminant analysis and k-nearest neighbor classifiers. Here, with an extended set of spectral features, a recognition rate of 88% is achieved for seven emotional states. Using vector-quantized MFCC features and discrete HMM classifiers, an average accuracy of 72% on six emotions is shown in Nwe et al. (2001). Their approach shows a comparable performance to similar work with pattern recognition, neural networks and nearest mean criteria. Polzin and Waibel (1998) use suprasegmental HMMs, where certain states within an HMM are combined, to detect four emotional states with an accuracy of 72%. We pick up their approach involving emotion-dependent models for our combined speech–emotion recognizer described in Section 4.6. The original approach is further described in Polzin and Waibel (2000), where a maximum likelihood regression technique is used to adapt a (neutral) speech-recognizer to different emotional states. Depending on the emotion set and the choice of features (verbal and non-verbal) recognition accuracies of up to 47% or 64% (compared to human recognition performance of 55% or 70%) are yielded with the recognition of three emotions. An overview on existing work and approaches using speech recognizer frameworks in combination with emotions is given in ten Bosch (2000), also including HMMs or neural networks.

An issue which is rather negligible in standard spoken dialogue applications, but still interesting to academia, is the classification of emotions in infants' cries as described in Matsunaga et al. (2006). They also use HMMs (three states, eight Gaussian mixtures) to distinguish five emotional states which are labeled in a hierarchical structure of three levels. With MFCC features, recognition rates of up to 75% are achieved.

Neural networks also enjoy a great popularity in the field of speech-based emotion recognition. The emotion recognizer in Tato et al. (2002) is based on feedforward neural networks implemented with the Stuttgart Neural Networks Simulator (see Zell et al., 1991) which are trained with a chunkwise-backprogragation learning algorithm with ten out of 37 acoustic features. Experiments are conducted for arousal classification (high, neutral, low) with 84% and 77% for speaker-dependent and speaker-independent models as well as for anger-happiness discrimination (74%) and bored-sad classification (66%) constituting the emotion pairs which are most commonly confused. In their experiments, Park and Si (2003) use a dynamic recurrent neural network with different acoustic feature sets and achieve relatively high recognition rates, however, with a very limited test corpus including four emotional states. A multi-layer perceptron and a probabilistic neural network are compared in Dan-Ning Jiang (2004) where recognition rates of up to 94% are achieved for six emotions with the probabilistic neural network based on acoustic features. Emotion recognition experiments on the SYMPAFLY corpus using a neural network are described in Batliner et al. (2004b). For the discrimination of six cover classes, a recognition rate of 51% is yielded, combining some of the negative classes the rate increases to 76%.

Particularly with regard to our proposed speech–emotion recognizer, we initially implement a plain emotion recognizer using prosodic and acoustic features and HMMs as commonly used in speech recognition (see Section 4.5). In order to follow the temporal aspects of the features within an utterance, we do not include support vector machines and (feedforward) neural networks in our considerations. Recurrent neural networks are also not considered due to their higher complexity. Instead, we employ HMMs for our acoustic model. In our plain emotion recognition approach, each emotion is represented by a three-state left-to-right HMM. Due to the relatively low number of HMMs in the acoustic model, we achieve a relatively low training and decoding complexity, even when considering different models for female and male speakers. An assessment of our approach's performance compared to existing work proves to be rather difficult due to different evaluation criteria (e.g., number of emotions, how to determine whether an emotion is correctly recognized or not, use of word or utterance level measures, etc.). On the one hand, there exist systems which, according to the respective publications, yield a higher emotion accuracy for the classification of more emotional states. On the other hand, there are also systems which exhibit lower recognition rates for the classification of fewer emotions. Leaving the plain emotion recognizer's performance as it is, in our research, we concentrate our attention to the development and improvement of a speech–emotion recognizer as described in Section 4.6 which is silhouetted against the above approaches by the combination of speech and emotion recognition. The idea behind our hybrid recognizer is to consolidate the (normally) separate but similar processes of speech and emotion recognition into one process taking advantage of the same signal preprocessing and classification methods. This is also implemented on the basis of HMMs – now the acoustic model contains one HMM per phoneme and its attached emotional state. Despite the high complexity of such a model, we achieve a satisfactory recognition performance with this approach outperforming our plain emotion recognizer.

For the recognition of emotions based on the feature vectors described in the previous section there exist several approaches. The "simple database" approach, storing all possible feature values for each emotional state and comparing these to the features to be recognized, is not implementable due to its calculating effort and the required memory. Thus, we limit our considerations to statistical methods, namely artificial neural networks and Hidden Markov Models. Both methods have in common that a large amount of training data, i.e., a labeled corpus, is used to train the classifier(s).

Neural networks, as described in Section 4.2.2, are quite popular in several classification applications, including emotion recognition based on different modalities. As this approach typically does not account for temporal aspects, one feature vector $[x_1, \ldots, x_n]$ is extracted from the biosignals, video caption or speech signal serving as the input of the trained network. The output indicates the emotional state, either as a number or as a vector the elements of which are assigned to the emotions as depicted in Fig. 4.17. In this exemplary system the set of emotions consists of "angry", "neutral" and "happy". The output can be therefore, e.g., [1 0 0] for an angry utterance or [0 0 1] for a happy utterance. Such an approach also allows "soft" decisions like [0.5 0.4 0.1] in cases of ambiguity which may occur when different people express their emotions in different ways or which may be due to insufficient or faulty training data. Focusing on the temporal aspects, i.e., on how the emotional expression evolves during an utterance, we favor HMMs over neural networks in our work. This is particularly relevant for the combination of speech and emotion recognition where the speech recognizer already employs HMMs.

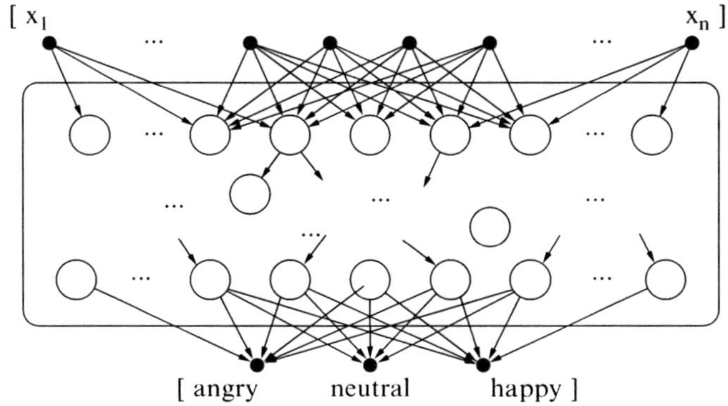

Fig. 4.17 Emotion recognition using a neural network

In the remainder of this chapter we firstly describe plain speech recognition using HMMs, then we specify plain emotion recognition using HMMs with respect to its parallels to speech recognition and, finally, we characterize an approach merging both methods to obtain a combined speech–emotion recognizer.

4.4 HMM-Based Speech Recognition

The goal of speech recognition can be formulated as a probabilistic problem to find a sequence of words W which is most probable given a sequence of acoustic observations X:

$$W = \arg\max_{W} P(W|X), \tag{4.20}$$

which can be rewritten applying Bayes' rule and which can be simplified as follows:

$$W = \arg\max_{W} P(X|W) \cdot P(W), \tag{4.21}$$

Here $P(X|W)$ represents an acoustic model and $P(W)$ is determined by a language model.

HMMs commonly apply in acoustic models of most state-of-the-art speech recognition systems, e.g., as plain HMMs as well as HMMs with Gaussian mixture models or neural networks (Rabiner 1989). In speech recognition, each phoneme is represented by one model, so that in the recognition process a Viterbi search is performed over all models resulting in a string of phonemes. The likelihood $P(X|M)$ of a sequence of features X given a model M is calculated with the aid of the forward part of the Forward-Backward algorithm.

As not all phoneme combinations are meaningful in most languages, the search is constrained with the aid of a dictionary containing all words to be recognized plus the respective phoneme transcriptions as shown in Fig. 4.18. The exemplary dictionary entries are extracted from the British English Example Pronunciations dictionary (BEEP, Robinson 1997b). Unlisted words like, e.g., the word "ULM" are

```
FOUR          f ao
FOUR          f ao r
GO            g ow
I             ay
NEW_YORK      n y uw y ao k
PLEASE        p l iy z
TO            t uw
ULM           uh l m
WEDNESDAY     w eh n z d ey
```

Fig. 4.18 Dictionary for a speech recognizer adapted from the BEEP dictionary (see Robinson, 1997b)

```
$city     = ( LONDON | NEW_YORK | PARIS | ULM );
$weekday = ( MONDAY | TUESDAY | WEDNESDAY | THURSDAY );
$date     = ( TODAY | TOMORROW | ON $weekday );

( [ I ( WANT | WOULD LIKE ) TO ( GO | TRAVEL) ] TO $city
  [ $date ] [ PLEASE ] )
```

Fig. 4.19 EBNF of a grammar for speech recognition

transcribed manually by regarding listed words which sound similarly: "uh (short 'u' like in 'good') l m". Including such a dictionary in the recognition, the recognition result is a string of words. However, as not all word combinations necessarily make sense, also restrictions on a higher level, i.e., a grammar or language model providing $P(W)$, need to be included in the Viterbi search.

For smaller speech applications like a travel information system, it is normally sufficient to create a rule-based grammar as shown in Fig. 4.19. In this example grammar the Extended Backus-Naur Form (EBNF) is used to define allowed sequences of words: the non-terminal symbol $city defines a sub-rule allowing one of the cities separated by | characters. In the main rule at the bottom "to $city" is mandatory, the words before or after this sequence surrounded by [ ] brackets are optional. Round brackets ( ) are used to group items, angle brackets < > (not contained in the example) surround arbitrary repetitions which may occur, e.g., in the recognition of phone numbers or in simple word loops.

Alternatively, depending on the application, instead of the EBNF or similar forms like the Java Speech Grammar Format (JSGF), etc., an Extensible Markup Language (XML) form as proposed in the Speech Recognition Grammar Specification (SRGS) by the W3C can be used to define grammar rules (Hunt and McGlashan 2004). In either cases, to facilitate the recognition process, the grammar is typically converted to a word network containing all allowed word transitions. For more complex applications such as dictation systems, it is rather tedious to define grammar rules that cover all possible word combinations in one or multiple languages. In these cases, typically, stochastic language models apply considering the probability of a word given adjacent words. The most accurate models can be obtained by including the complete context, i.e., calculating a word's probability given all previous words. As the model's complexity increases exponentially with the number of previous words to be considered, only short-term dependencies are included in state-of-the-art speech recognizers: bigrams include the relation between the previous word and the current word, trigrams include two previous words and the current words, n-grams include $n - 1$ previous words and the current word (Roark et al. 2007).

In simple "monophone" recognizers, each of the phoneme models is trained separately without allowing for dependencies on adjacent phonemes. Effects like coarticulation, where phonemes are differently pronounced depending on adjacent phonemes occurring in either the same word or in neighboring words, are included using "triphone" combinations consisting of a phoneme and the precedent and succeeding phonemes. E.g., according to the dictionary shown in Fig. 4.18, the isolated word "WEDNESDAY" consists of the following triphones: "?-w-eh", "w-eh-n", "eh-n-z", "n-z-d", "z-d-ey", and "e-ey-?". The combinations at word boundaries are actually "biphones" as words are typically separated by short pauses not permitting dependencies between words. Assuming approximately 50+ phonemes in an English large-vocabulary, there exists $(50+)^3 > 125{,}000$ triphone combinations including all of these in speech recognition would be rather impractical. Fortunately, not all of these combinations actually occur in the respective language, and, moreover, most of the triphone combinations with similar properties can be pooled by "tying" states, i.e., triphones share model parameters like transition probabilities or emission probability distributions (Young et al. 2006).

The training of the models consists of multiple re-estimation procedures iteratively updating the HMM parameters as well as pruning algorithms to sort out unreliable training data. As described in Section 4.2.1, the model re-estimation is accomplished with the aid of the Baum-Welch algorithm (Welch 2003; Young et al. 2006). The tying of states is performed on the basis of scripts and macros which need to be adapted to the language. In these macros also the context of the tied-state triphones can be included. The pruning of the training data is typically performed by aligning the utterances and the labels. At this, the trained models are used to perform a recognition on the training utterances. These recognition results are compared to the original labels of the training data so that the timing information, which phoneme occurs at which time, can be corrected and the suitable pronunciation variant can be selected for words which have multiple phonetical transcriptions, like "FOUR" in the dictionary shown in Fig. 4.18. Also, utterances are removed from the training set in cases where the results differ significantly from the labels.

In straightforward approaches covering small vocabularies and simple grammars, it is normally sufficient to use HMMs, the observation vector probability distribution functions of which are Gaussian functions. In these cases, the model parameters of the observations of an HMM state are combined in a vector of mean μ and variance ς^2, assuming that the real values actually distribute similar to these Gaussian functions. A prototype of such an HMM, described in the notation used by HTK, is shown in Fig. 4.20.

The model consists of 5 states, which is commonly used in automatic speech recognition. States 1 and 5 represent the start and the end states, observations occur only in the non-emitting states 2, 3 and 4. According to the transition matrix (TransP), the model is a left-to-right model as only transitions to the same or the

```
<BeginHMM>
  <NumStates> 5
  <State> 2 <NumMixes> 1
      <Mean> 39
        1.0 1.0 1.0 1.0 1.0 1.0 1.0 1.0 ...
      <Variance> 39
        1.0 1.0 1.0 1.0 1.0 1.0 1.0 1.0 1.0 1.0 ...
  <State> 3 <NumMixes> 1
      <Mean> 39
        1.0 1.0 1.0 1.0 1.0 1.0 1.0 1.0 1.0 1.0 1.0 ...
      <Variance> 39
        1.0 1.0 1.0 1.0 1.0 1.0 1.0 1.0 1.0 1.0 1.0 1.0 ...
  <State> 4 <NumMixes> 1
      <Mean> 39
        1.0 1.0 1.0 1.0 1.0 1.0 1.0 1.0 1.0 1.0 1.0 1.0 ...
      <Variance> 39
        1.0 1.0 1.0 1.0 1.0 1.0 1.0 1.0 1.0 1.0 1.0 1.0 ...
  <TransP> 5
    0.0 1.0 0.0 0.0 0.0
    0.0 0.6 0.4 0.0 0.0
    0.0 0.0 0.6 0.4 0.0
    0.0 0.0 0.0 0.7 0.3
    0.0 0.0 0.0 0.0 0.0
<EndHMM>
```

Fig. 4.20 Prototype of an HMM with Gaussian observation probability density distribution functions

following state (to the right) are permitted. During the training, the non-zero transition probabilities are adapted with respect to the training data, however, the probabilities of the non-existing transitions remain zero. In the prototype, the mean values and variances of all observations default to 1.

In order to obtain more "realistic" models, Gaussian Mixture Models are commonly used in large-vocabulary continuous speech recognition. As described in Section 4.2.1, almost any shape of a probability density function can be generated by linearly combining Gaussian functions. One of the few constraints is that the sum of all weighting factors, the Gaussian functions are multiplied with, is 1. In the training of phoneme models for automatic speech recognition, firstly, simple HMMs with only one mixture are trained up to a certain degree of accuracy. Then, each model is converted to an HMM-GMM with two mixtures by splitting up each Gaussian function into a sum of two Gaussian functions with similar parameters but different weighting factors. In the subsequent re-estimation steps, the parameters of the mixture model are refined and the number of mixtures can be gradually increased. Depending on the complexity and the number of Gaussian mixtures, the training of the HMMs involves between seven and 20+ re-estimation steps (Young et al. 2006).

The decoding or recognition includes the HMMs, the dictionary and a word network generated from the grammar, all combined to a joint network which, in the

```
      0  1500000 I        -1992.69
1500000  4800000 WANT     -4219.77
4800000  6100000 TO       -1736.19
6100000  7900000 GO       -2366.27
7900000  9400000 TO       -2054.00
9400000 12800000 PARIS    -4707.06
```

Fig. 4.21 Output of a speech recognizer using the grammar shown in Fig. 4.19 (Young et al. 2006)

end, consists of HMM states connected to each other according to the grammar, dictionary and HMM specifications. On this large network a Viterbi search is performed providing the most probable path(s) through the network given a sequence of input feature vectors. These paths represent word sequences which are typically combined to a word transition network (WTN), also referred to as word hypothesis graph (Gibbon and Lüngen 1999), the most probable path of which is the preferred recognition result. Optionally the second best or third best word sequence can also be extracted from the WTN. Using HTK, a typical speech recognizer output looks as shown in Fig. 4.21.

The first and second column contain the beginning and the end of a time period. In HTK, the time unit used for labeling, recognition and frame parameters is predefined as 100 ns. The third column contains the word which was recognized in the respective time period, and the numbers given in the fourth column show acoustic and linguistic likelihood scores. These scores are calculated on the basis of path metrics which are accumulated or maximized when traversing through the recognition network during the recognition. Typically, logarithmic measures are used to maintain a reasonable preciseness in the range of values during the calculation, so that a measure of 0 represents the highest possible score, (corresponding to a probability or confidence of 1), whereas a measure of $-\infty$ denotes a probability of 0, i.e., the lowest possible score.

4.5 HMM-Based Emotion Recognition

For speech-based emotion recognition, the principles described above can be adopted for the most part (ten Bosch 2000; Pittermann and Pittermann 2006a). The respective emotion recognizer includes a simplified dictionary, a language model and acoustic models, training and recognition are performed in the same manner. Several approaches exist for the labeling of emotional speech data: In cases where one single feature vector is extracted from the waveform it is correspondingly sufficient to label each utterance with one emotion without regarding pauses or other changes in the waveform (cf. also Fig. 5.1 in Section 5.1).

In order to account for temporal changes, i.e., in cases where feature vectors are extracted regularly, e.g., every 10 ms like in speech recognition, the emotional labeling is similar to word labeling. Each part of the waveform is assigned an emotional label and, optionally, pauses are also labeled in order to obtain a more accurate emotion recognition. In order to increase the annotators' flexibility, we allow multiple emotional states in one utterance.

The labeling process, which may be quite tiresome especially in cases where a large amount of data needs to be labeled, can be facilitated by several means: Instead of labeling each utterance manually, it can be prelabeled with the aid of a speech-pause detector which already inserts the pause labels. Such a detector is realized on the basis of energy measures calculated from the speech signal. The average power $\bar{P}_i$ of a signal region or time frame i is calculated as

$$\bar{P}_i = \frac{1}{i_{end} - i_{begin}} \cdot \sum_{j=i_{begin}}^{i_{end}} |s_j|^2, \tag{4.22}$$

where s_j is the value of the signal's jth sample. Frame i begins with the i_{begin}th sample and ends with the i_{end}th sample. For speech-pause detection a threshold ϑ is defined such that a frame i is labeled as pause if $\bar{P}_i < \vartheta \cdot \bar{P}_s$, where $\bar{P}_s$ is the average power of the utterance. The optimal threshold is determined by experiments by comparing the speech-pause detectors output to the actual shape of waveform: the higher ϑ is, the more frames are considered as pause. Depending on the quality of the speech signal, i.e., signal-to-noise ratio ϑ ranges from 0.0001 to 0.05 when comparing the average power as described above. When comparing signal amplitudes and the signal's maximum amplitude, the threshold $\hat{\vartheta}$ ranges from 0.001 to 0.1 (Pittermann and Pittermann 2006d).

Further labeling assistance is provided by the iterative "bootstrapping" method. It takes advantage of a recognizer, trained on a small set of labeled data, which is used to label the remaining data semi-automatically. The functionality of the bootstrap algorithm is illustrated in Fig. 4.22. Given a corpus size of several thousand utterances, the annotator (also referred to as "human expert") selects a subset of a few hundred arbitrary utterances and labels them manually according to predefined

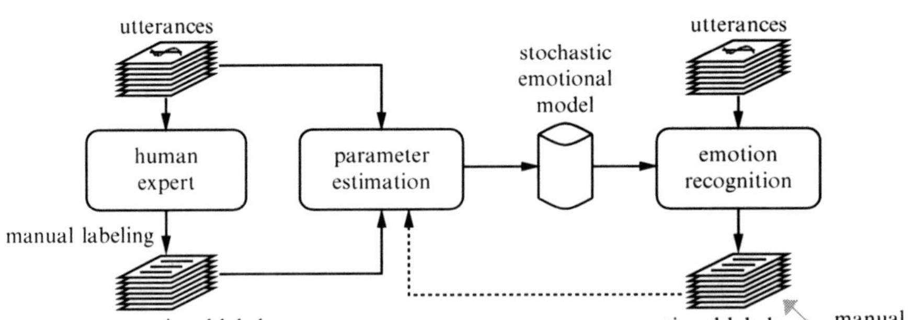

Fig. 4.22 Bootstrap algorithm for labeling (emotional) speech data (Pittermann and Pittermann 2006d)

constraints. These utterances and the respective labels are then used to estimate the parameters of preliminary stochastic (emotional) models and these models are used to perform a recognition on another subset of a few hundred utterances. The recognition result is then reviewed by the annotator and manual corrections are performed where necessary. Then both manually labeled and corrected utterances plus their labels are used to train new models which, in turn, are used to label unlabeled data, etc., until the whole corpus is labeled. The major improvement provided by this method is due to the fact that the stochastic models are supposed to become more accurate with an increasing number of training utterances. By that, the reviewing and correction effort of the annotator decreases significantly after the first few iterations. Moreover, the number of new unlabeled utterances per subset can be increased to speed up the process. In short, technically speaking, given a corpus $\mathcal{C}$, the algorithm can be described by the procedure in Fig. 4.23.

In order to distribute the workload, there exist approaches to enable multiple annotators to work on the same corpus (Ma et al. 2002). Although such an approach does not necessarily facilitate a single annotator's work, it speeds up the labeling process as annotators are able to work in parallel. Moreover, an annotator may even employ other annotators' results in the training process to obtain more accurate models and, by that, to reduce the own correction effort. A distributed system is typically based on a client-server architecture accessing one central database storing the utterances, labels and preliminary results. Such an approach is described in Abdennaher et al. (2007): A client software provides a graphical user interface allowing the annotator to select subsets to be used for parameter estimation and recognition, to perform the actual training and recognition process as previously specified, and to review and correct the recognition results.

An exemplary dictionary is given in Fig. 4.24. Each emotion and the pause is assigned one "emotioneme" (in analogy to phonemes in speech recognition), e.g., "DISGUST" $\leftrightarrow$ "dis". Such a simple dictionary, however, does not contribute to any significant constraints in the recognition network, as each emotion is only assigned one emotioneme. However, when deriving emotion recognition from speech recognition, it is practical to set up a dictionary as described in this example. Optionally, an emotion can be transcribed by a set of emotionemes in order to model acoustic effects at the beginning, in the middle and at the end of the respective emotion, e.g., "DISGUST" evolves to "dis-begin dis-middle dis-end". This, on the one hand,

1. let $i = 0$; choose subset $\mathcal{L}_0 \subseteq \mathcal{C}$ and annotate utterances in $\mathcal{L}_0$

while $\mathcal{L}_i \neq \mathcal{C}$ **do** {

2. perform stochastic model parameter estimation using the data in $\mathcal{L}_i$
3. select a subset $\mathcal{T} \in \mathcal{C}, \mathcal{T} \cap \mathcal{L}_i = \emptyset$
4. perform recognition on data in $\mathcal{T}$ using the stochastic models
5. review and correct the recognition results; let $\mathcal{L}_{i+1} = \mathcal{L}_i \cup \mathcal{T}, i = i + 1$

}.

Fig. 4.23 Technical description of the bootstrap algorithm for labeling (emotional) speech data

```
ANGER        ang
BOREDOM      bor
DISGUST      dis
FEAR         fea
HAPPINESS    hap
NEUTRAL      neu
PAUSE        pau
SADNESS      sad
```

Fig. 4.24 Dictionary for an emotion recognizer

```
$emotion = ( ANGER | BOREDOM | DISGUST | FEAR | HAPPINESS
           | NEUTRAL | SADNESS );

( < $emotion | PAUSE > )
```

Fig. 4.25 EBNF of a flexible grammar for emotion recognition

```
( PAUSE $emotion PAUSE )
```

(a)

```
( PAUSE < $emotion | PAUSE > )
```

(b)

```
( PAUSE ( < ANGER [ PAUSE ] > | < BOREDOM [ PAUSE ] > |
          < DISGUST [ PAUSE ] > | .... ) PAUSE )
```

(c)

Fig. 4.26 EBNF of alternative grammars for emotion recognition. The emotions in `$emotion` are the same like in Fig. 4.25

provides more accurate models, especially in cases where one emotion stretches over a longer time span, but, on the other hand, requires three times more models than for the one-emotioneme-per-emotion case. This in turn requires more training data and may lead to a lower recognition accuracy as there are more models to choose from in the recognition (Pittermann and Pittermann 2006a).

The "emotional grammar" shown in Fig. 4.25 represents the most general and flexible type of a language model for emotion recognition. In analogy to a "word loop" which is commonly used in speech recognition, every emotion or pause is allowed before or after any other emotion. Also, no restriction on the number of emotions per utterance is defined. Alternatively, more restrictive grammars as shown in Fig. 4.26 can be employed, assuming certain knowledge how the emotion(s) is/are temporally arranged in the utterance or forcing pauses between single emotional states.

Like in plain speech recognition, also statistical language models can apply by using an emotion loop grammar like in Fig. 4.25 and introducing restrictions such as the conditional probability $P(x|y)$ of emotion x given emotion y as the preceding emotion. For use in an actual recognizer, these probabilities can be derived from speech corpora, however, it is comprehensible that, e.g., the probability of changing from sad to happy is considerably lower than the probability of changing from neutral to disgusted.

In analogy to the phonemes in speech recognition, each emotioneme is represented by one HMM. With respect to the labeling and therefore the grammar used in the recognition process, an emotioneme tends to span a longer time period than a phoneme. This can either be considered by using HMMs with more than five states, e.g., ten or 20 states, by increasing the self-transition probabilities in a five-state HMM, or by using more general models also allowing right-to-left transitions. An increase of the number of states involves more training effort and allows more changes of the feature vectors during one emotion(eme). Thus, it has shown to be more practical using five-state HMMs when subdividing an utterance in many short emotion periods as opposed to using larger HMMs when, e.g., labeling one complete utterance with one emotion.

Representing emotionemes, the length of which exceeds the length of phonemes, the respective HMMs are required to be less restrictive than the HMMs used in speech recognition. Apart from the fact that the number of features varies from recognizer to recognizer, also the shapes of the probability density functions differ from feature parameter to feature parameter. This, like in speech recognition, is taken into consideration by using GMMs. However, as the training of emotioneme models does not include tied-state triphones (or tri-emotions, respectively) and special pause models, the appropriate number of Gaussian mixtures is already trained right away from the first (re-)estimation step. Thus, GMM prototypes, like the exemplary model shown in Fig. 4.27 are used in the training procedure.

As opposed to the prototype in Fig. 4.20, the observation probability density functions part of the states in the GMM prototype is subdivided into multiple mixture functions which are differently weighted. Still assuming that the resulting probability density function is at least somehow similar to a Gaussian function, there exists one "dominant" mixture, and the remaining mixtures have significantly smaller weights (see also Fig. 4.14).

The training of the emotioneme models is accomplished by repeating re-estimation steps and an alignment procedure to sort out unreliable training data. With respect to the number of Gaussian mixtures, in our system ranging from 1 to 8, a higher number of re-estimation steps, compared to speech recognition, is required to obtain accurate models. The "ideal" number of iterations, representing a trade-off between good performance and smallest possible number of iterations, has been determined in experimental series; it ranges from 25 for 1 mixture to 35 for 8 mixtures (Pittermann and Pittermann 2006a).

Using the dictionary shown in Fig. 4.24 and the grammar in Fig. 4.26 (b), the recognition network is structured as illustrated in Fig. 4.28.

```
<BeginHMM>
  <NumStates> 5
  <State> 2 <NumMixes> 3
      <Mixture> 1 0.5
      <Mean> 63
        1.0 1.0 1.0 1.0 1.0 1.0 1.0 1.0 1.0 ...
      <Variance> 63
        1.0 1.0 1.0 1.0 1.0 1.0 1.0 1.0 1.0 1.0 ...
      <Mixture> 2 0.3
      <Mean> 63
        1.0 1.0 1.0 1.0 1.0 1.0 1.0 1.0 1.0 1.0 1.0 ...
      <Variance> 63
        1.0 1.0 1.0 1.0 1.0 1.0 1.0 1.0 1.0 1.0 1.0 1.0 ...
      <Mixture> 3 0.2
      <Mean> 63
        1.0 1.0 1.0 1.0 1.0 1.0 1.0 1.0 1.0 1.0 1.0 1.0 ...
      <Variance> 63
        1.0 1.0 1.0 1.0 1.0 1.0 1.0 1.0 1.0 1.0 1.0 1.0 ...
  <State> 3 <NumMixes> 3
      [...]
  <State> 4 <NumMixes> 3
      [...]
  <TransP> 5
    0.0 1.0 0.0 0.0 0.0
    0.0 0.6 0.4 0.0 0.0
    0.0 0.0 0.6 0.4 0.0
    0.0 0.0 0.0 0.7 0.3
    0.0 0.0 0.0 0.0 0.0
<EndHMM>
```

Fig. 4.27 Prototype of a Gaussian mixture HMM

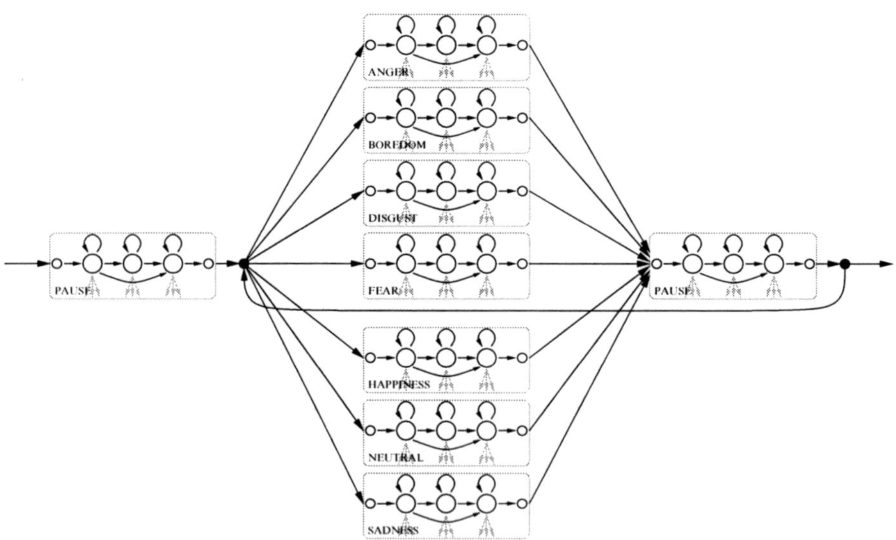

Fig. 4.28 Emotion recognition network based on the dictionary and the grammar shown in Fig. 4.24 and 4.26 (**b**)

```
       0   300000 PAUSE     -365.44
  300000   800000 NEUTRAL   -487.93
  800000  1100000 PAUSE     -321.48
 1100000  4400000 FEAR     -3101.73
 4400000  4700000 PAUSE     -323.36
 4700000 20500000 SADNESS -9212.50
20500000 21400000 PAUSE     -909.05
21400000 26100000 FEAR     -4656.30
26100000 27200000 PAUSE    -1183.06
```

vs.

```
       0   400000 PAUSE     -432.65
  400000  1200000 NEUTRAL -1389.36
 1200000  1400000 PAUSE     -317.85
 1400000  4100000 NEUTRAL -9102.45
 4100000  4700000 PAUSE     -524.46
```

Fig. 4.29 Emotion recognition results (ETNs) for different utterances and model types

Performing the same Viterbi search method on this network like in speech recognition, an emotion transition network (ETN) representing the most probable path in the network is obtained. Examples of ETNs for different utterances using different types of emotioneme models are shown in Fig. 4.29. Each HMM represents one emotioneme which is, according to the dictionary, equivalent to one emotional state. With the grammar requiring pauses at the beginning, between emotions and at the end, the loop closes between the respective nodes.

As opposed to speech recognition, where the correctness of the resulting word transition networks can be directly evaluated by comparing these to the reference labels and counting insertions, substitutions or deletions, the "correctness" of an ETN cannot be directly determined by objective measures. Especially in cases where the labels of the reference do not cover the complete utterance, the recognizer inserts arbitrary emotions which do not necessarily have to influence the result in a negative way, but which make up an error in terms of objective error measures. Thus, a goal for the recognition needs to be defined, e.g., the predominant emotion occurring in an utterance has to be correctly recognized (see Section 6.3). By that, only utterance-level errors are measured, in analogy to sentence errors, instead of emotion-level errors, in analogy to word errors. I.e., if the defined goal is not achieved, the utterance is considered as not correctly recognized. Such an approach entails that, even if the ETN of an anger-dominated utterance contains "ANGER" parts, it is considered as wrongly recognized as soon as, e.g., "NEUTRAL" parts occur more often. Alternatively, either the evaluation is completely subjective, with a human reviewer judging whether each recognition result is "acceptable" or not, or soft measures can be used to calculate a likelihood between result and reference: Assuming an utterance is labeled as 100% "ANGER" and the ETN contains 80% "ANGER", 15% "NEUTRAL" and 5%"HAPPINESS", the likelihood, i.e., the correctness ratio is 0.8 (see also Section 6.3).

4.6 Combined Speech and Emotion Recognition

In the previous two sections, speech recognition and emotion recognition have been considered as separate modules or processes operating independently of each other. However, due to their similarities, we argue that it is more efficient in terms of complexity and more effective in terms of recognition performance either to combine both processes or to take advantage of mutual information exchange between both processes. In this section, we describe the setup of a combined speech–emotion recognizer which is simultaneously recognizing the textual content and the emotional state of an utterance. I.e., instead of a string of words or a sequence of emotions, a sequence of emotionally pronounced words is recognized. This is accomplished by attaching the emotional state to the words (word–emotions). E.g., "WEDNESDAY" evolves into "WEDNESDAY-ANGER", "WEDNESDAY-BOREDOM", "WEDNESDAY-DISGUST", etc.

An excerpt of a dictionary to be used in combined speech emotion recognition is shown in Fig. 4.30. In the dictionary there exist different transcriptions for each word depending on the emotional state attached to it. From the linguist's point of view, this does not look useful, however, this is necessary to distinguish the acoustic properties of phonemes in different emotional states. For the sake of clarity, the word-only transcriptions of the respective words without emotions are given as comments after "#"s. Accordingly, each phoneme now exists in multiple variations. I.e., considering the seven emotional states included in our experiments, the "uw" phoneme evolves into seven emotional phonemes ("emophonemes"): "uwa", "uwb", "uwd", "uwf", "uwh", "uwn" and "uws". In these transcriptions we assume that there is only one emotional state attached to each word. I.e., the speaker's emotional state does not change within one word, so that there is, e.g., no "WEDNESDAY-ANGER-NEUTRAL" consisting of "wa eha nwa za don eyn".

Within one utterance, the emotional state may theoretically change at any point of time. Consideration can be given to this actuality by extending the speech recognition grammar shown in Fig. 4.19 to a speech–emotion grammar as shown in Fig. 4.31.

```
GO-ANGER                ga owa                # g ow
GO-BOREDOM              gb owb
...
I-DISGUST               ayd                   # ay
I-FEAR                  ayf
...
NEW_YORK-HAPPINESS      nh yh uwh yh aoh kh   # n y uw y ao k
NEW_YORK-NEUTRAL        nn yn uwn yn aon kn
NEW_YORK-SADNESS        ns ys uws ys aos ks
...
PLEASE-FEAR             pf lf iyf zf          # p l iy z
...
WEDNESDAY-SADNESS       ws ehs ns zs ds eys   # w eh n z d ey
```

Fig. 4.30 Dictionary for combined speech–emotion recognition

```
$city    = ( LONDON-ANGER | LONDON-BOREDOM | LONDON-DISGUST
           | LONDON-FEAR | LONDON-HAPPINESS | LONDON-NEUTRAL
           | LONDON-SADNESS | NEW_YORK-ANGER | NEW_YORK-BOREDOM
           | ... | PARIS-NEUTRAL | ... | ULM-SADNESS );
$weekday = ( MONDAY-ANGER | MONDAY-BOREDOM | ...
           | WEDNESDAY-DISGUST | ... | THURSDAY-NEUTRAL
           | THURSDAY-SADNESS );
date     = ( TODAY-ANGER | ... | TOMORROW-DISGUST
           | ... | ( ON-ANGER | ... | ON-SADNESS ) $weekday );

( [ ( I-ANGER | ... | I-SADNESS )
    ( ( WANT-ANGER | ...| WANT-SADNESS )
      | ( WOULD-ANGER | ... | WOULD-SADNESS )
        ( LIKE-ANGER | ... | LIKE-SADNESS ) )
      ( TO-ANGER | ... | TO-SADNESS )
      ( GO-ANGER | GO-BOREDOM | ... | TRAVEL-SADNESS )
  ]
  ( TO-ANGER | ... | TO-SADNESS ) $city [ $date ]
  [ PLEASE-ANGER | ... | PLEASE-SADNESS ] )
```

Fig. 4.31 Speech–emotion grammar adapted from the grammar shown in Fig. 4.19

```
$wordemotion = ( GO-ANGER | GO-BOREDOM | ... | I-FEAR | ...
               | NEW_YORK-HAPPINESS | ... | WOULD-SADNESS );

( [ PAUSE ] < $wordemotion > [ PAUSE ] )
```

Fig. 4.32 Word–emotion loop as a language model for flexible combined speech–emotion recognition

Technically, every word "X" is replaced by "(X-ANGER | X-BOREDOM | ...| X-SADNESS)", increasing the complexity of the grammar from $5 \cdot 4 \cdot 7 \cdot 2 = 280$ valid word-only sequences to 12,710,187,616 valid word–emotion sequences. In cases where an exact order of the words can not be predetermined, a more flexible language model allowing for these changes is required. The simplest approach to that is a word–emotion loop as shown in Fig. 4.32. The complexity of such a model, however, increases with the length of the utterance: E.g., allowing up to 5 consecutive word–emotions, there exist $133 + 133^2 + 133^3 + 133^4 + 133^5 = 41,931,067,073$ valid word–emotion sequences for a dictionary of 19 words evolving to $7 \cdot 19 = 133$ word–emotions. It can be determined in experiments that, like in speech recognition, the recognizer performance decreases when using more flexible language models and larger vocabularies. A reduction of the language models' complexity is described in Section 5.1.2 (Pittermann et al. 2007b).

A trade-off between high flexibility and low complexity is obtained when using n-gram language models which, typically employed in speech recognition, can also be adapted to speech–emotion recognition. For an accurate recognition, however, a large text corpus for training is required to account for all transitions between

```
      0  1400000 I-ANGER       -3972.94
1400000  4900000 WANT-ANGER    -9244.79
4900000  6200000 TO-NEUTRAL    -2937.85
6200000  7700000 GO-NEUTRAL    -4965.31
7700000  9500000 TO-NEUTRAL    -3959.03
9500000 12100000 PARIS-ANGER  -10854.37
```

Fig. 4.33 Output of a speech–emotion recognizer using the grammar shown in Fig. 4.31

both words and the respective emotional states. E.g., for a dictionary containing 19 words, there exist $19^2 = 361$ bi-gram combinations, evolving to $(19 \cdot 7)^2 = 17689$ word–emotion bi-gram combinations.

Introducing emophonemes instead of phonemes, the number of models is multiplied by seven, as each emophoneme is represented by one HMM. This not only requires more training effort, but also leads to a lower recognition performance due to the higher complexity of the recognition network unless any optimizations are included. Performing a Viterbi search like in plain speech or emotion recognition, we obtain a word–emotion transition network (WETN) like the example shown in Fig. 4.33.

The evaluation of a speech–emotion recognizer can be accomplished in a large variety of methods – either word–emotions are treated like words in plain speech recognition or word recognition performance and emotion recognition performance are considered separately applying different measures. Whereas for the word part "regular" criteria like insertions, deletions or substitutions can be applied, the emotion part, as described in Section 4.5, requires a more sophisticated evaluation.

4.7 Emotion Recognition by Linguistic Analysis

Alternatively, or in addition to the speech signal based emotion recognition methods, the emotional state of an utterance can be extracted by considering its textual content. On the one hand, as all operations are performed on text, this approach itself does not require complex signal analysis and classification methods, but, on the other hand, presumes that the text has been correctly recognized, i.e., the preceding speech recognizer performs reliably.

A neutral sentence, e.g., ""*I want to return on Monday*""s may be extended such that it represents happiness (positive mood) ""*Oh* great, *I'd* like *to return on Monday*"" or anger (negative mood) ""Damn, *I* have to *return on Monday*"". As indicated by the emphasized text, most of the emotional information is strongly related to keywords that need to be spotted. Such a list of keywords that contain emotional cues is provided by the BEEV (Cowie et al., 1999a, 2001).

Based on the BEEV, we have compiled a keyword dictionary in which, similar to a speech recognizer dictionary, words in the first column are transcribed to an emotion (anger, happiness, ...) in the second column as well as to a valence "++"

```
acceptable          happiness   ++
...
awful               anger       --
awful               disgust     -
...
boredom             boredom     0
...
cool                happiness   ++
...
funny               happiness   ++
...
irritated           fear        -
irritated           sadness     -
...
neutral             neutral     0
...
ok                  neutral     0
...
pissed [off]        anger       --
...
ridiculous          anger       --
...
sorry               sadness     -
...
wow                 happiness   ++
yawn                boredom     0
yuk                 disgust     -
```

Fig. 4.34 Emotional dictionary for linguistic emotion recognition. The first column contains the keywords, the second column contains an emotional state and the third column indicates the valence (positive/neutral/negative)

(very positive), “0” (neutral), “–” (negative) or “–” (very negative) in the third column, see Fig. 4.34. The valence is used to estimate an overall tendency of a sentence or utterance. This is useful in cases, where multiple different emotional keywords occur in the utterance, so that the tendency is determined by summing up the “+” and “–” as “+1” and “−1” resulting in a positive or negative number. Some words like, e.g., “awful” or “irritated”, have multiple emotion labels which is due to the different contexts in which they occur. Further aspects like irony which typically emanate from the whole sentence may not be covered by simple keywords, but require a more sophisticated grammar in combination with a speech signal analysis (Tepperman et al. 2006). This affective keyword dictionary contains a total of 376 keywords: 72 for anger, 30 for disgust, 62 for fear, 132 for happiness, 76 for sadness and 4 for neutral. The low number of neutral words in this compilation is not due to the fact that there are no neutral words. It is rather vice versa as we assume that all words are neutral except for those listed in this dictionary (Pittermann et al. 2008b).

The keyword list is integrated into the linguistic analysis such that the emotions of the occurring keywords of an utterance are concatenated. I.e., following the XML grammar notation described in Hunt and McGlashan (2004), the words are embedded as

```
<item> <tag>emotion+=',happiness';</tag> great </item>
```

in order to obtain emotion labels or as

```
<item> <tag>valence+=',++';</tag> great </item>
```

in order to obtain valence measures. Assuming that the user is in a neutral state if none of the emotion words occur in the utterance, each utterance is initialized as neutral, and a sentence like *""I want to return on Monday""* would be characterized as "neutral" or "0", whereas *""Oh great I'd like to return on Monday""* would be analyzed as "neutral, happiness, happiness" or "0, ++, ++". In the latter case, however, the initial "neutral" or "0" is typically ignored in the further processing. We refer to this grammar which constitutes the basis of the linguistic emotion recognizer as affective grammar.

Due to its autonomy from other components, the linguistic emotion recognition can be easily integrated into spoken language dialogue systems in different ways. The most prominent approaches are illustrated in Fig. 4.35.

In the first approach (a), the speech–emotion recognizer provides a word–emotion transition network which is optionally post-processed with the ROVER method (see Chapter 5.2) if multiple speech–emotion recognizers are involved. The word–emotions in the (processed) WETN are then separated into words and emotions and the resulting sentence is parsed by the linguistic analysis.

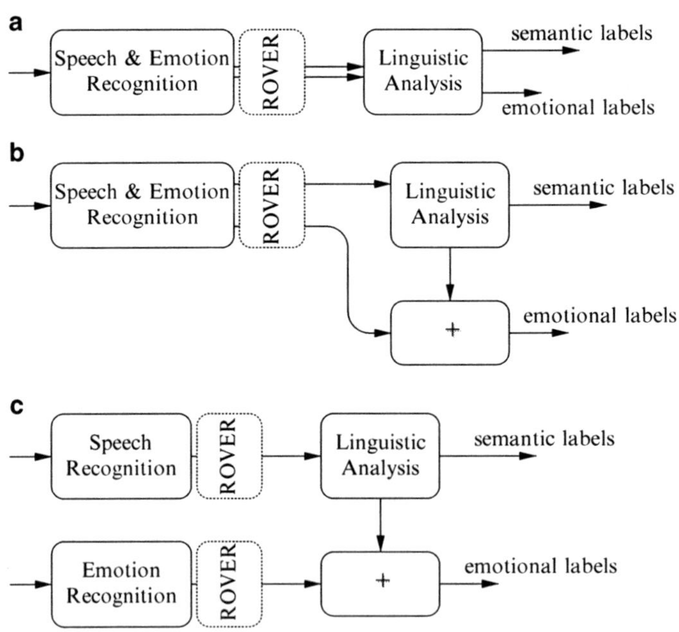

Fig. 4.35 Integrating linguistic emotion recognition and speech–emotion recognition into SLDSs (Pittermann et al. 2008b)

E.g., a sentence like "EXCELLENT-HAPPINESS I-NEUTRAL WOULD-HAPPINESS LIKE-NEUTRAL TO-NEUTRAL RETURN-HAPPINESS ON-NEUTRAL MONDAY-HAPPINESS" would be transformed into "EXCELLENT HAPPINESS I NEUTRAL WOULD HAPPINESS LIKE NEUTRAL TO NEUTRAL RETURN HAPPINESS ON NEUTRAL MONDAY HAPPINESS" and the emotions are also treated as words. Thus, the output of the linguistic analysis (omitting the initial "neutral") would be "happiness, happiness, neutral, happiness, happiness, neutral, neutral, happiness, neutral, happiness" as both "LIKE" and "EXCELLENT" are mapped to "happiness". Alternatively, the valence of the sentence would be "++, ++, 0, ++, ++, 0, 0, ++, 0,++". Comparing the influence of the speech–emotion recognizer and the linguistic emotion analysis, it is obvious, that the impact of the speech–emotion recognizer is significantly higher as all words by design are assigned one emotion whereas the number of emotional keywords in the sentence is much lower. In the above example, the ratio is 2 (emotional keywords): 8 (total number of word–emotions). In order to avoid such an unequal treatment, the emotional dictionary can be adapted to use different numbers of "+" or "−" in the valence for emotional words (like "great") and emotion words (like "anger"). The same idea can also be applied to the emotional labels, so that the keyword dictionary is modified, e.g., as shown in Fig. 4.36.

By that, a sentence like "GREAT-BOREDOM I-ANGER WOULD-ANGER LIKE-DISGUST TO-NEUTRAL RETURN-HAPPINESS ON-ANGER MONDAY-ANGER" which is regularly classified as "++, 0, –, –, ++, −, 0, ++, –, –" (total valence: −3) would be classified as "++++, 0, −, −, ++++, −, 0, +, −, −" (total valence: +4). Taking into account, that anger and happiness are quite likely to be confused, the latter approach turns up to be more practical.

A further approach for the integration of the emotional linguistic analysis is shown in Fig. 4.35 (b). This architecture also includes one speech–emotion recognizer, or optionally multiple speech–emotion recognizers applying the ROVER method, however, only the output word transition networks are passed to the linguistic analysis. The emotion transition networks or numerical emotion scores

```
...
 anger             anger               -
...
 boredom           boredom             0
...
 great             happiness happiness ++++
...
 pissed [off]      anger anger         ----
...
 sorry             sadness sadness     --
...
 wow               happiness happiness ++++
...
```

Fig. 4.36 Alternative emotional dictionary for linguistic emotion recognition. This approach uses different "weighting factors" for emotional keywords and emotion words

are passed to a further module which combines the emotion information from these ETNs or scores and the emotional output of the linguistic analysis into an adequate emotion representation. This combination can be performed by simply concatenating the emotional words and the emotions in the ETNs to obtain an output as described in the first approach. Additionally, more sophisticated combination methods on a numerical basis are feasible. For the example sentence "GREAT-NEUTRAL I-ANGER WOULD-ANGER LIKE-DISGUST TO-NEUTRAL RETURN-HAPPINESS ON-ANGER MONDAY-ANGER" we obtain two "happiness" words from the linguistic analysis and a soft emotion score of S_E ={ 0.554 ANGER, 0.188 NEUTRAL, 0.152 HAPPINESS, 0.106 DISGUST } from the voting module of the ROVER system. Based on these conditions, several methods are possible:

- The proportion of happiness in this example is 1.0, which can be added to the happiness proportion in the soft score: S'_E ={ 0.554 ANGER, 0.187 NEUTRAL, 1.152 HAPPINESS, 0.107 DISGUST } which then needs to be normalized (dividing by $1 + 1 = 2$) to $\bar{S}_E$ ={ 0.576 HAPPINESS, 0.277 ANGER, 0.094 NEUTRAL, 0.053 DISGUST }. This approach can also be referred to as averaging approach.
- Alternatively, a weighted average approach can apply, i.e., the final soft score $\bar{S}_E$ is calculated as

$$\bar{S}_E = (S_E + \gamma \cdot L_E)/(1 + \gamma),$$

  where $0 < \gamma < \infty$ is a weighting factor representing the importance of the linguistic emotion score L_E. E.g., for $\gamma = 0.5$, the final score for the numbers given above is

$$\begin{aligned}\bar{S}_E &= (\{0.554\text{A}, 0.188\text{N}, 0.152\text{H}, 0.106\text{D}\} + 0.5 \cdot \{0\text{A}, 0\text{N}, 1\text{H}, 0\text{D}\})/1.5 \\ &= \{0.369\text{A}, 0.125\text{N}, 0.435\text{H}, 0.071\text{D}\}.\end{aligned}$$

- The third method includes a statistical model, based on which the final score $\bar{S}_E$ is determined by maximizing the probability $P(\bar{S}_E|S_E, L_E)$ given the recognizer soft score S_E and the linguistic emotion score L_E as described above.

$$\bar{S}_E = \arg\max_{S^*} P(S^*|S_E, L_E).$$

  Such a model can be implemented as a Bayes classifier or a Markov Decision Process, trained on a corpus of annotated examples extracted from recorded human–computer or distilled human–human dialogues.

The third integration approach as illustrated in Fig. 4.35 (c) follows the ideas described for the second approach, except for the fact that words and emotions are recognized independently by separate modules using one or multiple recognizers for speech and emotion recognition at any time. This entails the effect that words and emotions are not aligned, so that calculation of emotional scores in the combination module needs to be adapted to the respective recognizers' properties.

E.g., for a sentence like "GREAT I WOULD LIKE TO RETURN ON MONDAY" the emotional labels may range from "HAPPINESS" to "BOREDOM ANGER HAPPINESS ANGER ANGER ... ANGER NEUTRAL DISGUST ... NEUTRAL ANGER ANGER" depending on the emotioneme models. To account for that discrepancy, the weighting factors need to be selected accordingly or the model for $\bar{S}_E$ needs to be trained on suitable data (Pittermann et al. 2008b).

4.8 Discussion

In this chapter, we have described our approaches to speech-based emotion recognition. We pay particular attention to what we refer to as speech–emotion recognition, i.e., the combined recognition of speech (text) and emotions. Accordingly, our selection of features and the properties are strongly geared to regular speech recognition. I.e., we use Mel-frequency cepstral coefficients for robust phonetic classification plus prosodic and acoustic features for the detection of the emotional state. The acoustic model consists of Hidden Markov Models each of which represents one phoneme in one emotional state. By that, the complexity of the acoustic model grows linearly with the number of emotions to be recognized, which initially leads to an increasing decoding complexity and a lower recognition performance. An evaluation of the proposed methods is given in Section 6.3. Optimizations to our combined speech–emotion recognizer as well as an approach to fusion multiple speech–emotion recognizers are described in Chapter 5.

Additionally, we propose an approach to emotion recognition by linguistic analysis by spotting emotionally relevant keywords as described in Section 4.7. Such an approach at first requires a reliable speech recognizer output. Due to the fact that explicit emotional keywords occur relatively infrequently, the linguistic analysis is rather useful in combination with the emotion recognizers described above. We describe the evaluation of our affective grammar approach in Section 6.3.4.

Chapter 5
Implementation

The ideas about adaptive dialogue management and speech-based emotion recognition as described in Chapters 3 and 4 constitute a firm groundwork as for theoretical aspects of the integration of emotions into adaptive SLDSs. This groundwork, however, features a large potential for improvement as well as a high degree of flexibility concerning an implementation. In this chapter, we identify approaches to improve the performance of our emotion and speech–emotion recognizers and we describe the implementation of our adaptive dialogue manager.

There exist a large variety of parameters which can be altered to increase the performance and robustness of speech-based emotion recognizers. In the following section, we address the optimization of our plain emotion recognizer and our combined speech-emotion recognizer. Optimizations of the recognition performance do not necessarily require a change of the recognizers. Instead, a post processing algorithm can be applied to reduce the recognition errors after the recognition process. Our approach to combining multiple plain emotion recognizers or combined speech-emotion recognizers to reduce the overall error rate is described in Section 5.2. In Section 5.3, we present an adaptive dialogue manager which is based on VoiceXML. This dialogue manager integrates the semi-stochastic emotional dialogue model as described in Section 3.8 to adapt the dialogue flow to the user's emotional state.

5.1 Emotion Recognizer Optimizations

In Sections 4.5 and 4.6, we have described two approaches to (speech–)emotion recognition using Hidden Markov Models. Evaluations of implementations in the described standard setup and configuration (see Chapter 6) have shown that these approaches provide a relatively useful performance, which, however, can be expanded by several means (Pittermann and Pittermann 2006a; Pittermann et al. 2007b). In this section we address problems observed in the experiments and we propose approaches to improve the recognizers' performance, for both plain emotion recognition and speech–emotion recognition.

J. Pittermann et al., *Handling Emotions in Human-Computer Dialogues*,
DOI 10.1007/978-90-481-3129-7_5,

5.1.1 Plain Emotion Recognition

For test purposes, a plain emotion recognizer including an HMM with one Gaussian mixture has been trained on arbitrary speech data taken from the Berlin Database of Emotional Speech (see also Section 6.3) made publicly available by the Technical University of Berlin (Burkhardt et al. 2005). For the labeling, an automatic speech–pause detector has been employed and the 'speech' parts have been assigned the predominant emotional state given for the respective utterance. Tests with this setup have, on the one hand, unearthed a rather low performance of the system, but, on the other hand, provided indications about the weaknesses or where optimizations can be taken up.

The first kind of optimization, which actually does not directly relate to the HMM classification, is applied to the labeling of the data. On the one hand, it is mandatory that the labels reliably match the actual emotional or textual content of the utterances. On the other hand, the labels should be designed and arranged such that they go with the employed model type. For speech recognition, both criteria are fulfilled by default – each word in the utterance is assigned the respective text which is subdivided into phonemes which, in turn, are represented by one model with the appropriate parameters. For plain emotion recognition, however, the labels of an utterance can be subdivided arbitrarily without modifying the actual emotional content. Exemplary labeling methods of one utterance are illustrated in Fig. 5.1. Apart

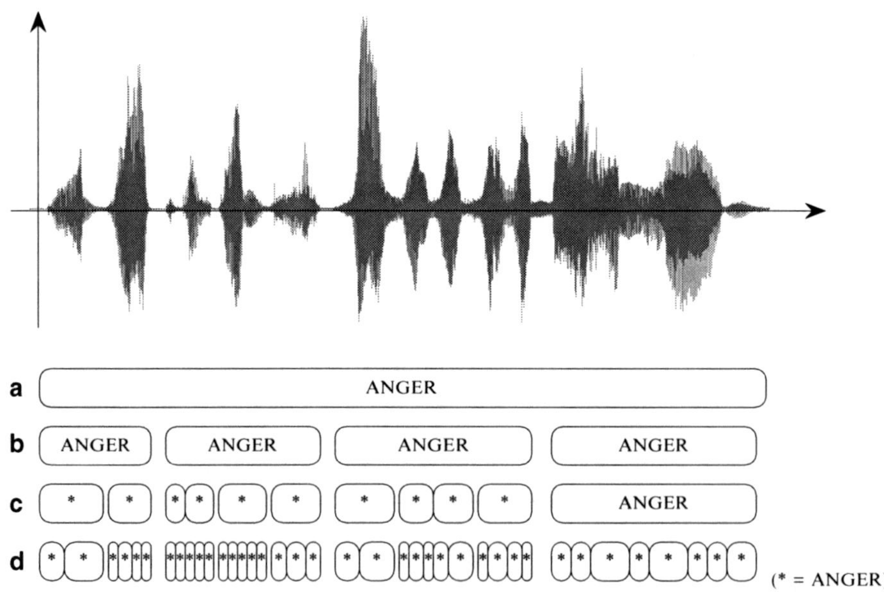

Fig. 5.1 Different labeling levels for emotional labeling: (**a**) utterance level, (**b**) group level, (**c**) word level and (**d**) phoneme level

from further possible variations, we distinguish four basic labeling levels (cf. the three labeling stages described by Campbell et al. 2006):

1. Utterance level, i.e., labeling the whole utterance with one single emotion as shown in Fig. 5.1(a).
2. Speech–pause/group level, i.e., groups of words, separated by pauses, are labeled with the respective emotion, Fig. 5.1(b).
3. Word level, i.e., like in speech recognition, the utterance is subdivided into words and each word is assigned one emotion as illustrated in Fig. 5.1(c).
4. Sub-word/phoneme level, i.e., each word is again subdivided into its phonemes or syllables and each of these is assigned one emotion, Fig. 5.1(d).

Labeling the whole utterance at once (level 1.) is the most trivial method, but does not make allowance for the models' properties well enough, as most utterances last a few seconds or longer, overstraining the range of an HMM. Labeling each phoneme separately (level 4.) might suit the HMMs best due to the temporal structure. However, for the annotator, it is quite tedious to subdivide an utterance into a large number of phonemes and label these individually. E.g., a short sentence like "I want to travel to New York" already contains 20 phonemes. In cases where the word boundaries are known, or where the words are already labeled, level 3. or, with the aid of a modified dictionary, even level 4. may be applied.

Further fine-tuning can be applied on the speech–pause level (level 2). Varying the speech–pause threshold, different segmentation characteristics can be procured and an optimal threshold (for certain model parameters) can be determined by simulation series: For each threshold value, automatic labeling is applied and emotional models are trained on the basis of these labels and the utterances. Then, recognition tests are performed on test utterances and the accuracy of the output is evaluated by means of word-level measures as used in speech recognition, i.e., substitutions as well as insertions and deletions count as errors. The optimal threshold is defined as the threshold for which the accuracy is maximal. E.g., for a five-state left-to-right HMM, the threshold is at around 1% of the maximum amplitude of the speech signal(s) (Pittermann and Pittermann 2006a).

A further approach to improve the emotion recognizer's performance also affects the emotion models and the functionality of the recognizer only marginally and, thus, actually cannot intrinsically be considered as an optimization. However, for the sake of completeness and with respect to the typical scenario in which the emotion recognizer operates, we will shortly describe the application of a reduced emotion set (Pittermann and Pittermann 2006d).

On the one hand, it is obvious and comprehensible that a reduction of the set of emotions automatically leads to an improvement of the recognition performance: Even if the recognizer just randomly selected its output, a reduction from seven to five emotions increases the recognition rate from $1/7 \approx 14.3\%$ to $1/5 = 20\%$. And one may wonder why certain emotions shall be excluded from the recognizer. On the other hand, the goal of the emotion recognizer is not only to recognize distinct emotions correctly but also to supply the dialogue management with useful information in order to adapt the dialogue according to the user's emotional state. Thus, we argue that there exist emotions that do not only show very similar acoustic and prosodic properties but which also evoke similar reactions towards the user. And, moreover, certain emotions do not necessarily influence the dialogue flow at all.

For the latter reason, disgust is omitted from the set of emotions. We argue that, e.g., in a travel information system we can assume that users are not disgusted when querying the departure time of a flight, and, moreover, there wouldn't be an appropriate way to respond to a disgusted user other than responding in a neutral or appeasing way. A similar argumentation holds for boredom. How shall the reaction towards a bored user differ from the reaction towards a neutral user?

Comparing the acoustic and prosodic features of the respective utterances (see the pitch distributions illustrated in Figs. 4.4 and 4.5 or the statistical computations in Fig. 4.7), it can be observed that neutral and boredom show a very strong similarity and that disgust, exhibits a certain degree of identicalness with any other emotion, is more difficult to distinguish from other emotions. These observations hold not only for pitch, but also for other features which can be verified by means of recognition results. For a regular emotion recognizer as described in Section 4.5 using all emotions, an analysis of the recognition results shows that there is a significantly high degree of confusion between neutral and boredom and that disgust equally interferes with the other emotions. Further details of the recognition results are given in Chapter 6.

Applying this idea, "disgust" utterances are omitted completely and "boredom" utterances are re-labeled as neutral utterances contributing to a common "neutral" model. By that, the number of emotional models reduces from seven to five. The models for anger, fear, happiness and sadness remain unchanged. The model for disgust is removed and the new model for neutral becomes a more general model with broader distributions than the original neutral model. By that, the likelihood of confusion between neutral and other emotions increases.

A third optimization approach, also visibly improving the recognition performance, is the discrimination of female and male speakers for all emotional states (Pittermann and Pittermann, 2006a). To accomplish that, labels, dictionary and language models need to be adapted. Each emotion evolves into a female and a male version, e.g., the label "ANGER" becomes "ANGER-F" and "ANGER-M". The respective dictionary is shown in Fig. 5.2. By that, the number of models increases from eight (seven emotions plus pause) to 15 (two times seven emotions plus pause). This, on the one hand, provides a larger choice for the recognizer which could lead to a lower recognition performance. However, on the other hand, these 15 models are stronger silhouetted against each other than the original eight models so that the recognizer's robustness is visibly increased. As opposed to speaker-independent speech recognition, where the (general) phoneme models shall suit as many different speaker types as possible, this approach focuses on highly constricted models. These differences between female and male speakers can, again, be observed when comparing the characteristics of the features like pitch (Figs. 4.4 and 4.5), intensity and their statistical computations (Fig. 4.7).

A suitable grammar is shown in Fig. 5.3. It is adapted from the most flexible grammar described in Fig. 5.3. Emotions are subdivided into female and male

```
ANGER-F          angf
ANGER-M          angm
BOREDOM-F        borf
BOREDOM-M        borm
DISGUST-F        disf
DISGUST-M        dism
...              ...
NEUTRAL-F        neuf
NEUTRAL-M        neum
PAUSE            pau
SADNESS-F        sadf
SADNESS-M        sadm
```

Fig. 5.2 Dictionary for a gender-dependent emotion recognizer

```
$emotionf = ( ANGER-F | BOREDOM-F | DISGUST-F | FEAR-F |
              HAPPINESS-F | NEUTRAL-F | SADNESS-F );
$emotionm = ( ANGER-M | BOREDOM-M | DISGUST-M | FEAR-M |
              HAPPINESS-M | NEUTRAL-M | SADNESS-M );

( < $emotionf | PAUSE > | < $emotionm | PAUSE > )
```

Fig. 5.3 A grammar for emotion recognition distinguishing female and male speakers

variations. However, only one of these two sets is allowed in an utterance, making allowances for the fact that the user does not change from female to male or vice versa while speaking. As mentioned above, the respective models differ strongly enough to improve the emotion recognition performance, and, moreover and secondary, the recognizer is also able to determine the speaker's gender.

5.1.2 Speech–Emotion Recognition

For combined speech–emotion recognition, in principle, the approaches described in the previous section can also apply, except for the labeling. As word–emotions are assigned to the respective parts of an utterance where these words are uttered, one cannot change their positions significantly. Thus, the labels may only vary by plus or minus a few milliseconds to achieve a more optimal segmentation of the phonemes.

The reduction of the emotion set also leads to an improvement of the recognition performance. With respect to the same feature characteristics as described in the previous section, it appears to be practical, again, to omit disgust and to merge boredom and neutral into a common neutral model. For combined speech–emotion recognition, also separate approaches, one only omitting disgust and one only merging boredom and neutral have been evaluated leading to slightly different results compared to plain emotion recognition (Pittermann et al. 2007b).

```
GO-ANGER-F              gaf owaf
GO-ANGER-M              gam owam
GO-BOREDOM-F            gbf owbf
GO-BOREDOM-M            gbm owbm
...
PLEASE-FEAR-F           pff lff iyff zff
...
WEDNESDAY-SADNESS-M     wsm ehsm nsm zsm dsm eysm
```

Fig. 5.4 Dictionary for combined speech–emotion recognition distinguishing female and male speakers

When applying the differentiation of female and male speakers in plain emotion recognition, the number of emotioneme models increases from 8 to 15, still constituting a manageable amount in terms of complexity and robustness. For speech–emotion recognition, however, where approximately 50 phonemes are already expanded to 350 emophonemes, the implementation of this approach leads to several practical problems. As shown in Fig. 5.4, e.g., a word like "GO" which has already evolved to word–emotion "GO-ANGER" now becomes word–emotion-gender "GO-ANGER-F".

This large amount of approximately 700 female/male emophonemes not only requires a larger training corpus to obtain equally accurate models but may also exceed (default) limitations of the recognition software. However, tests with a reduced emophoneme set have shown that, like in plain emotion recognition, the recognition performance can be improved when training different emophoneme models for female and male speakers and when using an adapted language model.

Up to now we have considered speech–emotion recognition as a one-step process searching the most probable path within a huge recognition network consisting of emophoneme models connected by a large variety of paths. With respect to the different evaluation criteria for speech–emotion recognition, optimizations can apply to either the speech recognition part or the emotion recognition part or both. From the technical point of view, a recognizer becomes more robust the fewer paths are allowed in the recognition network. In the one-step approach, however, such a reduction decreases the recognizer's flexibility significantly.

Thus, we propose a two-step approach involving two recognizers using (two) different recognition networks for speech and emotion recognition without requiring a noticeably higher effort than regular speech–emotion recognition. The architecture of this two-step approach is illustrated in Fig. 5.5. It consists of a feature parameter extraction, a based speech recognizer using a bi-gram language model, a language model adaptation module and a combined speech–emotion recognizer using this adapted language model (Pittermann et al. 2007b). In order to distinguish between this approach and "regular" speech–emotion recognition as described in Section 4.6, we also refer to the regular approach as "one-step" approach.

Firstly, the desired features which are used in regular speech–emotion recognition are extracted from the speech signal. These features typically include MFCCs plus prosodic and acoustic features. The MFCCs are passed to a plain speech

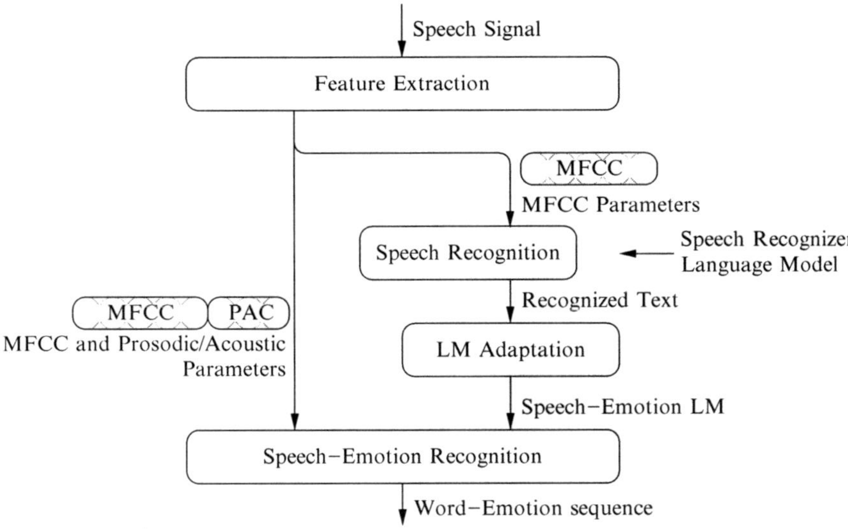

Fig. 5.5 Two-step approach for combined speech–emotion recognition (Pittermann et al., 2007b)

```
( I-ANGER WANT-ANGER TO-ANGER TRAVEL-ANGER TO-ANGER
  NEW_YORK-ANGER ON-ANGER MONDAY-ANGER |
  I-BOREDOM WANT-BOREDOM TO-BOREDOM TRAVEL-BOREDOM TO-BOREDOM
  NEW_YORK-BOREDOM ON-BOREDOM MONDAY-BOREDOM |
  I-DISGUST WANT-DISGUST TO-DISGUST TRAVEL-DISGUST TO-DISGUST
  NEW_YORK-DISGUST ON-DISGUST MONDAY-DISGUST |
  ... |
  I-SADNESS WANT-SADNESS TO-SADNESS TRAVEL-SADNESS TO-SADNESS
  NEW_YORK-SADNESS ON-SADNESS MONDAY-SADNESS )
```

Fig. 5.6 A simple grammar for two-step speech–emotion recognition

recognizer which involves an acoustic model consisting of standard phoneme models and a word-level language model. Depending on the application, this language model may be a domain-specific grammar like the one shown in Fig. 4.19 or a general statistical language model on a bi-gram or tri-gram basis. Considering this recognizer's output, a word–emotion grammar is created consisting of the recognized word sequence plus emotions attached to each word. E.g., if "I WANT TO TRAVEL TO NEW_YORK ON MONDAY" is recognized, the new grammar appears as shown in Fig. 5.6 or Fig. 5.7, alternatively.

The simple grammar in Fig. 5.6 leads to the most restrictive recognition network, consisting of only seven paths, as it assumes that the whole utterance is pronounced in the same emotional state. The alternative grammar (Fig. 5.7) provides $7^8 = 5,764,801$ possible paths (which is still less than 12 billion paths in the original word–emotion recognition network), allowing individual emotional states for each word.

```
$i         = ( I-ANGER | I-BOREDOM | I-DISGUST | I-FEAR |
               I-HAPPINESS | I-NEUTRAL | I-SADNESS );
$monday    = ( MONDAY-ANGER | MONDAY-BOREDOM | MONDAY-DISGUST |
               ... | MONDAY-SADNESS );
$new_york  = ( NEW_YORK-ANGER | NEW_YORK-BOREDOM | ... |
               NEW_YORK-NEUTRAL | NEW_YORK-SADNESS );
$on        = ( ON-ANGER | ... | ON-DISGUST | ON-FEAR |
               ON-HAPPINESS | ON-NEUTRAL | ON-SADNESS );
$to        = ( TO-ANGER | TO-BOREDOM | ... | TO-FEAR |
               TO-HAPPINESS | TO-NEUTRAL | TO-SADNESS );
$travel    = ( TRAVEL-ANGER | TRAVEL-BOREDOM | ... |
               TRAVEL-NEUTRAL | TRAVEL-SADNESS );
$want      = ( WANT-ANGER | ... | WANT-FEAR | WANT-HAPPINESS |
               WANT-NEUTRAL | WANT-SADNESS );

( $i $want $to $travel $to $new_york $on $monday )
```

Fig. 5.7 An alternative grammar for two-step speech–emotion recognition

Regarding this two-step recognition process, it can be determined how speech recognition can help to improve the performance of the associated emotion recognizer. Experiments have shown (see Chapter 6) that perfect knowledge of the textual content, i.e., a speech recognition rate of 100%, leads to significantly better emotion recognition rates than no knowledge of the textual content (as assumed in the one-step speech–emotion recognition approach). One of the reasons for this effect is illustrated in Fig. 5.8: The one-step recognizer can not use any information about the positions or the properties of the phonemes and, thus, first needs to segment the utterance appropriately and then find the most probable emophoneme (out of all emophonemes) or word–emotion sequence. The speech–emotion recognizer in the two-step approach has the advantage that it can avail itself of a mostly reasonable recognition result based on which it can not only subdivide the utterance appropriately but also assign correct phonemes to the respective positions, in this case the words "ON WEDNESDAY" are given. Then, choosing the suitable emophoneme, i.e., the emotional state among seven or five emotions, for each phoneme is less complex and provides more reliable results.

Summarizing, the two-step approach described above, the emotion recognition performance is improved by taking into account the (hopefully correct) output of the speech recognizer. Referring to the speech recognition process which is described by Eq. 4.21, the speech–emotion recognition process of a word–emotion sequence WE can be described by

$$WE = \arg\max_{WE} P(X|WE) \cdot P(WE). \tag{5.1}$$

Already knowing the output of the speech recognizer, in the two-step approach, the equation changes to

$$WE = \arg\max_{WE} P(X|WE, W) \cdot P(WE|W), \tag{5.2}$$

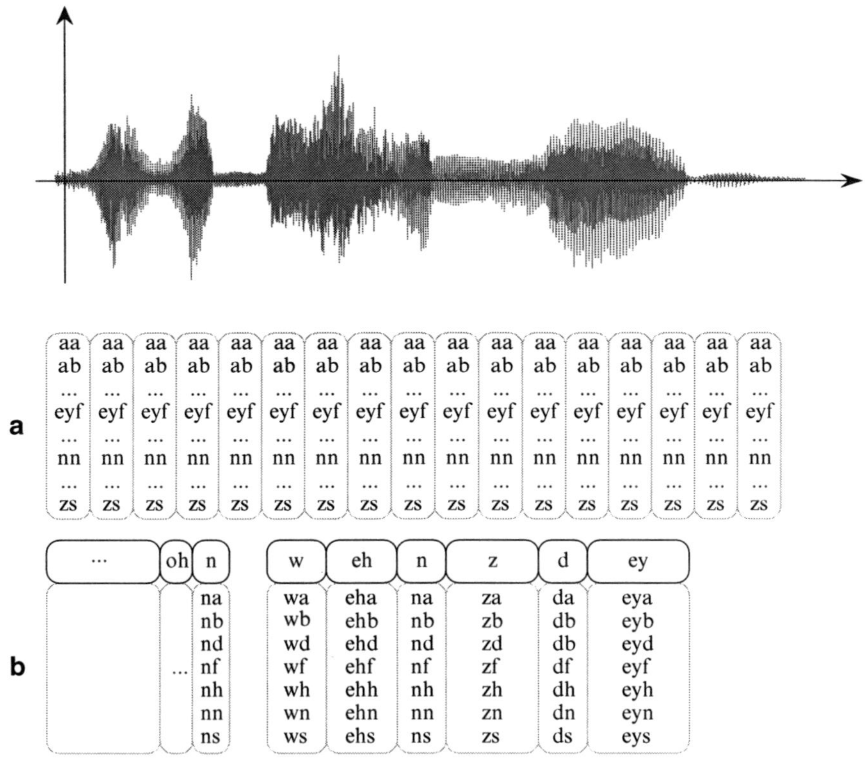

Fig. 5.8 (Emo-)phoneme alignment in one-step (**a**) and two-step (**b**) speech–emotion recognition

where W is the word part in the sequence of word–emotions. This can be summarized to

$$E = \arg\max_{E} P(X\,|\,WE, W) \cdot P(E), \tag{5.3}$$

where E is the emotion part in the sequence of word–emotions. This means, the recognizer can now "concentrate" on the recognition of emotions given W. Vice versa, it is also imaginable to use the output of an emotion recognizer to improve the speech recognition performance of the combined system. However, due to the fact that the emotion recognition performance is typically below the speech recognition performance, it does not seem sensible to apply this approach in actual systems as the speech recognizer will not benefit significantly from the extra information.

5.2 Using Multiple (Speech–)Emotion Recognizers

In systems where transmission or recognition errors occur stochastically, i.e., not predictably, it has been shown, that when slightly changing the system's parameters, different types of errors occur. Thus, a combination of multiple (different)

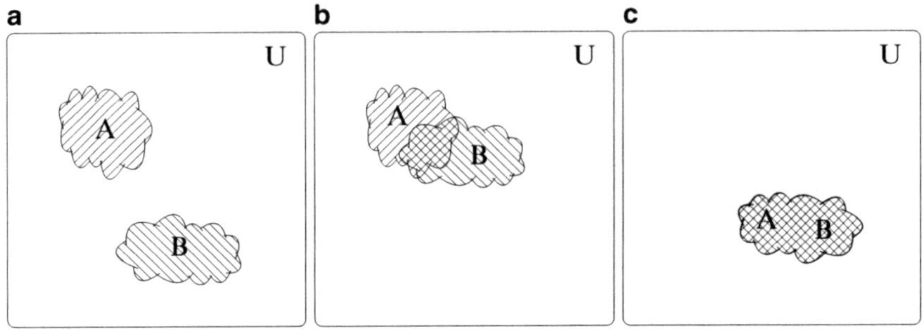

Fig. 5.9 Different scenarios for error distributions when using multiple recognition systems

systems may lead to a better detection or recognition performance if the parameters are appropriately chosen. E.g., in information transmission, the use of multiple antennas (multiple-input-multiple-output, MIMO, see Telatar 1999) helps to achieve a higher efficiency and reliability. Also the combination of sensory data (sensor fusion) which applies in, e.g., image processing or control engineering is employed to achieve a "better" data in terms of accuracy or completeness (Wagner and Dieckmann 1995; Kalman 1960).

An approach how to take advantage of the combination of multiple speech recognizers has been proposed by Fiscus (1997). The ROVER post-processing system exploits differences in the way how errors occur in multiple recognition systems. Figure 5.9 illustrates the idea behind this combinations: In these three scenarios, the entirety of all utterances U is represented by the squares and the misrecognitions of recognizers A and B are represented by striped clouds with the respective character (A or B).

In the first scenario (a), the A and B clouds do not overlap, i.e., utterances which are incorrectly recognized by the recognizer A are correctly recognized by recognizer B and vice versa. Here, an appropriate combination may theoretically lead to an error-free overall recognition. In the second scenario (b), there is some small overlap of both clouds, i.e., the utterances in the $A \cap B$ part certainly will not be recognized correctly by the combined system, whereas the utterances in $A \triangle B = (A \cup B) \setminus (A \cap B)$ could be corrected in a combined setup. Finally, in the third scenario (c), where $A = B$, there is no chance correcting errors as both recognizers "agree" on the incorrect recognitions. Alternatively, instead of considering utterances, the error distributions can also be determined on a word level.

The architecture of a ROVER system is outlined in Fig. 5.10. It is subdivided into two modules – an alignment and a voting module. In the alignment module, the word transition networks, i.e., the output word strings of the different speech recognizers are merged to one overall WTN. Based on this WTN, the best scoring output WTN is selected in the voting module.

The alignment process is illustrated in Fig. 5.11. In the dynamic programming approach, which is typically used in the ROVER system, the WTNs are aligned according to the sequence of words, regardless of the durations of the single words.

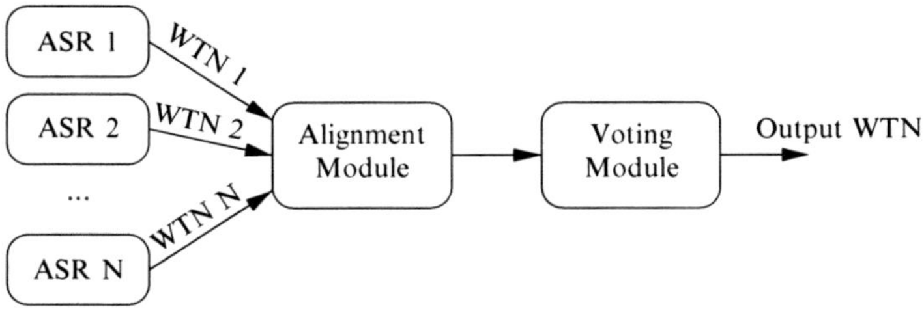

Fig. 5.10 Architecture of the ROVER system for speech recognition (Fiscus 1997)

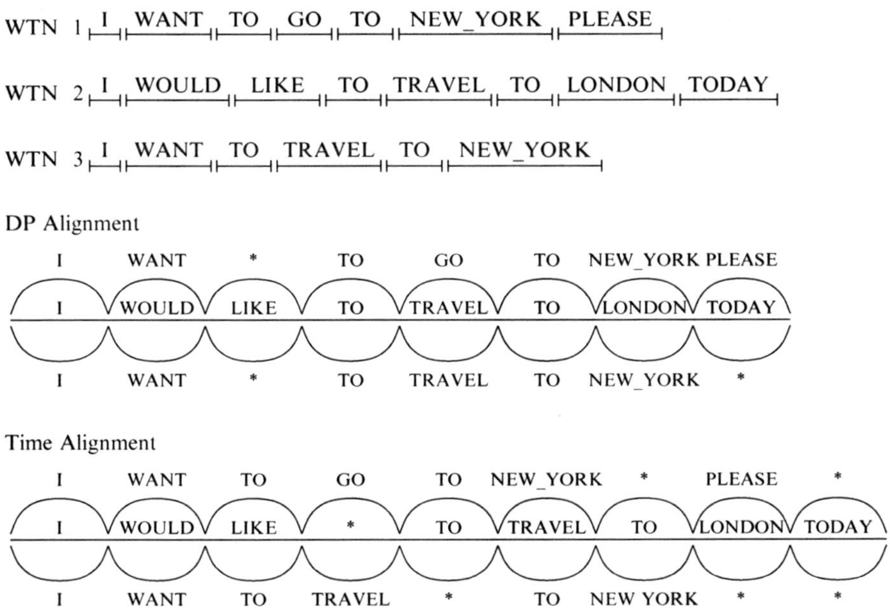

Fig. 5.11 Alignment of three word transition networks in the alignment module of a ROVER system (Fiscus 1997). One approach involves dynamic programming, the other approach aligns the WTNs on a time basis. The "*" represent null word transitions which are included in the WTNs when words are inserted or deleted

The WTN of the first recognizer serves as the basic WTN, so that the WTNs of the second, third, ...Nth recognizers are individually aligned to the basic WTN. This alignment is performed according to the minimal edit distance, which can be defined by the Levenshtein distance as illustrated in Fig. 5.12 (Levenshtein 1966; Fiscus 1997).

In this figure we align string x (I WANT TO GO TO NEW_YORK PLEASE) and string y (I WOULD LIKE TO TRAVEL TO LONDON TODAY). Both word strings are spread in a table and the 0th column ($d_{i,0}$) and the 0th row ($d_{0,j}$) of the distance matrix are initialized with 0, 1, 2,..., where $d_{i,0} = i$, $d_{0,j} = j$ and

	I WOULD LIKE TO TRAVEL TO LONDON TODAY								
	0	*1*	*2*	*3*	*4*	*5*	*6*	*7*	*8*
I	*1*	**0**	**1**	2	3	4	5	6	7
WANT	*2*	1	(1)	**2**	3	4	5	6	7
TO	*3*	2	2	(2)	**2**	3	4	5	6
GO	*4*	3	3	3	3	**3**	4	5	6
TO	*5*	4	4	4	3	4	**3**	4	5
NEW_YORK	*6*	5	5	5	4	4	4	**4**	5
PLEASE	*7*	6	6	6	5	5	5	5	**5**

Fig. 5.12 Calculation of the Levenshtein distance between two word transition networks

$0 \leq i \leq \text{length}(x)$, $0 \leq j \leq \text{length}(y)$ (see italicized numbers in second row and second column of the table). Then, for each i, $0 \leq i \leq \text{length}(x)$ and j, $0 \leq j \leq \text{length}(y)$ the distance metrics $d_{i,j}$ are calculated as

$$d_{i,j} = \min\left(d_{i-1,j} + 1;\; d_{i,j-1} + 1;\; d_{i-1,j-1} + \begin{cases} 0 & x_i = y_j \\ 1 & x_i \neq y_j \end{cases}\right). \tag{5.4}$$

Once all elements $d_{i,j}$ are calculated, the lower right element $d_{\text{length}(x),\text{length}(y)}$ denotes the Levenshtein distance between x and y, in this example, the distance is 5. The modifications ("edits") which are necessary to transform x into y can be reconstructed by traversing the distance matrix along the minimal values of $d_{i,j}$, starting from the final element $d_{\text{length}(x),\text{length}(y)}$, as emphasized by the bold numbers. In this example, putting the cart before the horse, "TODAY" is a substitution for "PLEASE" as $d_{7,8} = d_{6,7} + 1$. Analogously, "LONDON" (instead of "NEW_YORK") and "TRAVEL" (instead of "GO") are also substitutions. The "TO"s coincide in both cases. Continuing from "TO" ($d_{3,4}$), there exist two possible paths – either the bold path stating that "LIKE" is a substitution of "WANT" and "WOULD" is an insertion, or the other path (numbers in round brackets) stating that "LIKE" is an insertion and "WOULD" is a substitution of "WANT".

Alternatively, the word transition networks can also be aligned with respect to their chronological coincidence (time alignment), again with WTN 1 serving as the reference. In Fig. 5.10, this alignment has been accomplished according to a simple algorithm calculating and including the maximal overlap of words. Comparing WTNs 1 and 2, the "I"s match perfectly, "WOULD" and "WANT" show a significant overlap. The overlaps "TO" ↔ "LIKE" and "GO" ↔ "LIKE" are equal, so that the deletion can be either with "TO" or with "GO". The following "TO"s as well as "NEW_YORK" and "TRAVEL" coincide, however, the following "TO" in WTN 2 is an insertion similar to "TODAY" at the end.

Regardless of which approach has been applied, an overall ("composite") WTN like the examples shown in Fig. 5.11 is generated in the alignment module. From this WTN, the best scoring (most probable/likely) word sequence is selected in the voting module. In this module, for each word w in the composite WTN, a score

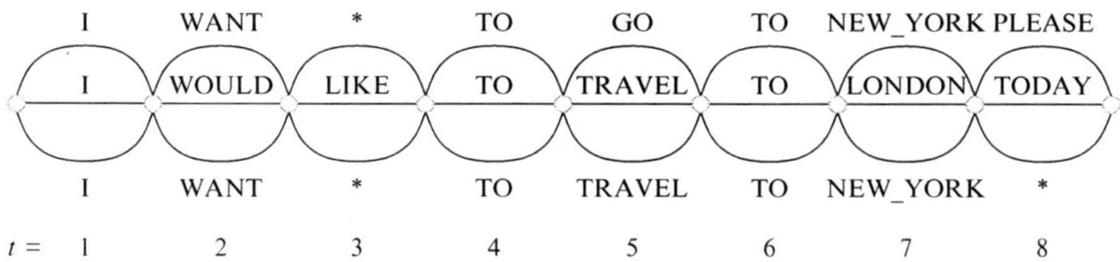

Fig. 5.13 Exemplary composite WTN for the scoring module of the ROVER system (see also Fiscus 1997)

$S(w)$ is calculated on the basis of its frequency of occurrence and its speech recognizer confidence measure. Figure 5.13 shows a composite WTN based on which this scoring shall be exemplified.

The frequencies of occurrence are determined individually at each (discrete) time i. Initially, a set of unique words is determined for each t and the frequency of each word w from this set is calculated as $N(w,t)/N_s$, where N_s is the number of input systems and $N(w,t)$ is the number of occurrence of w at time t. The confidence measures $C(w,t,r)$ of word w at time t in WTN r can be directly obtained from the speech recognizer output as shown in Fig. 4.21 which, however, in order to conform with the score calculation, need to be converted to non-log measures by exponentiating and multiplying with a suitable constant. Then, the score $S(w)$ is calculated as

$$S(w) = \alpha \cdot \frac{N(w,t)}{N_s} + (1-\alpha) \cdot C(w,t), \tag{5.5}$$

where $C(w,t)$ is the confidence of word w at time t determined from the confidence measures $C(w,t,r)$ of all involved WTNs and where α is a weighting factor determining the proportions of frequency of occurrence and confidence measure in the scoring (Fiscus 1997; Pittermann and Pittermann 2006b). The values for α as well as for the confidence $C(*)$ of a null word are trained with the aid of an exhaustive search on representative material for the lowest word error rate. On the basis of the above formula, Fiscus (1997) proposes three voting approaches:

1. Plain frequency of occurrence ($\alpha = 1$),
2. Calculating the confidence score as the average of all involved confidence measures ($\alpha < 1$)

$$C(w,t) = \frac{1}{N_s} \sum_{r=1}^{N_s} C(w,t,r),$$

3. Using the maximum confidence measure as the confidence score ($\alpha < 1$)

$$C(w,t) = \max_r C(w,t,r).$$

With respect to the fact that the confidence measures to some extent constitute indicators about the output's correctness, approaches 2 and 3 lead to better results than the first approach. In the third approach, only the confidence measures of the "most confident" recognizer are considered by what mistakes by other recognizers are ignored and by what the third approach also performs better than the second one.

Based on the scores $S(w)$ an output WTN is constructed. In cases where no knowledge about the sentence structure is available, this WTN contains a plain sequence of words with the best score regardless of their context. I.e., the output is a sequence W of words w_i such that

$$W = \arg \max_{W=w_1,w_2,\ldots} \sum_i S(w_i). \tag{5.6}$$

Optionally, a WTN with the N best words can be produced, based on which a Viterbi search including a language model can be performed.

In order to estimate the performance of a ROVER system, upper and lower bounds on the overall recognition rate can be defined (Pittermann and Pittermann, 2006b). The theoretical limit R^+ is the maximal recognition rate that may be achieved on predefined input conditions, i.e., on given WTNs. Considering the error distributions illustrated in Fig. 5.9, the upper bound is determined by the uncorrectable utterances $\mathrm{A} \cap \mathrm{B}$, i.e., in the first scenario, R^+ would be 100%, in the second scenario, R^+ would be 100% minus the percentage of $\mathrm{A} \cap \mathrm{B}$ in U, and in the third scenario R^+ would be equal to the recognition rate of one of the recognizers. Extending the considerations on a ROVER system combining N recognizers, the theoretical limit is calculated as

$$R^+ = \left(1 - \frac{|\mathrm{E}_1 \cap \mathrm{E}_2 \cap \ldots \cap \mathrm{E}_N|}{|\mathrm{U}|}\right) \cdot 100\%, \tag{5.7}$$

where $\mathrm{E}_1, \ldots, \mathrm{E}_N$ are the error distributions of recognizers $1, \ldots, N$, U is the entirety of all utterances or words, and $|\cdot|$ is the cardinality of a set. This definition of the upper bound implicitly assumes perfect knowledge about the correctness of the single recognizers' recognition results. Moreover, it also supports the claim that error-free recognition is theoretically possible for $N \to \infty$ when combining an infinite number of recognizers. Taking into account these circumstances, it is understandable that an actual overall recognition rate of R^+ is not very likely to be achieved.

Analogously, a lower bound R_- for the overall recognition rate can be defined. In this bound, the worst case, i.e., the fact that ROVER always decides on the wrong word/utterance no matter how often it is correctly recognized by the other recognizers, is assumed. E.g., in the example scenarios, the lower bound is defined by all errors as 100% minus the percentage of $\mathrm{A} \cup \mathrm{B}$. Generally, given the same conditions like for the upper bound, for N recognizers, R_- is defined as

$$R_- = \left(1 - \frac{|\mathrm{E}_1 \cup \mathrm{E}_2 \cup \ldots \cup \mathrm{E}_N|}{|\mathrm{U}|}\right) \cdot 100\%. \tag{5.8}$$

Table 5.1 Determination of upper and lower limits for a ROVER system on the basis of four exemplary input word transition networks

#	WTN 1	WTN 2	WTN 3	WTN 4	Reference
001	TO	TO	TO	TO	TO
	NEW_YORK	NEW_YORK	NEW_YORK	NEW_YORK	NEW_YORK
002	TO	TO	TO	TO	TO
	ULM	~~LONDON~~	~~NEW_YORK~~	ULM	ULM
	TODAY	TODAY	~~PLEASE~~	TODAY	TODAY
				~~PLEASE~~	
003	I	I	I	I	I
	WANT	~~WOULD~~	~~WOULD~~	WANT	WANT
	TO	~~LIKE~~	~~LIKE~~	TO	TO
	TRAVEL	TO	TO	TRAVEL	TRAVEL
	TO	~~GO~~	TRAVEL	TO	TO
	~~LONDON~~	TO	TO	~~NEW_YORK~~	PARIS
	TODAY	~~LONDON~~	~~ULM~~	~~PLEASE~~	TODAY
		-	-		
004	TO	TO	TO	TO	TO
	~~NEW_YORK~~	~~LONDON~~	PARIS	PARIS	PARIS
	PLEASE	PLEASE	PLEASE	~~TODAY~~	PLEASE
005	TO	TO	TO	TO	TO
	PARIS	PARIS	~~LONDON~~	~~LONDON~~	PARIS
	TODAY	TODAY	-	TODAY	TODAY
	PLEASE	PLEASE	PLEASE	PLEASE	PLEASE

This bound, to some extent, also assumes perfect knowledge about the recognition results' correctness, and for $N \to \infty$ recognizers with sufficiently different error distributions this lower bound converges to 0.

In order to determine the upper and lower bound for the combination of actual recognizers, their WTNs need to be compared as shown in Table 5.1. In this small example, word errors including substitutions, deletions and insertions are indicated as ~~deleted~~ text. The single recognizers achieve word recognition accuracies between $11/19 \approx 58\%$ (WTN 3) and $17/19 \approx 89\%$ (WTN 1) as well as sentence recognition rates between $1/5 = 20\%$ (WTN 4) and $3/5 = 60\%$ (WTN 1). Having a closer look at the particular utterances, it can be noticed that the first utterance is correctly recognized by all recognizers, utterances 002, 004 and 005 are correctly recognized by at least one recognizer and utterance 003 is not correctly recognized by any recognizer. Thus, the upper and lower bounds on the utterance level are calculated as follows:

$$\begin{aligned} R_{-,(\mathrm{u})} &= \left(1 - \frac{|\{002, 003, 004, 005\}|}{|\{001, 002, 003, 004, 005\}|}\right) \cdot 100\% = \left(1 - \frac{4}{5}\right) \cdot 100\% = 20\%, \\ R^{+,(\mathrm{u})} &= \left(1 - \frac{|\{003\}|}{|\{001, 002, 003, 004, 005\}|}\right) \cdot 100\% = 80\%. \end{aligned} \tag{5.9}$$

Looking at the single words, most words are correctly recognized by at least one recognizer (e.g., "ULM" in utterance 002) or even by all recognizers (e.g., "NEW_YORK" in utterance 001). Only "PARIS" in utterance 003 is not correctly recognized by any recognizer. The upper and lower bounds on the word level are calculated as:

$$\begin{aligned} R_{-,(\mathrm{w})} &= \left(1 - \frac{|\{\mathrm{ULM}, \mathrm{TODAY}, \mathrm{WANT}, \mathrm{TRAVEL}, \ldots, \mathrm{TODAY}\}|}{|\{\mathrm{TO}, \ldots, \mathrm{PLEASE}\}|}\right) \cdot 100\% \\ &= (1 - 10/19) \cdot 100\% \approx 47\%, \\ R^{+,(\mathrm{w})} &= \left(1 - \frac{|\{\mathrm{PARIS}\}|}{|\{\mathrm{TO}, \ldots, \mathrm{PLEASE}\}|}\right) \cdot 100\% \approx 95\%. \end{aligned} \tag{5.10}$$

Comparing these numbers, it can be seen that the utterance level recognition rate may be improved from 60% to 80% but, in the worst case, may also decrease to 20%. Also, the word accuracy has the potential to increase from 89% to 95% or to drop to 47%. Based on these bounds, the developer can roughly estimate whether it is worth using and combining multiple recognizers.

5.2.1 ROVER for Emotion Recognition

Using an emotion recognizer as described in Section 4.5, the ROVER idea may also apply for emotion recognition (Pittermann and Pittermann 2006b). On the one hand, in these systems the emotions are recognized on a time basis so that they can be aligned and treated like words. On the other hand, these emotion recognizers provide information about confidence measures so that a scoring can be performed.

The architecture of an emotional ROVER system is similar to the architecture shown in Fig. 5.10. Instead of the automatic speech recognition systems providing word transition networks, this approach combines multiple emotion recognition systems providing ETNs. In the alignment module, these ETNs are combined to a composite ETN based on which the most probable sequence of emotions is selected in the voting module. As illustrated in Fig. 5.14, the alignment can be performed according to time measures or edit distance measures. As opposed words which have to be considered individually, it may be useful to merge adjacent emotions if they represent the same emotional state.

In this figure, it can be seen that the different recognizers' output may differ significantly. E.g., the ETN of the third recognizer consists of many emotions with short durations whereas ETN 1 consists of only three emotions although the overall duration of the ETNs is approximately the same. Taking into account the temporal overlap(s) of the emotions in the ETNs, the dynamic programming alignment approach which treats emotions like words (word-like alignment) is a rather inappropriate representation. Thus, instead, we propose a time-based alignment summarizing emotions and omitting pauses which do not represent emotional states. In this approach, contiguous emotions of the same state are merged as shown in

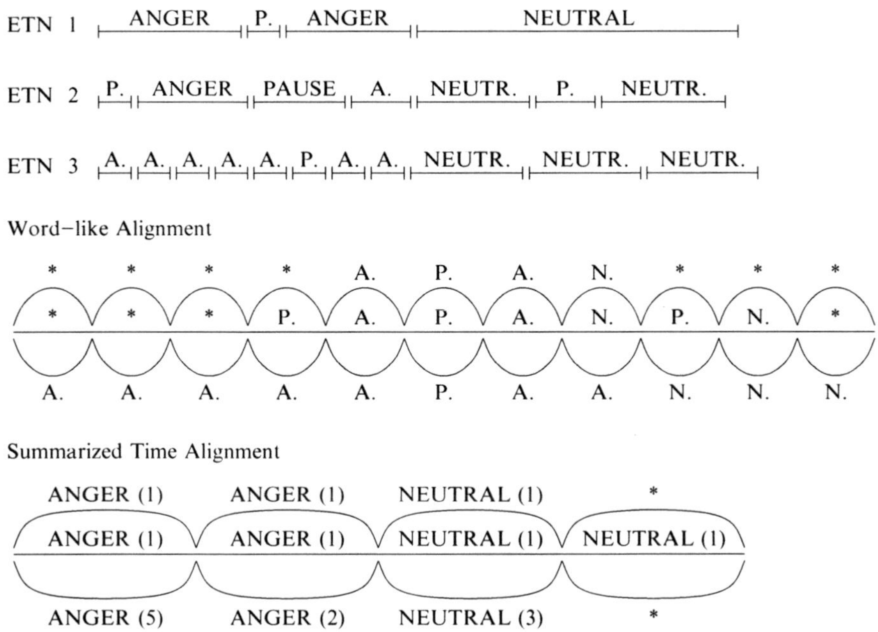

Fig. 5.14 Alignment of three ETNs in the alignment module of a ROVER system. One approach involves dynamic programming (DP) considering the individual emotions, the other approach summarizes the emotions in the ETNs and aligns these on a time basis. The "*"s represent null emotion transitions which are included in the ETNs when words are inserted or deleted

the lower part of Fig. 5.14. The emotions in the first two ETNs are separated by pauses or adjacent emotions do not represent the same state (e.g., "ANGER" ⇔ "NEUTRAL"). Thus, in these cases, only the pauses are removed from the ETNs. In the third ETN, there are five consecutive "ANGER"s, another two consecutive "ANGER"s and three consecutive "NEUTRAL"s that may be combined to "ANGER ANGER NEUTRAL". However, when combining emotions, their individual confidence measures also need to be combined somehow to make them comparable to the other confidence measures. This is accomplished with an integrative approach, calculating the average value $\bar{C}(e,t',r)$ for all combined emotions as

$$\bar{C}(e,t',r) = \frac{1}{T_2 - T_1} \cdot \int_{T_1}^{T_2} C(e,\theta,r)d\theta = \frac{1}{T_2 - T_1} \cdot \sum_{t \in \mathcal{T}_e} C(e,t,r) \cdot l(e,t,r), \tag{5.11}$$

where T_1 and T_2 are start and end time of the combined emotion interval, θ is the continuous time, t is the discrete time as used in the speech recognizer ROVER description, t' is the new time index after the combinations, $\mathcal{T}_e$ is the set of all t of the emotion e that are combined and $l(e,t,r)$ is the duration of emotion e at t in ETN r. E.g., for the combination of "ANGER"s in the following emotion recognizer output (the logarithmic confidence measures have already been converted to positive values)

```
    ...
 800000    1100000   ANGER      0.5678
1100000    4400000   ANGER      0.8546
4400000    8700000   ANGER      0.4289
8700000   15500000   SADNESS    0.7584,
    ...
```

the average confidence measure would be

$$\bar{C}(\text{ANGER}, t_0', r_0) = \frac{1}{8700000 - 800000} \cdot (0.5678 \cdot (1100000 - 800000)$$
$$+0.8546 \cdot (4400000 - 1100000)$$
$$+0.4289 \cdot (8700000 - 4400000)) = 0.6120. \tag{5.12}$$

In the voting module, the scores for the emotions in the composite ETN are determined. In analogy to the ROVER for speech recognition, especially if the ETNs are aligned according to the edit distance, the score of an emotion can be calculated according to Eq. 5.5:

$$S(e) = \alpha \cdot \frac{N(e,t)}{N_s} + (1-\alpha) \cdot C(e,t), \tag{5.13}$$

where $N(e,t)$ is the number of occurrences of emotion e at time t and where $C(e,t)$ is the representative (average or maximum) confidence measure of emotion e at t. Applying this score, the most probable sequence of emotions can be determined.

For most applications, however, it is sufficient to assign a single emotion to each utterance instead of determining a sequence of emotions. Thus, we propose an alternative voting approach, defining the predominant emotion E^* of an utterance as

$$E^* = \arg\max_e S'(e), \tag{5.14}$$

where $S'(e)$ is the modified score of an emotion e calculated as

$$S'(e) = \alpha \cdot N'(e) + (1-\alpha) \cdot C'(e). \tag{5.15}$$

Here, $N'(e)$ is the normalized overall length of emotion e:

$$N'(e) = \left(\sum_{\{t,r\}|_{e'=e}} C(e',t,r) \right) / \left(\sum_{t,r,e'} C(e',t,r) \right), \tag{5.16}$$

where $l(e',t,r)$ is the duration (length) of emotion e' in ETN r at time t. Due to the fact that, as illustrated in Fig. 5.14, depending on the recognizer, one emotion may occur more often in one recognizer but span the same time period, we consider the overall duration of an emotion instead of counting its frequency of occurrence. Similarly, $C'(e)$ is the normalized overall confidence of emotion e calculated as

$$C'(e) = \left(\sum_{\{t,r\}|_{e'=e}} C(e',t,r) \right) \Bigg/ \left(\sum_{t,r,e'} C(e',t,r) \right), \tag{5.17}$$

where $C(e',t,r)$ is the adapted or average confidence measure of emotion e' in ETN r at time t.

The performance of a ROVER system for emotion recognition can also be estimated with the aid of upper and lower bounds as described by Eqs. 5.7 and 5.8. Only determining the predominant emotion of an utterance, the word-level measures become superseded by utterance-level measures. Analogously, $R_{-,(\mathrm{u})}$ and $R_{+,(\mathrm{u})}$ can be calculated on the basis of the ETNs provided by the emotion recognizers. Example ETNs to be used in limit considerations are shown in Table 5.2. Regarding the utterances' predominant emotions (in bold letters), the individual utterance recognition rates range from 40% (recognizer 4) to 80% (recognizer 1). Based on these measures, the lower bound $R_{-,(\mathrm{u})}$ is at 20% (utterance 002 is very likely to be

Table 5.2 Determination of upper and lower limits for a ROVER system on the basis of four exemplary input emotion transition networks. Pauses are omitted, emotions are summarized where applicable

#	ETN 1	ETN 2	ETN 3	ETN 4	Reference
001	ANGER	SADNESS	ANGER	NEUTRAL	
	ANGER	DISGUST	NEUTRAL	ANGER	
	ANGER	**~~SADNESS~~**	**ANGER**	**~~NEUTRAL~~**	**ANGER**
002	FEAR	FEAR	HAPPINESS	NEUTRAL	
	NEUTRAL	NEUTRAL	NEUTRAL	NEUTRAL	
	BOREDOM	NEUTRAL	BOREDOM	BOREDOM	
	NEUTRAL	**NEUTRAL**	**NEUTRAL**	**NEUTRAL**	**NEUTRAL**
003	HAPPINESS	NEUTRAL	HAPPINESS	BOREDOM	
	BOREDOM	NEUTRAL	HAPPINESS	SADNESS	
	HAPPINESS	BOREDOM	NEUTRAL	BOREDOM	
	NEUTRAL	HAPPINESS	NEUTRAL	NEUTRAL	
		BOREDOM	NEUTRAL	BOREDOM	
			HAPPINESS		
	HAPPINESS	**HAPPINESS**	**~~NEUTRAL~~**	**~~NEUTRAL~~**	**HAPPINESS**
004	ANGER	NEUTRAL	SADNESS	NEUTRAL	
	NEUTRAL	DISGUST	NEUTRAL	SADNESS	
	~~NEUTRAL~~	**DISGUST**	**~~SADNESS~~**	**~~NEUTRAL~~**	**DISGUST**
005	NEUTRAL	DISGUST	BOREDOM	NEUTRAL	
	BOREDOM	NEUTRAL	NEUTRAL	DISGUST	
	NEUTRAL	BOREDOM			
	NEUTRAL	**~~DISGUST~~**	**NEUTRAL**	**NEUTRAL**	**NEUTRAL**

recognized correctly, even in the worst case), the upper bound $R_{+,(\mathrm{u})}$ would be at 100%. These measures include most of eventualities that may influence the decision in the voting module.

For three existing emotion recognizers which achieve individual utterance emotion recognition rates of 72%, 70.7% and 70.7%, an upper bound $R_{+,(\mathrm{u})}$ has been determined as 77.3%, standing for a possible absolute improvement of 5.3%. In a different scenario, combining emotion recognizers with 64%, 62.7% and 60%, an upper bound of 74.7%, constituting a possible improvement of 10.7%, has been found (Pittermann and Pittermann 2006b).

5.2.2 ROVER for Speech–Emotion Recognition

On the basis of the ROVER approaches described above, it is also possible to implement a method to combine the output of multiple speech–emotion recognizers as described in Section 4.6. In principle, word–emotions we can be considered and treated like plain words, so that an architecture like the one illustrated in Fig. 5.10 can be employed. I.e., N_s speech–emotion recognizers provide N_s WETNs which are aligned to a composite WETN in the alignment module. In the voting module a score $S(we)$ is calculated for each word–emotion we in the composite WETN (see Eq. 5.5) as

$$S(we) = \alpha \cdot \frac{N(we,t)}{N_s} + (1-\alpha) \cdot C(we,t), \tag{5.18}$$

where $N(we,t)$ and $C(we,t)$ are frequency of occurrence and confidence measure of we. Based on these scores, the most probable sequence of word–emotions is determined.

However, due to the fact that the number of possible word–emotions is seven times the number of possible words, it is very likely that no majorities can be found for word–emotions in the composite WETN making the frequency of occurrence considerations ($N(we,t)$) useless in the voting. A further problem is the evaluation of a word–emotion sequence. From the speech recognizer point of view, a word–emotion is incorrectly recognized when at least one of word or emotion differ from the reference word–emotion, which is a very strict measure. A constructed worst-case example for the straightforward ROVER approach is shown in Table 5.3. Considering the individual WETNs, it is obvious that the word "I" has been correctly recognized by all four recognizers, but the associated emotional state has not been recognized by any of the recognizers. Furthermore, the reference sentence "I WANT TO TRAVEL TO PARIS" has been correctly recognized by recognizer 1, but, again, none of the associated emotions have been recognized. Vice versa, the suitable emotional states (BOREDOM, NEUTRAL) have been recognized by the other three recognizers, however, most of the words have not been correctly recognized. In brief, the ROVER approach would probably fail, voting for a word string which does not approximate the reference sentence. An analysis of word–emotion error distributions with similar behavior is illustrated in Fig. 5.15.

Table 5.3 Example word–emotion transition networks bringing out limitations of a straight-forward ROVER system for speech–emotion recognition. For the sake of overview he emotional states are abbreviated (A = ANGER, B = BOREDOM, etc.)

#	WETN 1	WETN 2	WETN 3	WETN 4	Reference
...					
092	I-A	I-H	I-A	I-N	I-B
	WANT-N	WOULD-B	WANT-D	WOULD-B	WANT-B
	TO-A	LIKE-N	TO-D	LIKE-B	TO-N
	TRAVEL-A	TO-B	GO-B	TO-B	TRAVEL-B
	TO-N	GO-B	TO-N	TRAVEL	TO-D
	PARIS-N	TO-A	NEW_YORK-B	TO-B	PARIS-B
		ULM-B		PARIS-D	
...					

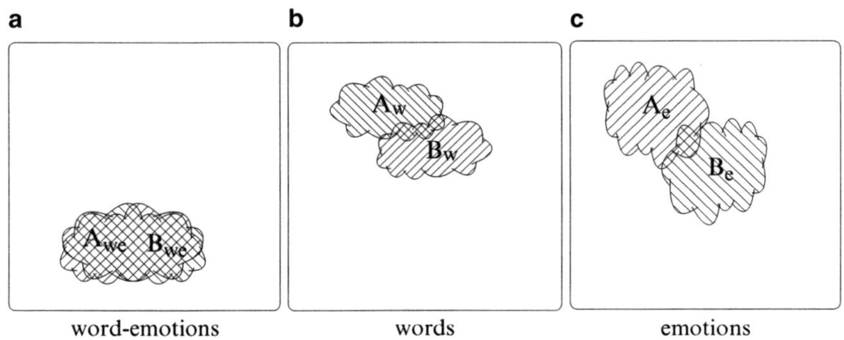

Fig. 5.15 Word–emotion error distributions for two speech–emotion recognizers and the respective word-only and emotion-only error distributions

Figure 5.15 shows the error distributions in the word–emotion (we) space as well as the respective error distributions in the plain word space and the emotion space. The error clouds in the we-space show a significant overlap $|A \cap B| \gg |A \triangle B|$, which means that the regular ROVER approach has no potential to correct a large number of errors on the WETNs. However, looking at the error clouds in the w-space or in the e-space, the overlap $|A_w \cap B_w|$ or $|A_e \cap B_e|$ is considerably smaller independent of the cardinality of the sets. Typically, there are more emotion recognition errors than word recognition errors, so that $|A_w| \leq |A| \leq |A_e|$ or $|B_w| \leq |B| \leq |B_e|$, respectively.

In order to account for the larger word–emotion variety and the independence between words and emotions, we propose a modified ROVER method as illustrated in Fig. 5.16.

Here, the input WETNs are separated into WTNs and ETNs by splitting each word–emotion we into word w and emotion e (see Fig. 5.17) (Pittermann et al. 2007b).

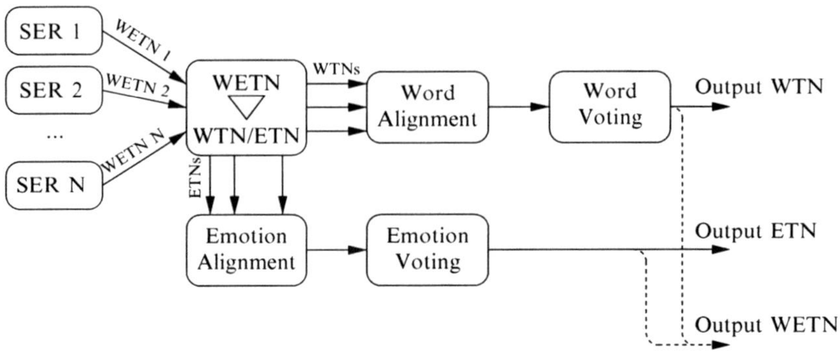

Fig. 5.16 Architecture of a ROVER system for speech–emotion recognition

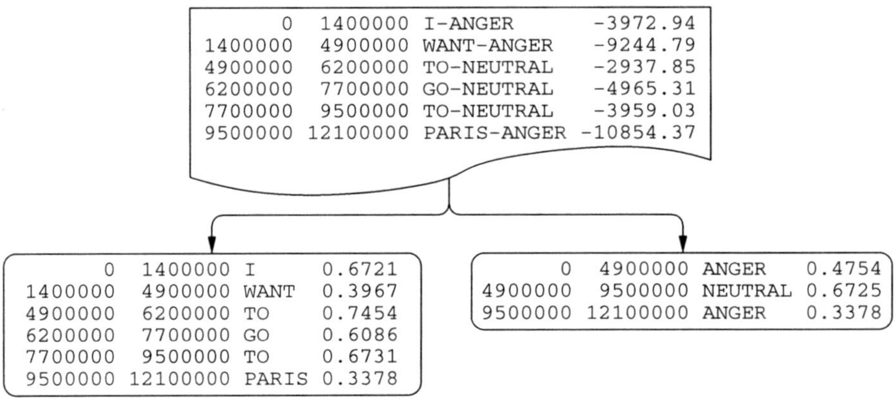

Fig. 5.17 Transformation of a word–emotion transition network into the respective word transition network and emotion transition network (Pittermann et al. 2007b). The emotions in the ETNs are summarized according to Section 5.2.1 before transferred to the alignment module

I.e., the w part is contained in the WTN, and the e part goes into the ETN. Whereas the WTN remains untouched, the emotions in the ETN are summarized as described in Section 5.2.1. The recognizer confidence measures are contained in both networks for further use in the respective ROVER subsystems.

In the word subsystem, the WTNs are aligned according to edit distance or time measures as described in the beginning of this section (see also Fig. 5.11). Also, the voting is based on word scores $S(w)$ calculated according to Eq. 5.5.

In the emotion subsystem, the summarized ETNs are also aligned on a time basis. Depending on the application, the scores are calculated in two ways. Either output of the ROVER system is required to be a time-dependent sequence of emotions (in analogy to the output WTNs), then the scores $S(e)$ are calculated according to Eq. 5.13. This approach has the advantage that once the output WTN and ETN are determined, these can be merged into a common output WETN where the we consist words w and the aligned emotions e. Alternatively, it is sufficient to determine one predominant emotion for each utterance. Then, the scores $S(e)$ are calculated as shown in Eq. 5.15 and the emotion with the maximum score is selected as the predominant emotion E^*. This approach can also be extended to "soft" emotional labels which account for the mixture of emotions due to the categorization and perceptual uncertainties. The "hard" decision (i.e., the reduction to one emotion) involves a significant loss of information for further processing.

Thus, the soft values help to improve the robustness of, e.g., the adaptive dialogue management. Firstly, for each utterance u, a set $\mathcal{E}_u$ of emotions which occur in the composite ETN for u is defined. Then, for each $e \in \mathcal{E}_u$, the score $S'(e)$ is calculated as

$$S'(e) = \alpha \cdot N'(e) + (1 - \alpha) \cdot C'(e), \tag{5.19}$$

where $N'(e)$ and $C'(e)$ are the normalized frequency of occurrence and confidence measures as described in Eqs. 5.16 and 5.17. In the soft emotional values either all emotions in $\mathcal{E}_u$ may be included or, alternatively, a limited number of best emotions. I.e., a set $\mathcal{E}_u^* \subseteq \mathcal{E}_u$ is found and the share of an emotion $e \in \mathcal{E}_u^*$ is calculated as

$$\bar{S}'(e) = S'(e) / \left(\sum_{e' \in \mathcal{E}_u^*} S'(e') \right), \tag{5.20}$$

so that the soft output is $\mathcal{E}_u^*$ and the respective scores. This may be, e.g., "{ 0.654 ANGER, 0.239 NEUTRAL, 0.107 DISGUST }".

5.3 Implementation of Our Dialogue Manager

In this section, we describe the implementation of an adaptive dialogue manager integrating the ideas shown in the previous section. For our implementation, we choose VoiceXML as a basis, as it is commonly used in research and provides a high flexibility in terms of platform independence and modularity of the surrounding modules such as speech recognition, linguistic analysis, text-to-speech synthesis, etc. Moreover, VoiceXML supports the integration of ECMAScript code (also known as JavaScript, see Ecma International 1999, 2005) for dynamic calculations and, thus, parameter adaptations during the dialogue.

Extensively using ECMAScript, it is virtually possible to create one's own dialogue manager within the VoiceXML framework, yielding the advantage, that all existing dialogue system components can be used and the behavior of the dialogue manager can be personalized as desired (Pittermann et al. 2005). The success of such an implementation, however, stands or falls depending on the capabilities of the utilized VoiceXML interpreter. Unfortunately, it has shown that the interpretation of ECMAScript code is not consistent among different VoiceXML interpreters. Then, in the best case, it happens that the ECMAScript dialogue manager works properly, in the worse cases, this dialogue manager either shows a strange behavior or even refuses to work at all, e.g., due to interpretation errors. Further errors can occur due to an ambiguous variable declaration (e.g., var i=1) which does not include a type declaration. With this, the addition of two integer variables can lead to a string concatenation (e.g., 1+1=11) if one of the variables happens to be of string type.

To obtain a robust (in terms of functionality) dialogue manager within the dialogue system, we choose the use of compiled VoiceXML as proposed by Bühler and Hamerich (2005), where the VoiceXML form is translated into ECMAScript code consisting of very basic functions so that it can be smoothly processed by standard VoiceXML interpreters. E.g., using the compiled VoiceXML output, it is even possible to run a simple dialogue manager in the web-browser of a standard PDA. Apart from its platform-independence, the compiled VoiceXML approach also has the advantage that it can be run in the Java-based Rhino ECMAScript implementation provided by the Mozilla Foundation (see http://www.mozilla.org/rhino/). Using this ECMAScript implementation, it is then possible to include all kinds of Java classes and their methods in the VoiceXML code, which in turn contributes to a more robust implementation featuring more functionality and flexibility. Thus, apart from the interfaces between the VoiceXML interpreter and the manager which are realized in ECMAScript, the entire functionality of the dialogue manager is implemented in Java.

Our dialogue manager basically adopts the ideas of storing relevant dialogue parameters in a dialogue history array $\underline{\mathbf{H}}_D$ as defined in Chapter 3. This includes the implementation of the ideas described in Section 3.5, i.e., the adaptation of the dialogue flow and the stylistic realization of the system output to the user's input and the recognized emotional state of the user.

The simplified structure of the dialogue manager, embedded in a dialogue system architecture, is illustrated in Fig. 5.18. In this diagram and in the further system description, we refer to the term "context", represented by the grey box, as the entirety of all information, variables, modules and methods which are involved in the functionality of the dialogue manager. Furthermore, we paraphrase what we call "system reaction" in the previous section by the term "field under discussion" (FUD). Parts of this structure and the approach accounting for an FUD are adopted from the TRINDI dialogue move engine as described in Larsson and Traum (2000) and Larsson (2000). As opposed to standard VoiceXML, here, like in the generalized dialogue model, the term "field" also includes all possible attached dialogue control parameters like, e.g., the user's emotional state. Among these fields, we distinguish between pre-defined fields and dynamically generated fields which can be included at run time.

In our implementation, the context is represented by one Java class which embraces the following sub-classes as shown in Fig. 5.18:

- "fields" is an array containing all dialogue fields regardless of whether these are pre-defined beforehand by the designer or dynamically generated during the course of the dialogue. The pre-defined fields are described with the aid of XML in a customized variation of VoiceXML.

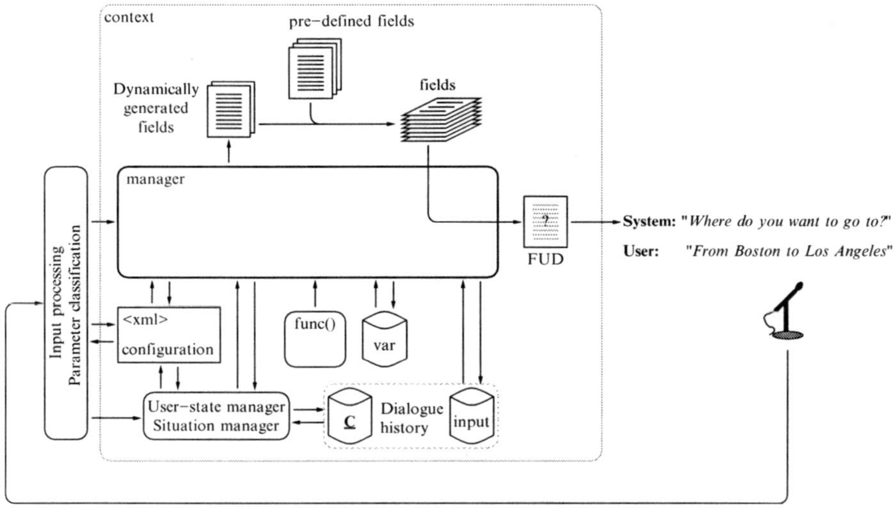

Fig. 5.18 Simplified architecture of the adaptive dialogue manager embedded in a standard dialogue system environment

- The configuration, implemented as XML, contains all setup parameters which determine the basic behavior of the dialogue manager. This includes, e.g., the selection of the dialogue strategy (rule-based vs. semi-stochastic) or confirmation strategy, how the dialogue system shall overanswering cases where the user provides more information than required (flexible vs. pure system initiative), etc.
- The user-defined functions ("func()") enable the dialogue developer to include further functionality to the system. These functions can be called and accessed from any class within the context and from the field and configuration descriptions. By that, e.g., prompts for certain fields can be adapted or an external database can be accessed during the dialogue.
- "var" serves as a collecting basin for user-defined variables which can also be accessed from any class or dialogue field within the context. These variables, e.g., allow the calculation and storage of further parameters which are not contained in the dialogue history.
- For the sake of overview, the dialogue history is subdivided into two parts – the actual user "input" which is required to fill the dialogue fields, and the respective control parameters which are contained in $\underline{\mathbf{C}}$.

In this setup, one dialogue turn takes place as follows: The user replies to a certain system prompt, e.g., *"From Boston to Los Angeles."*. This utterance is processed by the speech recognizer and linguistic analysis, emotion recognizer, etc. and the semantic representations as well as further dialogue control parameters are extracted according to the specifications of the system configuration. These representations and parameters are passed to the dialogue manager and (optionally) to the user-state and situation manager which assists the dialogue manager in adapting the dialogue

```
<?xml version="1.0"?>
<!DOCTYPE vxml PUBLIC "vxml"
            "http://www.w3.org/TR/voicexml20/vxml.dtd">

<vxml version="2.0">
 <script src="ds_main_system.js"/>
 <var name="context"
           expr="new DSContext('dialog.xml','config.xml')"/>

 <form id="mydialog">
   <grammar src="ds_grammar.grxml"></grammar>

   <block name="block0">
    <value expr="context.dialog.getDI('ds-start').getText()"/>
   </block>

   <field name="ds_input" expr="''"/>
   <field name="ds_control" expr="''"/>

   <field name="exit">
    <prompt bargein="true">
      <value expr="context.FUD.getQuestion()"/>
    </prompt>
   </field>

   <filled mode="any" namelist="ds_input">
    <value
     expr="context.manager.fullAnalysis(ds_input,ds_control)"/>
    <value expr="update()"/>
   </filled>

   <block name="block1">
    <value expr="context.dialog.getDI('ds-start').getText()"/>
   </block>
 </form>
</vxml>
```

Fig. 5.19 VoiceXML form used to integrate the functionality of our dialogue manager into the dialogue framework (Pittermann et al. 2007c)

flow. For further use, the input and the parameters are stored in the dialogue history based on which the dialogue flow and realization is adapted. After an alignment with all previous user turns, the manager selects a new field under discussion (FUD) which is then prompted according to the other dialogue control parameters as suggested by the user-state and situation manager.

The integration of our implementation of the dialogue manager is accomplished with the aid of a consistent VoiceXML form as shown in Fig. 5.19. At the beginning, the context class is initialized on the basis of the dialogue and configuration XML files and an external ECMAScript file containing the interface functionality is loaded. The dialogue form contains two blocks (ds-start and ds-end) which can include a personalized greeting or goodbye message like *"Welcome to the University of Ulm's experimental dialogue system!"*. These messages will be outputted at the very beginning and at the very end of the dialogue. The fields called ds_input and ds_control are solely used to receive the user input and the respective control parameters from the linguistic analysis and the signal processing module. As these fields do not include any prompts, these are pre-assigned an empty string as value.

Being initialized before the dialogue form, our dialogue manager's main functionality is called from the field exit and the filled node. In the exit field, the prompt of the current FUD is outputted and in the filled node, the user's input and the control parameters are processed and the further progress of the dialogue is determined in accordance with the employed dialogue model and the system configuration. The function update() is used to synchronize the dialogue manager and the VoiceXML interpreter. It should be noted that, despite that fact that the actual dialogue can contain an almost infinite number of fields, there is only one VoiceXML field (exit) containing a prompt statement. Here, the VoiceXML form interpretation algorithm (FIA) is outwitted in such a manner that only the fields ds_input and ds_control are filled during the dialogue, whereas exit is not filled until a stopping condition is fulfilled, e.g., when all the fields within the context of our dialogue manager are filled or when the dialogue shall be aborted. This is made possible by a grammar which actually does not know about the existence of the exit field.

The course of a dialogue description form is shown in Fig. 5.20. The FIA first executes block0 where the introductory text like *"Welcome to... "* is generated. As the fields ds_input and ds_control are already filled, the FIA then executes the field exit and prompts a text or question generated by the dialogue manager. The user's reply and the respective control parameters are temporarily stored in ds_input and ds_control before being processed by the dialogue manager. As exit could not be filled by the VoiceXML interpreter, the FIA prompts a newly generated text when executing exit again in the next turn. This process is repeated until the dialogue

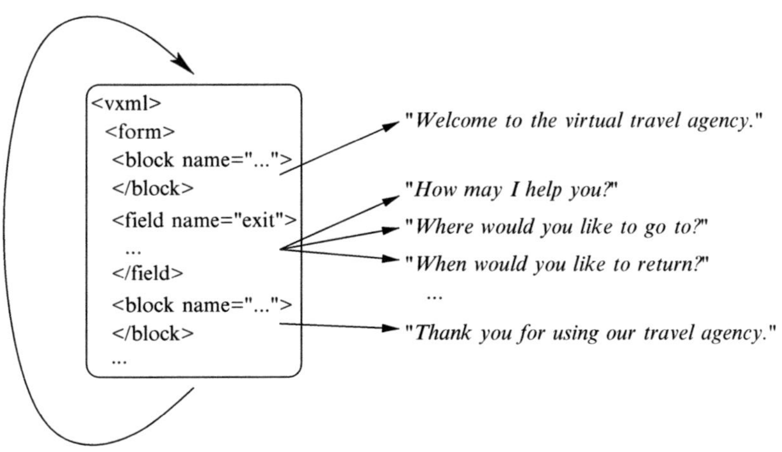

Fig. 5.20 Illustration of the cooperation between the VoiceXML form interpretation algorithm and the dialogue manager

manager fills exit with an arbitrary value (which will not be evaluated by the dialogue manager), then block1 is executed and a conclusion like *"Thank you for..."* is outputted.

Constituting a central part within the context, the manager ensures a proper course of the dialogue. In order to accomplish this it provides the following basic functionality:

- Generation and stylistic realization of prompts according to the control parameters and the current field under discussion. For the actual implementation of a dialogue system, the designer can choose between a rule-based strategy as described in Section 3.5 which imitates the standard behavior of a VoiceXML interpreter or a semi-stochastic approach as described in Section 3.8.
- Generation of confirmation prompts under certain conditions: If the user of a travel information system says, e.g., *"I want to travel to Munich to Paris."*, the linguistic analysis of a standard VoiceXML interpreter would choose *"Paris"* (i.e., the last item mentioned) as the destination whereas our system can allow for overanswering by temporarily saving both current and also previous answer(s). Using this feature it can prompt for confirmation such as *"Would you like to travel to Munich or to Paris?"*. Also if the predicted recognition rate calculated from the speech recognizer confidence measures is below a certain threshold, the system can prompt for an explicit confirmation like *"I understood you wanted to travel to Paris. Is this correct?"* or include an implicit confirmation like *"From where would you like to depart to Paris?"*. It should be noted that these confirmations are already included in each field and can be prompted arbitrarily during the dialogue, whereas in standard VoiceXML the dialogue designer needs to add an extra field for each kind of confirmation. Also, a slot called yesno is included in the grammar to pass the respective reply to the dialogue manager.
- Reception of the user's input and control parameters as well as the analysis and storage of these: The mapping of user input to the respective fields can be done in multiple different ways. Considering an automated travel information system, a city can either be a destination or a departure city. Typically, an utterance like *"I want to travel from Munich to Paris."* is analyzed as departure_city = 'Munich' and destination = 'Paris' whereas *"I want to travel Munich Paris."* does not necessarily lead to any results in standard VoiceXML. In the linguistic analysis, we avoid this problem by introducing an additional, more general semantic label called city resulting in city = 'Munich' and city = 'Paris' and leaving it to the dialogue manager how to proceed with this information. If in this case the configuration parameters are set accordingly and the order of the fields is 1. departure_city and 2. destination, the first city can be considered as the departure city and the second city is considered as the destination. Here, we assume that it is more probable in common speech that the first city of such a sentence is the departure city and the second city is the destination. For a better interpretation the system can also ask for confirmation in these cases.
- Dynamic generation of fields or explicit confirmations during the course of the dialogue. This feature can be useful in cases where not all dialogue conditions can be foreseen in the development phase or when certain dialogue conditions

are very unlikely to occur. In standard implementations, these cases are typically covered by pre-defined fields rendering the dialogue flow, i.e., the dialogue description rather unclear, especially when each confirmation requires its own field. Thus, in this system, the dialogue manager decides at run time whether a certain confirmation shall be prompted, making it easier to implement the ideas proposed in Litman and Pan (2002) when speech recognition problems occur. Alternatively, following the ideas in Section 3.5, when it is necessary to adapt the prompt style to certain conditions, e.g., when a bad message needs to be communicated to an angry user, this message can be tailored to the situation. E.g., if the desired flight is fully booked, the system can either ask for an alternative time, date or origin or it can offer an alternative itinerary and ask if the user accepts that suggestion: *"I'm sorry, but all flights from Stockholm to Paris on March 1st are fully booked. Would it be OK for you to travel on March 2nd?"*

Each dialogue item occurring in the dialogue holds a certain set of default properties which can be modified with the aid of an XML-based dialogue description as shown in Fig. 5.21. The first dialogue item is the ds-start item which solely contains the text which is outputted at the beginning of the dialogue. This text is put in quotation marks in order to allow the inclusion of variables and text concatenation. In this mixed initiative dialogue, the first field is of initial type allowing the user to provide as much information as flexibly as possible. Here, to properties of this field are set up: questions corresponds to the prompt tag in VoiceXML and, thus, contains the questions or prompts posed by the system. As opposed to VoiceXML, multiple questions can be defined and the system randomly selects one of these at runtime. These are separated by ///. Similarly, explanations can be defined which are outputted before the new prompt in cases where the user does not say anything ("noinput") or says something which does not match any item of the linguistic analysis ("nomatch"). These explanations are sorted according to the order in which they are outputted, making it possible to follow the incremental prompts idea proposed in Yankelovich (1996) where the length and the information content of a prompt increase every time the user input is not useful.

The departure_city field possesses the entire set of parameters that can be assigned to a field. The use of questions and explanation is equivalent to the initial field. Furthermore, there is a summary which, in combination with another field's prompt, can form an implicit confirmation (e.g., *"I understood you want to travel from Paris. Where would you like to go to?"*) or, in combination with the confirmation text, can form an explicit confirmation (e.g., *"I understood you want to travel to Munich. Is this correct?"*). The slots property provides the dialogue manager more flexibility to react on ambiguous user input like *"I want to travel Munich Paris."* as described above. I.e., if the user's input contains a city which matches neither departure_city nor destination, that city is labeled as city. Typically, a VoiceXML field can contain one value received from the user, e.g., destination = 'Paris'. In some cases, however, it may be more convenient to store multiple values, e.g., when asking *"From which cities would you be able to depart?"*. Setting the max_num_ans (maximum number of answers) parameter to 3, it is then possible to store three departure cities, e.g., departure_city = {'New York',

```
<dialog-xml>
  <ds-start>
    'Welcome to the UTA, the virtual Travel Agent in Ulm!'
  </ds-start>
  <initial name="initial1">
    <questions    value="'How may I help you?'///'What can I
do for you?'"/>
    <explanation  value="'You can inquire or book a trip'///
'You can say something like &quot;I want to book a flight to
Oslo&quot;.'"/>
  </initial>
  <field name="departure_city">
    <questions    value="'From where would you like to
travel?'"/>
    <explanation  value="'Please tell me your departure
city.'"/>
    <summary      value="'I understood you want to travel
from '+$$jsstring:dlg:departure_city:utterance$$+'.'"/>
    <confirmation value="'Is this correct?'"/>
    <slots        value="city"/>
    <max_num_ans  value="1"/>
    <cond         value="true"/>
    <text         value="''"/>
  </field>
  <field name="destination">
    <questions    value="'What is your destination?'///'Where
would you like to go to?'"/>
    <summary      value="'I understood &quot;'
+$$jsstring:dlg:destination:utterance$$+'&quot;.'"/>
    <confirmation value="'Is this correct?'"/>
    <slots        value="city"/>
  </field>
  <field name="date">
    <questions    value="'When would you like to depart?'"/>
    <summary      value="'I understood &quot;'
+$$jsstring:dlg:date:utterance$$+'&quot;.'"/>
    <confirmation value="'Is this correct?'"/>
  </field>
  ...
  <confirmation name="conf1">
    <questions    value="'Are you sure you want to travel
from '+...+' to '+...+' on '+...+'?'"/>
    <slots        value="departure_city///destination///date"/>
  </confirmation>
</dialog-xml>
```

Fig. 5.21 Standard dialogue description excerpt of our travel information system

'Philadelphia', 'Boston'}. As known from VoiceXML, fields can be assigned conditions about whether they are included in the dialogue or not. Similarly, the argument of cond can be an ECMAScript statement evaluating to true or false determining whether the field shall be included. Like in the other properties, values of other fields, variables or functions can also be accessed with the aid of the interface call $$jsstring:dlg:field_name:answer$$. Finally, the text parameter can contain any kind of text which cannot be assigned to the other parameters but which might be needed in certain dialogue conditions. Apart from that it can also contain comments or further information in the dialogue description file. Analogously, the fields destination and date are included, with date being defined by the minimal number of required parameters. Apart from the initial and field type dialogue items, there also exists the confirmation type which allows the system to prompt for special confirmations which are, like in this example, combining multiple fields. Here, the slots are the ones which are reset when the user disagrees. E.g., if the system asks *"Are you sure you want to travel from Munich to Paris on December 2nd?"* and the user replies *"No."*, the fields departure_city, destination and date are reset. For the sake of overview, the variable access is replaced by "..." in the dialogue description excerpt.

In order to provide the functionality of the dialogue manager as described up to now, a specialized grammar is required to, somehow, outwit the linguistic analysis of the VoiceXML interpreter. An excerpt of a grammar which could be used for the travel information system described by the code in Fig. 5.21 is shown in Fig. 5.22. In this example, the grammar is structured in order to provide a maximum flexibility allowing the user to say literally anything. This mixed initiative approach is manifested in the <count number="1+"> in the root rule allowing any combination of the items once or more often in the answer and confirm rules and further rules which are omitted in this example. In the answer rule, a garbage model is included to also allow words or phrases which are not required for the linguistic analysis like *"I want to ... "*, *"hm"*, *"please"*, etc. Furthermore, all cities appearing after *"from"* are classified as departure_city, all cities after *"to"* are considered as destination, all cities without any context are just treated as city, etc. The confirm rule is used to cover the various replies of a user to a confirmation prompt or to other prompts requiring *"yes"* or *"no"* answers.

The major difference compared to standard VoiceXML grammars is given by the += operator instead of =. This means that the repeated input for a field leads to the concatenation of the previous input and the current input in the linguistic analysis. In order to separate semantic labels from values and to distinguish between different fields within one utterance, different separator levels are used. Although the resulting strings are not evaluated by the ECMAScript interpreter, these separators are selected in such a way that they can not be confused with ECMAScript operators like, e.g., =. Thus, here, a semantic label is assigned a value with @. Multiple of these are concatenated with + and appended to the name of the rule using #. The results of multiple rules are concatenated with ||. Depending on the parsing tool, this += approach implicates the problem that at the beginning a result string is appended to a not initialized string leading to strange output like, e.g., undefined||answer#undefined+destination@Paris||... which after some preprocessing evaluates to answer#destination@Paris||.... Using such a grammar and the appropriate preprocessing, a user input like, e.g., *"Yes, I would like to travel from Munich to Paris or to London tomorrow."* would be

```
<?xml version="1.0"?>
<grammar xml:lang="en-US" version="1.0" root="ROOT">

<rule id="ROOT" scope="public" >
  <count number="1+">
    <item><ruleref uri="#answer"/> <tag>ds_input+='||answer#'
      +answer.val;</tag></item>
    <item><ruleref uri="#confirm"/> <tag>ds_input+='||confirm#'
      +confirm.val;</tag></item>
    ...
  </count>
</rule>

<rule id="answer">
  <one-of>
    <item>
      <ruleref import="garbage#base_garbage"/>
    </item>
    <item>
      from
      <ruleref uri="#cities"/>
      <tag>val+='+departure_city@'+cities.val;</tag>
    </item>
    <item>
      <ruleref uri="#cities"/>
      <tag>val+='+city@'+cities.val;</tag>
    </item>
    <item>
      <ruleref uri="#date"/>
      <tag>val+='+date@'+date.val;</tag>
    </item>
    ...
  </one-of>
</rule>

<rule id="confirm">
  <ruleref uri="#yesno"/> <tag>val+='+yesno@'+yesno.val;</tag>
</rule>

<rule id="cities">
<one-of>
  <item> london   <tag>val='London'</tag></item>
  ...
</one-of>
</rule>
...
</grammar>
```

Fig. 5.22 Excerpt of an XML grammar tailored to the travel information system shown in Fig. 5.21

```
<config-xml>
  <constants>
    <grammarseparator1 value="'\\|\\|'"/>
    ...
    <grammarconfirmationslot value="'yesno'"/>
    <grammarconfirmation_yes value="'yes'"/>
    ...
  </constants>
  <dialoghandling>
    <allow_multiple_answers value="true"/>
    <multiple_answers_same_slot value="true"/>
    <multiple_answers_choose_last value="false"/>
    ...
    <stochastic_dm value="true"/>
    ...
  </dialoghandling>
  ...
</config-xml>
```

Fig. 5.23 Excerpt of a configuration file for the dialogue manager

analyzed as ds_input = cond#yesno@yes||answer#departure_city@Munich+ destination@Paris+ destination@London+ date@tomorrow.

As mentioned above, the functionality and behavior of the dialogue manager is determined by the configuration parameters which are defined in a separate configuration XML file. An excerpt of such a configuration file including some of the significant parameters is shown in Fig. 5.23. The configuration is subdivided into constants and dialoghandling plus output where the verbosity of the output can be adjusted, e.g., for debugging reasons. The prompt style is influenced by prompts, userfunctions establishes the integration of task-related user-defined functions, ctrlparam defines the use of dialogue control parameters and files administrates the file names of dialogue description and dialogue model parameter files.

The constants section contains all important string and number constants which occur anywhere in the dialogue manager. Most of them concern the linguistic analysis where the developer can use arbitrary names and symbols as separators or slot names. The most significant parameters in terms of dialogue manager behavior are contained in the dialoghandling section. Here, the developer can intentionally constrain the flexibility of the system limiting the number of answers per user turn to one (allow_multiple_answers = true allows a flexible mixed-initiative behavior, ... = false expects the user to reply to the current FUD and nothing else like in plain system initiative). The parameters multiple_answers_same_slot and multiple_answers_choose_last define whether overanswering shall be allowed and, if no, how multiple replies to one field shall be processed. E.g., selecting multiple_answers_same_slot = true and multiple_answers_choose_last = true imitates the behavior of a standard VoiceXML interpreter analyzing *"... to Paris to London ... "* as destination = 'London' whereas multiple_answers_choose_last = false temporarily stores destination = {'Paris','London'} and leaves it up to the dialogue manager how to treat that case. Apart from further parameters which, among other things, allow or disallow the user to repeat or change the stored values of filled fields, the developer can also choose whether the semi-stochastic dialogue model shall be used (stochastic_dm = true) or whether the standard rule-based dialogue model shall be employed (... = false).

Up to now, we have only considered the plain dialogue capabilities of the dialogue manager implementation. As described in Section 3.2, however, a dialogue model contains more than only dialogue fields, but also dialogue control parameters like, e.g., the user's emotional state or the confidence measures provided by the automatic speech recognizer. The integration of (an arbitrary number of) control parameters is implicitly provided for in this implementation. To accomplish that, the definition of a field is extended with respect to the employed control parameters. Depending on the number of combinations, an arbitrary number of extended questionsX... statements are included. An example is given in the destination field in Fig. 5.24. Here, we presume three emotional states "happy", "neutral" and "angry" represented by their numerical values $E(U) = 0.3$, 1.0 and 1.7. Accordingly, three extended prompts questionsX0_3, questionsX1_0 and questionsX1_7 are defined. I.e., if the dialogue manager determines FUD=destination:0.3, the following system turn will be *"Excellent, and what is your destination?"*.

A prominent example of how the confidence measures provided by the automatic speech recognizer can be included is shown in the confirmation field in Fig. 5.24. Here, the condition on which the confirmation is requested is directly dependent on the value of the confidence measure which is assumed to be stored as a variable called confidence. If this value, which is typically the log-likelihood of the speech recognizer output ranging from $-\infty$ to 0, is below a threshold of, e.g., -4.0, the system includes this combined confirmation. Furthermore, by default, the confidence

```
...
<field name="destination">
  <questions       value="'What is your destination?'///
    'Where would you like to travel to?'"/>
  <summary         value="'I understood '
    +$$jsstring:dlg:destination:utterance$$+'.'"/>
  <confirmation   value="'Is this correct?'"/>
  <questionsX0_3  value="'Excellent, and what is your
    destination?'"/>
  <questionsX1_0  value="'Where would you like to travel to?'"/>
  <questionsX1_7  value="'I am sorry to bother you again.
    Where did you say you wanted to travel to?'"/>
</field>
...
<confirmation name="conf_dep_dest">
  <questions       value="'Are you sure you want to travel from '
    +...+' to '+...+'?'"/>
  <slots          value="departure_city///destination"/>
  <cond           value="$$float:var:confidence:value$$ < -4.0"/>
</confirmation>
```

Fig. 5.24 Extended dialogue description excerpt of our travel information system corresponding to the dialogue model illustrated in Fig. 3.34

measures have a direct influence whether or not single confirmations (implicit or explicit) for certain fields are included. To accomplish that, two threshold values θ_i and θ_e, $\theta_i > \theta_e$ are defined. If the confidence measure of a user utterance addressing a certain field is lower than θ_e, an explicit confirmation for this field is included. If the confidence measure of the same turn is between θ_e and θ_i, an implicit confirmation is included. Otherwise, no confirmation is requested.

If, following the considerations in Section 3.9, further control parameters shall be included in the prompt realization, the extended questions can be described as questionsXp1Xp2X...XN representing N further parameters which are described as destination : $p_1^{(j_1)}$: $p_2^{(j_2)}$: ... : $p_N^{(j_N)}$.

5.4 Discussion

In this chapter, we have described implementation aspects and improvements for our proposed speech-based emotion recognizers. The reduction of the emotion set as well as gender discrimination constitute just a modicum of practical aspects when dealing with plain emotion recognition. As discussed in Section 5.1.1, also data annotation methods influence the performance of the emotion recognizer.

Using appropriate features, selecting representative emotions and applying the two-step approach described in Section 5.1.2, we are able to keep the speech–emotion recognizer's complexity at a reasonable level resulting a robust recognition performance. In the two-step speech–emotion recognizer, we employ the output of an upstream speech recognizer as a-priori information for the actual speech–emotion recognizer, which particularly leads to a better emotion recognition performance. An assessment of our plain emotion recognizer as well as of our combined speech–emotion recognizer is given in Sections 6.3.1 and 6.3.2.

The combination of multiple speech–emotion recognizers in order to improve the recognition performance is described in Section 5.2. We exploit the dissimilarities in the output of different speech–emotion recognizers to reduce the overall error rate. With respect to the high complexity which is linearly increasing with the number of involved recognizers, it is important to be able to assess the performance gain beforehand in order to estimate whether it is worth the effort. Here, we consider theoretical limits which indicate whether the use of multiple recognizers actually increases or decreases the performance compared to a single recognizer. In Section 6.3.3, we discuss the actual performance of our approach to combining multiple speech–emotion recognizers.

In Chapter 3, we have described a theoretic foundation to semi-stochastic emotional dialogue modeling. Here the term "semi-stochastic" refers to the fact that all states (fields and emotions) including their properties are predefined whereas the transitions between the states are defined by probabilities. An implementation of this dialogue model is described in Section 5.3. Our proposed dialogue manager is integrated in the VoiceXML framework utilizing all of the framework's advantages while offering possibilities which exceed the framework's limitations.

Chapter 6
Evaluation

Identifying the weaknesses of a system as well as establishing test criteria and measures which make different systems and concepts comparable are one of the major challenges in the evaluation of human–computer interfaces. We have given a synopsis about the evaluation of SLDS in Section 1.5 (cf. also Fig. 1.3). In the following section, we outline further aspects of the evaluation of SLDSs and their components. In the remainder of this chapter, we present an in-depth evaluation of the performance of the emotion recognizers described in Chapters 4 and 5, and we describe our approach to measure the usability of the dialogue manager described in Chapter 5.3.

6.1 Description of Dialogue System Evaluation Paradigms

Over the years, the PARADISE approach proposed by Walker et al. (1997b) has emerged to a quasi-standard for the evaluation of human–computer interfaces. In this model (see Fig. 6.1), it is assumed that both task success and dialogue costs constitute relevant contributors to the top level objective (user satisfaction). The dialogue costs objective, in turn, can be further subdivided into efficiency and quality objectives. For each of the low level objectives, there exist performance measures such as the κ-factor for the task success, various time measures for the efficiency and utterance measures for the quality. In order to be able to determine a value for κ, here, attribute value matrices for the dialogue domain and the therein included tasks are defined, i.e., attributes like departure_city are assigned possible values like *"London"*, *"Paris"*, etc. Then, a confusion matrix for all attributes is determined on the basis of the dialogues to be evaluated and κ is calculated as described for the labeling process in Section 2.3 as

$$\kappa = \frac{P(A) - P(E)}{1 - P(E)}, \tag{6.1}$$

where $P(A)$ is the probability that the attribute value matrices for the evaluation dialogues agree with those of the reference dialogue(s) and $P(E)$ is the probability that

J. Pittermann et al., *Handling Emotions in Human-Computer Dialogues*,
DOI 10.1007/978-90-481-3129-7_6,

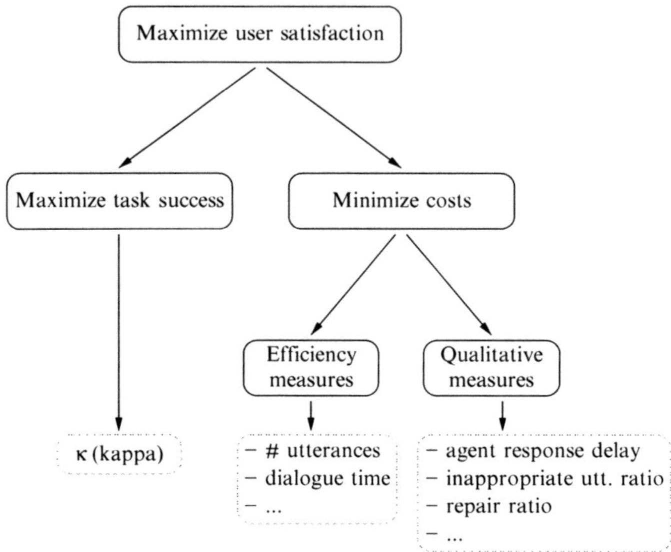

Fig. 6.1 Structure of objectives for the evaluation of SLDSs using the PARADISE approach (Walker et al. 1997b)

the attribute value matrices coincide by chance (see also Cohen 1960; Fleiss 1971; Carletta 1996). Taking into account a task's complexity by including $P(E)$, κ can be used to compare dialogue systems which even perform different tasks (Walker et al. 1998).

Dialogue costs, also referred to as cost functions c_i, such as number of dialogue turns, number of repair utterances, etc., can be also determined by comparing attribute value matrices in order to take into account the structure of dialogues (segments and subdialogues).

Given these measures of success and costs, the PARADISE approach calculates the overall performance $P_{\{D|S\}}$ of a dialogue D or a dialogue segment S as

$$P_{\{D|S\}} = \alpha \cdot \Pi(\kappa) - \sum_{i=1}^{n} (w_i \cdot \Pi(c_i)) , \qquad (6.2)$$

where α is a weighting factor setting up the influence of κ and w_i are weighting factors for the individual cost functions c_i, $i = 1, \ldots, n$. $\Pi(\cdot)$ is a normalization function which is used to compensate the problem that the values of c_i and κ are not on the same scale. It is defined as

$$\Pi(x) = \frac{x - \mu_x}{\sigma_x}, \qquad (6.3)$$

where μ_x and σ_x are mean value and standard deviation of x (Cohen 1995; Walker et al. 1997b).

The weighting factors α and w_i in Eq. 6.2 are calculated by determining the impact of κ and c_i on the user satisfaction which has been defined as the top level objective of the PARADISE approach. This is accomplished by putting values for user satisfaction, κ and c_i, which have been collected in experiments, into relation and applying a multivariate linear regression to obtain α and w_i. In this calculation, the statistical significance of c_i also plays an important role. I.e., it is not unlikely that certain cost measures finally do not contribute to the user satisfaction. It should also be noted that the values for α and w_i are different for different dialogue systems.

The application of the PARADISE approach to two different SLDSs including possible task specifications, κ and cost measures and the organization of a user survey is circumstantially described in Walker et al. (1998). Being originally designed for SLDSs and agents, the PARADISE model itself has been adapted to a huge variety of fields like the PROMISE approach to the evaluation of multimodal dialogue systems (Beringer et al. 2002) and has been amply analyzed (see, e.g., Hajdinjak and Mihelič 2006).

As an alternative to PARADISE, Dybkjær and Bernsen (2001) propose a framework to assess the usability of SLDSs. This framework embraces 15 measures to be evaluated ranging from the appropriateness of the modality (is speech the modality of choice for the given application/domain?), the system's understanding capabilities, the quality and adequacy of its output, the structure of the dialogue flow to the system's intelligence (including its reasoning or error handling capabilities), cooperativity and finally also the users' satisfaction.

Despite the fact that holistic evaluation schemes such as PARADISE provide a good basis for the comparison of different dialogue systems, their informative value about possible improvements of the system or its components is relatively low. Moreover, involving an extremely high effort for user studies and surveys, these evaluation schemes are not suitable at all to debug a system under development, i.e., to track possible faults in individual components or in the overall system.

Thus, we do not limit the term "evaluation" to the final assessment of an end-to-end system but also include preliminary tests to evaluate and optimize the performance of single components or of the overall system as already indicated in Fig. 1.3. Moreover, we argue that the PARADISE approach is too extensive for our needs in terms of parameters to be considered and overall user studies to be conducted. Accordingly, for the evaluation described in this chapter, we do not implement the complete PARADISE idea, but concentrate on selected details, particularly on performance measures of speech and emotion recognition as well as the user acceptance of our approach to adaptive dialogue management.

Regarding the architecture of adaptive SLDSs as described in the previous chapters, specific performance measures can be defined for each of the components as shown in Fig. 6.2. Focusing our consideration on the processing of the user's input, we primarily go into detail about the performance measures of the respective components. Thus, to simplify matters, we presume that the assessment of components like synthesis is, to some extent, included in the holistic evaluation.

For the linguistic analysis (parsing), the concept accuracy is determined on the basis of reference sentences and the respective labels containing the included

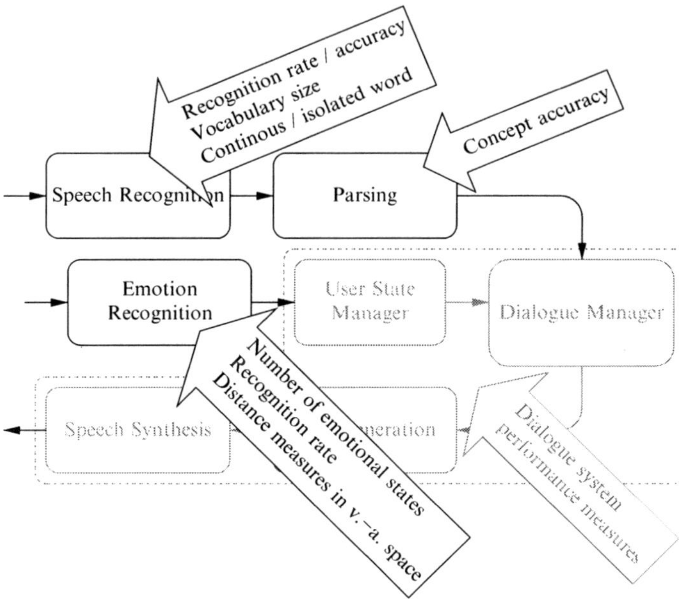

Fig. 6.2 Architecture of an adaptive SLDS and typical performance measures for its input processing components

semantic representations. Typically, the sentences are analyzed and the output is compared to the reference labels. The accuracy is calculated as the proportion of correctly recognized representations and analysis errors. These errors include misinterpretations (a keyword is assigned the wrong semantic meaning, e.g., *"To Paris."* is analyzed as departure_city='Paris'), additional interpretation (a word without meaning is assigned a semantic meaning, e.g., *"From London, please."* is analyzed as departure_city='London', destination='Please') or missing interpretations (e.g., *"From London to Paris."* is interpreted as departure_city='London').

The performance measures for speech and emotion recognition are quite similar. These typically include information about the complexity of the recognizer (vocabulary size or number of emotions, isolated words vs. continuous text) and the distance between the recognized text or emotion sequence and the reference sentences or emotional state sequence. With respect to the similarity of both recognizers, we will comprehensively describe these measures in the following section before evaluating the performance of the emotion recognizers described in Chapters 4 and 5.

6.2 Speech Data Used for the Emotion Recognizer Evaluation

In the experiments conducted in the course of the development and the performance evaluation of speech-based emotion recognizers, we use speech data from two corpora – the Berlin Database of Emotional Speech, presented in Burkhardt et al. (2005), and a smaller database containing recordings of spontaneous emotional speech, shortly described in Pittermann et al. (2007b).

The Berlin Database of Emotional Speech which is publicly available from the Technical University of Berlin includes seven emotions, namely anger, boredom, disgust, fear, happiness and sadness along with neutral recordings serving as references. Ten different everyday-speech sentences not containing any linguistic cues to certain emotional states like *"Der Lappen liegt auf dem Eisschrank."* (*"The cloth is lying on the fridge."*) or *"An den Wochenenden bin ich jetzt immer nach Hause gefahren und habe Agnes besucht."* (*"At the weekends I used to go home and visit Agnes."*) have been performed by different actors in all different emotional states and recorded in an anechoic chamber to minimize background noise and other disturbing effects like echoes. Ten actors, five female and five male speakers, have performed these sentences, each utterance in all seven emotions so that a high comparability across all emotions and speakers is ensured. With some sentences having been recorded more often than once by the same speaker in the emotional state, a total of 813 utterances is available for our experiments. The emotional quality of each utterance has been rated by 20 persons, who were asked to listen to the utterances only once in front of a computer monitor and then to choose the emotion which they think is acted in the respective utterance as well as to specify how convincing this emotion has been brought out (Burkhardt et al. 2005; Bartels et al. 2006).

For the experiments with natural speech data, the Ulm Small Database of English Spontaneous and Affective Speech (USDESAS) is used. The database consists of 586 utterances involving four emotions (anger, boredom, happiness and sadness) and neutral. Selected neutral utterances have also been assigned "certainty" and "doubt" as cognitive states. Twelve speakers (two female and ten male) have been recorded while interacting with a quiz and a personality test both designed to elicit certain emotional reactions from the candidates. These persons, mainly international students and staff members, have been novices in the field of human–computer interaction without any experience with such systems. The recording sessions have been performed in a Wizard-of-Oz environment as shown in Fig. 6.3 (see also Bernsen et al. 1994) and the persons have not been told beforehand that the interaction actually is between them and a simulated system controlled by a human "wizard". Nevertheless, they have given their consent that the recorded data may be used in experiments. The recordings have been performed in a secluded room without any background noise and the utterances have been labeled manually based on subjective

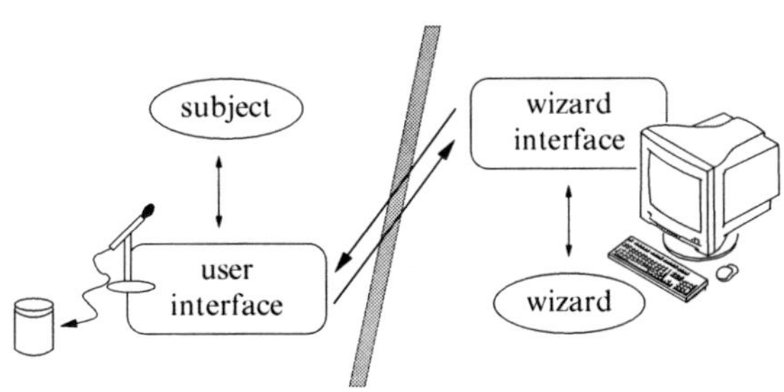

Fig. 6.3 Wizard-of-Oz setup for the recording of emotional speech data (Bernsen et al. 1994)

impressions. In order to achieve a higher credibleness of the data, the utterances have been relabeled iteratively by different annotators using the bootstrapping algorithm as described in Section 4.5. As opposed to the acted emotional speech data where the emotional state for an utterance is clearly predefined and where the utterances are "clean" in terms of additional sounds, the spontaneous speech data are often hard to categorize and also contain additional sounds like laughter or breath sounds making it easier to annotate these data but discombobulating the emotion recognizer during the training and recognition.

Despite the fact that certain utterances in the databases (particularly in the Berlin Database of Emotional Speech) have been judged as "not very convincing", all of these utterances are also included in our experiments to assess the systems' robustness against such outliers. In all experiments, disjoint training, development test and evaluation test sets are randomly selected from the speech database(s) used in the respective experiments. From the 813 utterances of the Berlin Database of Emotional Speech, we choose 600 utterances for the training, 63 utterances for development tests and 150 utterances for evaluation test. Constraints on the selection of the sets are, e.g., disjoint speaker and sentences sets for training and testing. All experiments are carried out repeatedly with different set combinations and the results of the individual test runs are averaged.

6.3 Performance of Our Emotion Recognizer

Within the architecture of an adaptive SLDS, automatic speech recognition constitutes the component the performance of which is the easiest to measure. As opposed to the evaluation of the dialogue manager or the quality assessment of text-to-speech synthesis which inherently require the subjective opinion of test users, the performance evaluation of a speech recognizer can be broken down into the comparison of two strings of words and the recognition error rate can be directly determined from the distance of the two strings.

The Levenshtein distance may also be involved in the calculation of the word accuracy of a speech recognizer on the basis of a set of test sentences. Given a reference sentence *"I would like to travel to Paris today."* and the respective recognizer output *"I want to go to Paris today, please."*, we determine the Levenshtein distance according to Eq. 5.4 (see also Fig. 5.12).

At a first glance, both sentences look similar, yet they contain the same information. I.e., the linguistic analysis in an SLDS would extract the same semantic representations from both sentences. Nevertheless, we can determine the Levenshtein distance as $L = 4$. Including the overall length N of the reference sentence (8 words), the accuracy for this sentence is only $(8 - 4)/8 = 50\%$. Regarding the details, we observe two substitutions (*"want"* instead of *"would"* and *"go"* instead of *"travel"*), one deletion (*"like"* is missing in the recognized sentence) and one insertion (*"please"* does not occur in the reference sentence).

Depending on the application, insertions are sometimes not considered as critical errors in the evaluation. Thus, there exists the strict performance measure of the accuracy which is defined as

$$\text{Acc.} = \frac{N - (S + D + I)}{N} \cdot 100\% = \frac{N - L}{N} \cdot 100\%, \tag{6.4}$$

and the more relaxed measure of the word recognition rate which is defined as

$$\text{WordCorr.} = \frac{N - (S + D)}{N} \cdot 100\%. \tag{6.5}$$

By definition, the word recognition rate ranges between 0 and 100% as the number of substitutions and deletions may not be larger than the number of words in the reference sentences. On the other hand, the accuracy may also take negative values (in the worst case $-\infty$) as the number of insertions is not limited. E.g., *"From Munich to Paris."* instead of *"To Munich."* leads to two insertions and one substitution resulting in an accuracy of $(2 - 3)/2 = -50\%$. Analogously, the sentence recognition rate is defined as

$$\text{Sent.Corr.} = \frac{C_S}{N_S} \cdot 100\%, \tag{6.6}$$

i.e., the number of totally correctly recognized sentences C_S divided by the number of sentences N_S. Accepting only correct sentences without any substitutions, deletions or insertions, and thus being lower than the word recognition rate, the sentence recognition rate constitutes the hardest criterion for the evaluation of speech recognizers.

Due to their similarity of our proposed emotion recognizers to standard speech recognition systems, the criteria described above can, to some extent, also be used to evaluate speech-based emotion recognizers as described in Chapter 4. Thus, it seems obvious to treat substitution, deletions and insertions as errors and to calculate the accuracy of the recognizer as described above. Having done so, one might notice that these numbers are frustratingly low although a manual visual comparison of recognition result and reference by rule of thumb looks rather positive.

A selection of results from our plain emotion recognizer as described in Section 4.5 and the reference labels are shown in Fig. 6.4. In the left column, the reference labels for three utterances are given and the respective recognition results are given in the right column. The numbers in this figure represent beginning and end time of the individual emotions in milliseconds.

In both reference and recognizer results it is presumed that there are (short) pauses between individual emotions representing eventual transitions which are difficult to classify. As these pauses do not actually contribute to the recognition of emotions, the pauses are not considered in the evaluation. Applying standard speech recognizer criteria, the first recognized utterance contains one substitution and one insertion, the second utterance contains one substitution and one deletion and the

0	120	PAUSE	0	30	PAUSE
120	760	SADNESS	30	350	ANGER
760	790	PAUSE	350	380	PAUSE
790	1810	ANGER	380	730	ANGER
1810	1870	PAUSE	730	760	PAUSE
			760	1840	ANGER
			1840	1870	PAUSE
0	30	PAUSE	0	30	PAUSE
30	50	NEUTRAL	30	2690	NEUTRAL
50	70	PAUSE	2690	2730	PAUSE
70	3080	NEUTRAL	2730	3610	BOREDOM
3080	3110	PAUSE	3610	3640	PAUSE
3110	3630	NEUTRAL			
3630	3640	PAUSE			
0	30	PAUSE	0	30	PAUSE
30	1710	HAPPINESS	30	590	HAPPINESS
1710	1780	PAUSE	590	620	PAUSE
			620	1750	ANGER
			1750	1780	PAUSE

Fig. 6.4 Reference emotional labels (*left column*) of three utterances and the respective output of a speech-based emotion recognizer (*right column*)

third utterance contains one insertion. Summarizing the errors and the number of reference emotions, we obtain an accuracy of $(6-5)/6 = 1/6$. Moreover, the sentence recognition rate is even 0%.

Taking into account the durations of the individual emotions, however, in this example, it is striking that the sadness period in the first utterance is shorter than the anger period making anger the predominant emotion in the reference. Comparing this to the predominant emotion in the recognizer output which is also anger, we could now declare the first utterance as correctly recognized. The same applies to the second utterance, where the boredom period in the recognizer output is significantly shorter than the neutral period and where the predominant reference emotion is also neutral. For the third utterance, however, the predominant emotion in the recognizer output is anger as opposed to happiness which is the only emotion in the reference.

All in all, comparing the predominant emotions, we obtain a recognition rate of $2/3$ for these examples. With respect to the further use of the emotion recognizer output in the adaptive SLDS (see Chapter 3), the application of this criterion is justifiable, last but not least as other approaches to emotion recognition, like artificial neural networks, are also not able to account for temporal aspects and, thus, only provide one emotion per utterance anyway.

For the determination of the predominant emotion in one utterance, we follow two methods: either by comparing the frequencies of occurrences or by comparing the relative durations of the individual emotions. In the first method, the predominant emotion E_0 is determined as

$$E_0 = \arg\max_E N(E), \tag{6.7}$$

where E are all emotions occurring in the utterance and $N(E)$ is the number of occurrences of emotion E in the utterance. Similarly, the second method determines E_0 as

$$E_0 = \arg\max_E D(E) = \arg\max_E \sum_i D_i(E), \tag{6.8}$$

where $D(E)$ is the overall duration of emotion E in the utterance calculated as the sum of the durations of the individual occurrences of the emotion $D_i(E)$.

Following the scoring idea in the ROVER approach as described by Eq. 5.5, it may seem self-evident to combine both methods, i.e., determining E_0 as

$$E_0 = \arg\max_E \alpha \cdot N(E) + (1-\alpha) \cdot D(E), \tag{6.9}$$

where $\alpha \in [0, 1]$ is a weighting factor defining the proportions of the first and the second method – $\alpha = 1$ corresponds to the first method, $\alpha = 0$ represents the second method and any value for α in between leads to a mixture of both methods. An exhaustive search for a representative value for α over all experiments conducted with different emotion recognizer setups has lead to the conclusion that both methods individually lead to similar results. I.e., it is left to the evaluator whether to determine E_0 using the first or second method. A combination of both methods, however, is breaking a butterfly on a wheel and therefore unnecessary.

As opposed to speech recognition where the words are explicitly defined and provide a reliable basis for comparison, the labeling of emotions involves a large degree of subjectivity and, thus, may lead to strong discrepancies among different annotators. In the Berlin Database of Emotional Speech (Burkhardt et al. 2005), which we also use for the evaluation of the emotion recognizers, this problem has been addressed by providing the labels of all 20 annotators plus the actual emotion the actor was asked to play for each utterance. E.g., if the actor was asked to say a sentence in an sad manner and nine annotators labeled the utterance as sad, eight chose bored, two chose neutral and one chose disgusted, the reliability of the utterance would be 45% and it would or should not be surprising if the utterance was not recognized "correctly" by any of the systems.

Considering seven emotions (anger, boredom, disgust, fear, happiness and sadness plus neutral) as contained in the database, one could now think of a vectorial representation in a seven-dimensional space, where the position is given by the proportion of the respective emotion among all annotators. E.g., the utterance discussed above would be represented by [0.0 anger, 0.4 boredom, 0.05 disgust, 0.0 fear, 0.0 happiness, 0.45 sadness, 0.1 neutral]. Theoretically, the HMMs in the recognizer could be trained accordingly and the recognition performance could be assessed on the basis of the distance between reference and recognizer output in the seven-dimensional space. Commonly used for calculations in vector spaces, here, the Euclidean distance is a possible distance measure. However, assuming seven emotions and 20 annotators, there exist a vast number of discrete points in the vector space and therefore models to be trained, which is not feasible unless an appropriate amount of training data is available. It should be noted, that the emotional space does not necessarily have to be seven-dimensional – depending on the number of emotions to be considered, it may also be of lower or higher dimension.

Nevertheless, the output of the recognizer as shown in the right column in Fig. 6.4 may also be transformed into the seven-dimensional space by considering the relative duration of each of the seven emotions in each utterance. E.g., in the output of the second utterance, the proportion of boredom is (36,100–27,300)/(26,900–300+36,100–27,300) $\approx par0.25$, so that the second recognized utterance can be described by [0.0 anger, 0.25 boredom, 0.0 disgust, 0.0 fear, 0.0 happiness, 0.0 sadness, 0.75 neutral]. It should be noted that in the first case, the values are averaged over all annotators and in the second case, the values are derived from the temporal trend of emotions within an utterance which, strictly speaking, does not allow us to compare the vectors of both cases. However, as it is not identifiable based on which point of time within the utterance each of the annotators bases his judgment, we presume that different annotators form their opinions at different points of time like the recognizer also does, so that we argue that both cases form a comparable basis as well.

Again considering the second utterance in Fig. 6.4 the Euclidean distance between reference and recognizer output is $\sqrt{(1-0.75)^2+(0-0.25)^2} \approx 0.35$. By definition, the minimum Euclidean distance in this space is 0 and the maximum distance is $\sqrt{2}$. Thus, for a better comparability, we normalize each distance dividing its value by $\sqrt{2}$, so that, e.g., the normalized distance for the second utterance is 0.25. Similarly, the distance for the first utterance is approximately $\sqrt{(0-0.39)^2+(1-0.61)^2}/\sqrt{2} = 0.39$ and 0.67 for the third utterance.

Using these numbers, the recognition performance of a plain emotion recognizer can be determined as

$$\text{EmotionCorr.} = \left(1 - \frac{1}{|\mathcal{U}|} \cdot \sum_{u \in \mathcal{U}} \bar{d}(u)\right) \cdot 100\%, \tag{6.10}$$

where $\mathcal{U}$ is the set of all test utterances, $|\mathcal{U}|$ is the cardinality of $\mathcal{U}$ and $\bar{d}(u)$ is the normalized Euclidean distance for utterance u in the emotional vector space. Summarizing the numbers calculated for the utterances in Fig. 6.4, we obtain a recognition performance of $1 - \frac{1}{3}(0.25 + 0.39 + 0.67) = 56.3\%$.

Instead of the complicated distance calculations in the multi-dimensional emotional space, the same considerations can also be made in the two-dimensional valence-arousal space as depicted in Fig. 6.5 (see also Fig. 3.14 in Section 3.5). Predefining fixed values for each emotion, each utterance is represented as the linear combination of its included emotions. For the emotions in this example, we presume the following coordinates: anger $(-0.9, 0.9)$, boredom $(-0.1, -0.3)$, happiness $(0.9, 0.8)$, neutral $(0.0, 0.0)$ and sadness $(-0.8, -0.2)$. The coordinates of reference and recognition result for each utterance are marked with black and grey circles and the distances are represented by black solid lines.

For example, for the first utterance, the reference value is $0.39 \cdot \text{sadness} + 0.61 \cdot \text{anger} \approx (-0.861, 0.471)$. Here the proportions are also derived from the duration

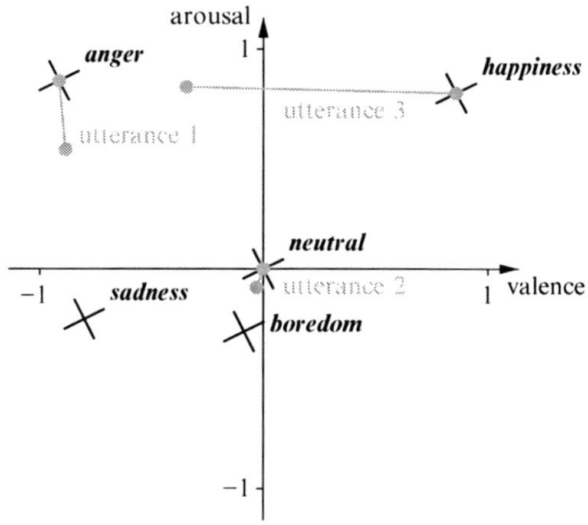

Fig. 6.5 Evaluating the emotion recognizer performance on the basis of distance measures in the valence-arousal space (Pittermann et al. 2008a)

of each emotion in the utterance. The coordinates of the recognition results coincide with anger and are $(-0.9, 0.9)$, so that the Euclidean distance is ≈ 0.43. The maximum Euclidean distance in the valence-arousal space is $2\sqrt{2}$, so that the normalized distance for the first utterance is approximately 0.152. Analogously, the distance for the second utterance is 0.028 and for the third utterance 0.427. Applying Eq. 6.10, we obtain a recognition performance of 79.8%.

Furthermore, as an alternative to distance calculations in the multi-dimensional emotional space or the valence-arousal space, we can also calculate the error of the resulting emotional value $E(U)$ which is described as a projection in the valence-arousal space as presented in Section 3.5. Typically, the values are distributed according to the curve shown in Fig. 3.16. Positing $E(U) = 0$ for extreme happiness (upper right corner of the valence-arousal space), $E(U) = 1$ for neutral and $E(U) = 2$ for extreme anger (upper left corner), a different continuous gradients can be defined following different types of functions. Given further predefined points in the space and their values $E(U)$, linear combinations of polynomials, exponential functions, harmonic or other functions may be found to satisfy the requirements. In Section 3.5, we are content with a relatively simple approach as described by Eq. 3.8:

$$E(U) = 1 - v \cdot (a + 2)/3, \tag{6.11}$$

where v and a are the values of valence and arousal.

Calculating the emotional values for both reference and recognizer output, we determine the distance as the absolute difference of both values. Inserting the valence-arousal values from above, we obtain for the first utterance $E_{ref}(U) = 1 - (-0.861) \cdot (0.471 + 2)/3 \approx 1.71$ and $E_{rec}(U) = 1 + 0.9 \cdot 2.9/3 \approx 1.87$. With a maximum possible distance of 2 among the values of $E(U)$, the normalized distance for the first utterance is $|1.71 - 1.87|/2 = 0.08$. Analogously, the normalized distance for the second utterance is 0.008 and for the third utterance 0.56. Again, applying Eq. 6.10, we obtain a further recognition performance measure of 78.2% for these utterance.

Comparing the different approaches to measuring the performance of an emotion recognizer, it is remarkable that the standard speech recognizer measures do not necessarily provide a useful basis for assessing the strengths and weaknesses of a system, whereas application-oriented measures help to rank recognition errors according to their actual impact on the further processing of the emotions. Advancing these ideas, it is absolutely imaginable that the definition of different performance measures as done above or in literature, is tempting to somehow gloss over one's own results by searching for some strange measure which leads to fabulous results. E.g., if, for whatever reason, we decided to use the mean square difference instead of the mean absolute difference when comparing the emotional values (by what the square difference needs to be normalized by factor 1/4) and to apply the compensatory square root at the right place, we could obtain a theoretical performance of "more than 90%" without any information value (Pittermann et al. 2008a).

Unless otherwise indicated, for the performance assessment of the emotion-recognizers and speech–emotion recognizers in the remainder of this section, we mainly apply the "predominant emotion"-criterion as this provides an application-independent basis for comparison and helps to point out and eventually to avoid model-conditioned weaknesses of the recognizers. Nevertheless, we argue that the three distance measures described above still constitute equitable and comparable measures. Thus, for our plain emotion recognizer, we also accomplish a performance evaluation on the basis of $E(U)$.

6.3.1 Plain Emotion Recognition

When considering only the emotional content of an utterance, we face the problem from which part(s) of the utterance, the relevant emotion(s) shall be extracted (as opposed to speech recognition where the words are clearly defined by their boundaries). Accordingly, we use a speech–pause detector to determine which parts of the utterance are relevant (speech) and which are not (pause). Thus, the performance of the emotion recognizer also stands or falls with the output of the speech–pause detector. A simple straightforward approach is to define a threshold factor $\theta_{sp} \ll 1$ and to label each contiguous block of samples (here: a frame of 10ms length) as pause if their amplitudes are smaller than θ_{sp} times the maximum signal amplitude. Alternatively, but requiring a lot more computational effort, an adaptive minimum-tracing method can be applied to speech-relevant bands in the frequency domain as described in Linhard and Haulick (1999). To keep things simple, this approach is not employed in our experiments.

As expected, applying standard speech recognizer evaluation criteria a relatively low emotion accuracy of approximately 20–30% is achieved for any of the values for the speech–pause threshold θ_{sp}. Among the features described in Section 4.1.4 we use exclusively prosodic and acoustic coefficients, which we refer to as "PAC-24", including pitch, three formants, intensity, jitter, harmonicity, pitch in voiced parts and the respective computational statistics including minimum, maximum, range, mean and standard deviation for selected features. These features are extracted from the speech signals with the aid of Praat (Boersma 2001, 2002).

In conformance with speech recognition, we choose a frame rate of 100 Hz, i.e., the features are calculated every 10 ms. As opposed to speech recognition where the state transitions in the phoneme models occur at such high rates, a lower frame rate would also be sufficient for emotion recognition. However, to enable a later exchange of features and to retain a higher consistency for the sake of a better comparability when evaluating speech–emotion recognizers, we afford the extravagance to actually operate the recognizer at that pace.

In addition to different speech–pause thresholds, we also use different numbers of Gaussian mixtures in the HMMs. As described in Section 4.2.1, mixtures are used to approximate more complex probability density functions with the aid of linear combinations of multiple Gaussian distributions with different mean values, variances and weights in the overall distribution.

For our experiments, we consider speech–pause thresholds of 0.002, 0.005, 0.01, 0.07 and 0.1 (of the maximal absolute signal amplitude, averaged of 10 ms windows) and the HMM prototypes include 1, 2, 3, 4 or 8 Gaussian mixtures. The recognition rates for these environment parameters are shown in Table 6.1. Here, as discussed in Section 5.1.1, only a reduced set of emotions, where disgust is omitted and where boredom and neutral are merged, is considered. I.e., the boredom parts in all utterances are relabeled as neutral contributing to a slightly more general acoustic model for neutral and should accordingly also be recognized as neutral after the training whereas all utterances containing disgust are removed from all training and test sets. Looking at the table, it can be observed that the recognition rates range from 49.3% to 65.3% variegatedly distributed across the table. Tendencies how the recognition rates depend on θ_{sp} or the number of mixtures are difficult to derive, except for a recommendation not to employ more than four Gaussian mixtures in the models. The maximum recognition rate is obtained with two mixtures – at a relatively low θ_{sp} of 0.005 but also at the highest θ_{sp} of 0.1.

As described in Sections 4.1.4 and 5.1.1, it is sensible to employ separate acoustic models for female and male speakers with respect to the different ways how women and men express emotions. By that, the number of models and thus the

Table 6.1 Plain emotion recognition rates (in %) for different speech–pause thresholds and numbers of Gaussian mixtures (Pittermann and Pittermann 2006a)

	1 mixt.	2	3 mixt.	4 mixt.	8 mixt.
$\theta_{sp}=0.002$	57.3	53.3	64.0	57.3	58.7
$\theta_{sp}=0.005$	56.0	**65.3**	58.7	60.0	58.7
$\theta_{sp}=0.01$	54.7	60.0	61.3	49.3	57.3
$\theta_{sp}=0.07$	61.3	62.7	60.0	54.7	54.7
$\theta_{sp}=0.1$	62.7	**65.3**	52.0	61.3	50.7

Table 6.2 Plain emotion recognition rates (in %) for different speech–pause thresholds and numbers of Gaussian mixtures with gender discrimination (different models for female and male speakers, Pittermann and Pittermann 2006a)

	1 mixt.	2 mixt.	3 mixt.	4 mixt.	8 mixt.
θ_{sp} =0.002	64.0	66.7	64.0	66.7	69.3
θ_{sp} =0.005	62.7	70.7	62.7	70.7	58.7
θ_{sp} =0.01	61.3	**72.0**	65.3	64.0	70.7
θ_{sp} =0.07	56.0	62.7	61.3	61.3	65.3
θ_{sp} =0.1	66.7	60.0	62.7	57.3	61.3

recognizer complexity are doubled. With a smaller amount of training data available for each model (when using the same training material), one might expect less accurate models and, thus, a lower recognition performance. Typically, a recognizer's performance tends to decrease when the recognition perplexity increases, e.g., when the domain becomes more complex, when the language model becomes more flexible or when the vocabulary size increases. This behavior has notably been observed by Lippmann (1997) who has conducted appropriate experiments with both humans and automatic speech recognizers.

We implement this idea of gender discrimination with the same parameters (emotion set, θ_{sp} and mixtures) as described above. The respective recognition rates are listed in Table 6.2. Looking at these results, it can be seen that the values still look untraceably distributed among the table but now range from 56% to 72% which is noticeably higher than the values in Table 6.1 although there are more HMMs in the acoustic model. This is mainly owed to the differences in pitch and formants in female and male speech. The maximum recognition rate is 72% at $\theta_{sp} = 0.01$ with two mixtures in the models.

It should be noted that, although there are different HMMs for female and male speakers, there is no extra "gender recognizer" required. I.e., the emotion recognizer returns both emotion and gender simultaneously like "ANGER-F" (female anger), where the gender information is neglected in our further analyses. Not yet further documented, in noncompetitive experiments, this emotion recognizer achieves a gender recognition rate of 90% and above, providing robust extra information for free which could be used, e.g., in combination with a speaker identification module in enhanced user modeling.

The influence of reducing the number of emotions can also be identified with the aid of confusion statistics, either in matrices or in a table listing the five most prominent types of errors which occur most often as shown in Table 6.3. In Table 6.3(a) the most frequent errors when considering all six emotions and neutral are listed. Looking at these numbers, it is striking that the most frequent error involves the confusion of boredom and neutral. It should be noted that in both tables, the confusion (emotion-a $\rightleftharpoons$ emotion-b) includes both cases where emotion-a is recognized as emotion-b and where emotion-b is recognized as emotion-a. Moreover, the numbers in the tables are averaged over all recognizers with different values for θ_{sp}

Table 6.3 Relative frequency of errors with the full (**a**) and the reduced (**b**) set of emotions (Pittermann and Pittermann 2006a)

All emotions:

error	freq.
Boredom $\rightleftharpoons$ neutral	10.4%
Anger $\rightleftharpoons$ happiness	7.0%
Fear $\rightleftharpoons$ neutral	5.4%
Neutral $\rightleftharpoons$ sadness	4.5%
Anger $\rightleftharpoons$ fear	3.8%

(**a**)

Reduced set of emotions:

error	freq.
Fear $\rightleftharpoons$ neutral	7.9%
Anger $\rightleftharpoons$ happiness	7.7%
Neutral $\rightleftharpoons$ sadness	6.5%
Anger $\rightleftharpoons$ fear	5.0%
Happiness $\rightleftharpoons$ neutral	2.7%

(**b**)

and the number of mixtures. The more typical confusion in speech-based emotion recognition, namely between anger and happiness, still constitutes a frequent error and is at the second position. Further "important" confusions involve also neutral and fear. In this ranking, confusions with disgust occur with relatively low frequencies, supporting the conjecture that the use or omission of disgust has no influence on the recognizers' performance.

Nevertheless, a reduced set of emotions without boredom (merged with neutral) and disgust (omitted) is considered for comparison in Table 6.3(b). By design, the confusion of boredom and disgust does not occur any more, however, it is remarkable that all confusions with neutral gain in importance, especially (fear $\rightleftharpoons$ neutral) which now constitutes the most prominent confusion, still sending (anger $\rightleftharpoons$ happiness) off to be second. The increase of almost all numbers is due to the fact that the proportion of the confusions with disgust is now missing in the statistics by what the proportion of the other errors increases. Moreover, the neutral models (female and male) now exhibit a larger variance as they are also trained with boredom utterances and the previous confusions between boredom and other emotions are now shoveled on to neutral $\rightleftharpoons$ other emotions.

All in all, the training of different models for female and male speakers leads to a significant improvement of the plain emotion recognizer performance. For five emotions, such a recognizer achieves a recognition rate of up to 72% on the Berlin Database of Emotional Speech (Pittermann and Pittermann 2006d). For these comparisons, we assume that the emotion accuracies are commonly determined by hard measures (match vs. no match) on the utterance level which we also apply with our predominant emotion criterion. Including soft distance measures on the values of $E(U)$ (see also the evaluation of the combined speech–emotion recognizer) for our best performing emotion recognizer, this "new" emotion accuracy is above 85%.

Calculating the average emotion classification recall ratio of all 20 annotators for all utterances of the Berlin Database Emotional Speech, we determine a human emotion recognition accuracy of 85.2% constituting an desirable upper bound (Meng et al. 2007). This number seems realistic compared to the observations by Schröder (2000) who observes a human emotion recognition rate of 81% for 10 emotional states.

In comparison to similar work by, e.g., Nwe et al. (2001) who use HMMs with MFCC features achieving an accuracy of 72%, the performance of our straightforward approach seems satisfactory. In contrast, Luengo et al. (2005) report an accuracy of 92% for the HMM-based classification of seven emotions, however, in a single-speaker scenario. Vogt and André (2006) report recognition rates of up to 86% on acted speech also taken from the Berlin corpus which actually constitutes a "better" performance than the reference labels by the annotators. Dating further back, a system similar to ours also achieves an emotion accuracy of 72% (Polzin and Waibel, 1998). Nevertheless, depending on the used speech data, performances of 60% or lower are also reported for similar systems (e.g., Kwon et al. 2003; Iwai et al. 2004; Ai et al. 2006). The accuracies mentioned in this paragraph may also serve as a basis for comparison with the emotion accuracies of our speech–emotion recognizers.

6.3.2 Speech–Emotion Recognition

The combination of speech and emotion recognition does not only reduce the recognition complexity by using similar or even the same features from the speech signals, but also add a new dimension or degree of freedom to the evaluation compared to plain speech or emotion recognizers. Due to the fact that emotions are now tied to words (actually to phonemes, but we evaluate only the recognizer output and not intermediate results like, e.g., emophoneme sequences) we can now read off and assign the exact time dependencies when a certain emotional state persists and when it should be recognized.

At a first glance, the use of standard speech recognizer measures on the basis of the edit distance makes sense again. In fact, these criteria provide a solid basis for performance comparison while constituting a strict measure. However, on closer examination, it turns out that the thereby determined recognition rates are not very meaningful about the nature of the errors and also about the actual impact of the errors in further processing steps of the overall dialogue system. E.g., if a (neutral) reference utterance like

> *"I-neutral want-neutral to-neutral travel-neutral to-neutral London-neutral tomorrow-neutral."*

is recognized as *"I-disgust want-boredom to-boredom travel-neutral to-fear London-fear tomorrow-boredom."* we obtain the same recognition rate of $1/7 \approx 14.3\%$ for this sentence as when recognizing *"We-neutral would-neutral like-neutral to-neutral go-neutral to-neutral Paris-neutral today-neutral.*" or "*We-disgust would-boredom like-boredom to-neutral go-fear to-neutral Paris-fear today-boredom*". Obtaining the same recognition rates for all three example cases, it becomes difficult to judge whether and to which extent errors occur on the word and/or emotion level.

Thus, in addition to these admittedly convenient and easy to determine measures, we also consider word and emotion recognition rates separately – the word recognition rates are determined on the basis of the edit distance and the emotion recognition rates are determined on the basis of the predominant emotion as described above.

Another important factor in the evaluation of speech recognizers is the language model used in the recognition. Accordingly, here, size and flexibility of the emotional language model have a strong influence on the output of the speech–emotion recognizer.

In our experiments we include three different types of emotional language models:

- A *word–emotion loop* the use of which leads to the lower bound on the recognition performance
- a *sentence* grammar strongly constrains the recognizer flexibility leading to rather unfair results which could be referred to as an upper bound on the performance
- a *bi-gram* emotional language model trained on the word–emotions occurring in the Berlin Database of Emotional Speech

As indirectly suggested in the HTK Book (Young et al. 2006) for the evaluation of plain speech recognizers, we evaluate the speech–emotion recognizers at three different stages of the training process of the HMMs:

- Stage-1 (corresponding to hmm7 of the tutorial in the HTK Book): the monophone HMMs have been iteratively re-estimated for five times and a short pause model has been derived from the silence model.
- Stage-2 (corresponding to hmm9): based on stage-1, a realignment of the data using the Viterbi recognizer has been performed and the monophone HMMs have been re-estimated twice more.
- Stage-3 (corresponding to hmm15): based on stage-2, tied-state triphones have been derived and the respective HMMs have been re-estimated four more times.

Concerning the feature sets we employ most of the sets described in Section 4.1.4. These sets range from 39 plain MFCCs and 24 plain prosodic and acoustic coefficients to various combinations of these features. Here, the prosodic and acoustic features are again extracted with the aid of Praat (Boersma 2001, 2002), the MFCCs are calculated by HTK's HCopy tool (Young 1994; Young et al. 2006). For the sake of consistency we use the same notation (MFCCDA-39, PAC-24, etc.) in our experiments as well.

Table 6.4 Evaluation of the speech–emotion recognizer using prosodic and acoustic features (PAC-24) by means of word–emotion (**a**), word (**b**) and emotion (**c**) recognition rates

(**a**) Word–emotion recognition rates in %

	W.-e. loop	Sentence	Bi-gram
Stage-1	17.8	23.0	18.7
Stage-2	17.1	22.3	19.5

(**b**) Word recognition rates in %

	W.-e. loop	Sentence	Bi-gram
Stage-1	19.6	34.5	21.9
Stage-2	22.3	34.7	22.6

(**c**) Emotion recognition rates in %

	W.-e. loop	Sentence	Bi-gram
Stage-1	46.2	46.2	45.3
Stage-2	45.3	48.1	47.2

An overview on the recognizer performance when using only the 24 prosodic and acoustic features is shown in Table 6.4. For this feature type, there are no results available for the third training stage (tied-state triphones) as the plain prosodic features cause numerical instabilities in the speech recognizer training. The word–emotion recognition rates (a) range from 17% (word–emotion loop) to 23% (sentence grammar), the plain word recognition rates (b) range from a little more than 19% (word–emotion loop) to 34.7% (sentence grammar) and the emotion recognition rates (c) are relatively dense between 45% and 48.1% independent of the used emotional language model. Concerning the word recognition performance, it should be noted that the significant difference between the word–emotion loop and the sentence grammar is due to the fact that with the sentence grammar the recognizer is able to include the (known) length of the utterances in its calculations. I.e., if all sentences were known and of different lengths, the recognizer still would have good chances, guessing the correct sentence even if the features were useless. Apart from the fact, that in all respects the recognition performance is unacceptably low, and that, as expected, the word–emotion recognition rates are lower than plain word or emotion recognition rates, it is remarkable that the emotion recognition rates are noticeably higher than the word recognition rates. This is due to the fact that prosodic parameters alone do not suffice for speech recognition but can be employed for emotion recognition as done in the previous section.

The recognition rates for the same scenarios as described above, now with plain MFCC features including delta and acceleration coefficients (MFCCDA-39), are listed in Table 6.5. It is not exaggerated to state that MFCCs outperform the plain prosodic and acoustic features in any discipline – the recognition rates for word–emotions, words and emotions are significantly better. It is noticeable that for almost all measures, the recognition rates decrease after the third stage. This is partly due to the fact that the use of triphones leads to a higher model complexity which in turn requires more training data (and then also more iterations). Being originally designed to optimize the performance of speech recognizers, the MFCCs lead to

Table 6.5 Evaluation of the speech–emotion recognizer using Mel-frequency cepstral coefficients and their first and second order regression coefficients (MFCCDA-39) by means of word–emotion (**a**), word (**b**) and emotion (**c**) recognition rates

(**a**) Word–emotion recognition rates in %

	W.-e. loop	Sentence	Bi-gram
Stage-1	34.5	74.2	54.5
Stage-2	28.2	75.8	58.1
Stage-3	26.9	70.7	56.1

(**b**) Word recognition rates in %

	W.-e. loop	Sentence	Bi-gram
Stage-1	53.3	>98	80.7
Stage-2	50.5	>98	87.3
Stage-3	49.6	>98	87.7

(**c**) Emotion recognition rates in %

	W.-e. loop	Sentence	Bi-gram
Stage-1	65.1	67.0	61.3
Stage-2	64.2	67.9	63.2
Stage-3	62.3	65.1	62.3

unsurprisingly good word recognition rates. When using the sentences grammar, constituting the most restrictive language model, we achieve the highest word recognition rates – up to 100% in our simulation with a limited number of utterances. Taking into account possible statistical deviations, we can assume that recognition rates of at least 98% are achievable with these settings. Such rather unusual performance outliers are indicated by ">98%". On the other hand, the recognition rates of approximately 87% for the bi-gram language model appear more realistic with respect to the flexibility of a bi-gram model. Kim (2006) discusses the performance of commercially available speech recognition software, for which most developers claim accuracies of up to 99% in their product descriptions. According to an assessment conducted by Zhou et al. (2005), the baseline word accuracy of a system like Nuance's IBM ViaVoice (see http://www.nuance.com/viavoice/) is approximately 88% for their evaluation scenario.

The highest emotion recognition rate of 67.9% (and also the highest word–emotion recognition rate of 75.8%) is obtained with the sentences grammar. Compared to the plain emotion recognizer (Tables 6.1 and 6.2) for five emotions, we now achieve similar recognition rates for seven emotions without even distinguishing female and male speakers.

The idea of using Gaussian Mixture Models is applied for mixed MFCCs and prosodic and acoustic features and we consider two further training stages:

- Stage-4: based on stage-3, the HMMs are extended to GMMs based on two Gaussian mixtures. These models are re-estimated two more times.
- Stage-5: based on stage-4, the two-mixture GMMs are extended to three-mixture GMMs and re-estimated two more times.

With respect to the fact that the listing of all recognition rates for all combinations of mixed-features would be rather disproportionate, we illustrate an idea of the results

Table 6.6 Evaluation of the speech–emotion recognizer using Mel-frequency cepstral coefficients and five additional prosodic and acoustic coefficients (MFCPAC-44) by means of word–emotion (**a**), word (**b**) and emotion (**c**) recognition rates

(**a**) Word–emotion recognition rates in %

	W.-e. loop	Sentence	Bi-gram
Stage-1	31.9	72.7	55.4
Stage-2	30.2	74.1	56.5
Stage-3	28.2	69.8	52.6
Stage-4	30.8	74.0	56.5
Stage-5	34.4	72.7	58.4

(**b**) Word recognition rates in %

	W.-e. loop	Sentence	Bi-gram
Stage-1	52.3	>98	83.2
Stage-2	52.3	>98	85.1
Stage-3	50.9	>98	83.3
Stage-4	54.2	>98	84.6
Stage-5	56.5	>98	84.7

(**c**) Emotion recognition rates in %

	W.-e. loop	Sentence	Bi-gram
Stage-1	64.2	67.0	67.0
Stage-2	63.2	68.9	66.0
Stage-3	61.3	65.1	61.3
Stage-4	63.2	69.8	64.2
Stage-5	68.9	67.9	67.0

at the example of the MFCPAC-44 feature set consisting of the 39 MFCCDA coefficients plus pitch, intensity and three formants. These results are shown in Table 6.6. Comparing these recognition rates to the ones in Table 6.5, it is noticeable that the numbers do not differ significantly but that there are cases where the combined features perform better and there are cases where the MFCCs perform better. First of all, it can be seen that the maximum emotion recognition rate of 69.8% is better than with plain MFCCs although the corresponding word–emotion recognition rate is lower than with MFCCs. Even with the word–emotion loop or the bi-gram language model, higher emotion recognition rates are achieved compared to MFCCs. Although the word recognition rates using the sentences grammar are still at 100%, most of the word–emotion and word recognition rates are lower than those with MFCCs. Accordingly, the highest word–emotion recognition rate of 75.8% is achieved with plain MFCCs (see Table 6.5). The highest emotion recognition rate of 71.7% is achieved with the MFCPAC-48 feature set using the word–emotion loop after stage-5 of the training.

The average error frequencies between emotion pairs are illustrated by the confusion matrix in Table 6.7. In this table the numbers are calculated from the recognition results for MFCPAC-40, MFCPAC-41, MFCPAC-44 (see Table 6.6), MFCPAC-46, MFCPAC-48, MFCPAC-52 and MFCPAC-56. For these seven recognizers the average emotion error rate is 36.2%. The left numbers are the absolute error frequencies in % and the right numbers in parentheses are the relative

Table 6.7 Emotion confusion matrix (average values in %) for speech–emotion recognizers with different feature sets combining MFCCs and prosodic and acoustic features (Meng et al. 2007)

	Sadness	Neutral	Happiness	Fear	Disgust	Boredom
Anger	0.00 (0.00)	0.13 (0.35)	6.66 (18.4)	1.60 (4.43)	1.12 (3.08)	0.32 (0.89)
Boredom	3.22 (8.88)	5.03 (13.9)	0.17 (0.46)	0.83 (2.28)	0.50 (1.39)	
Disgust	0.34 (0.93)	1.37 (3.78)	1.60 (4.41)	0.87 (2.41)		
Fear	0.00 (0.00)	2.56 (7.06)	6.04 (16.7)			
Happiness	0.71 (1.96)	1.06 (2.93)				
Neutral	2.08 (5.76)					

error frequencies (in relation to the 36.2%). The three most frequent errors occur between anger and happiness, fear and happiness as well as boredom and neutral. The total error rate of these three emotion pairs amounts to 49% of the overall error rate, i.e., almost half of the errors are due to these error sources whereas errors between anger and sadness or fear and sadness are not very likely to occur. It should be noted that these numbers are all based on the predominant emotion criterion as described in the beginning of this section. Applying distance measures in the valence-arousal space or along the gradient of $E(U)$, we can obtain more application-specific performance measures.

Assuming the emotions are distributed in the valence-arousal space according to anger $\rightarrow (-0.9, 0.9)$, boredom $\rightarrow (-0.1, -0.3)$, disgust $\rightarrow (-0.2, -0.8)$, fear $\rightarrow (-0.1, 0.4)$, happiness $\rightarrow (0.9, 0.8)$, neutral $\rightarrow (0.0, 0.0)$ and sadness $\rightarrow (-0.8, -0.2)$ and calculating $E(U)$ as described by Eq. 6.11, we obtain $E_{\text{anger}}(U) = 1.87$, $E_{\text{bored.}}(U) = 1.06$, $E_{\text{disg.}}(U) = 1.08$, $E_{\text{fear}}(U) = 1.08$, $E_{\text{happ.}}(U) = 0.16$, $E_{\text{neut.}}(U) = 1$ and $E_{\text{sadn.}}(U) = 1.59$. If the emotions were more or less equally distributed along the scale of $E(U)$, one could argue to use the maximum distance (here $|E_{\text{anger}}(U) - E_{\text{happ.}}(U)| = 1.71$) to normalize the distances. In this case, with most of the values above 1, we use the average distance of all 21 emotion pairs which can be determined as 0.602. By that, we obtain, e.g., $\Delta_{anger-happ.} = 1.71/0.602 = 2.84$ or $\Delta_{bored.-fear} = |1.06 - 1.08|/0.602 = 0.03$. The absolute error frequencies in Table 6.7 can then be directly multiplied with the normalized distances and we obtain the "new" error frequencies as shown in Table 6.8. Here, again, the numbers on the left denote the absolute error frequencies and the numbers in parentheses on the right denote the relative error frequencies which are based on the new overall error rate of 44.4% (the sum of all numbers on the left side). It should be noticed that this overall error rate is directly dependent on the reference distance which we here define as the average distance. I.e., increasing this reference distance by defining it differently would make it possible to tune the new error rate to a more attractive value than 44.4%. However, as these modifications have no influence on the relative error frequencies, we leave it at that rather unattractive value. The most frequent errors again occur between happiness and anger as well as between happiness and fear. These two errors already constitute more than 60% of all errors. Due to the new distance measures, their proportion in the overall has increased significantly compared to the old measures. On the third place, we find

Table 6.8 Weighted emotion confusion matrix (average values in %) based on the values in Table 6.7 with respect to the normalized distance

	Sadness	Neutral	Happiness	Fear	Disgust	Boredom
Anger	0.00 (0.00)	0.19 (0.43)	18.9 (42.6)	2.10 (4.73)	1.47 (3.31)	0.43 (0.97)
Boredom	2.83 (6.38)	0.50 (1.13)	0.25 (0.56)	0.03 (0.07)	0.02 (0.05)	
Disgust	0.29 (0.65)	0.18 (0.41)	2.44 (5.50)	0.00 (0.00)		
Fear	0.00 (0.00)	0.34 (0.77)	9.22 (20.8)			
Happiness	1.68 (3.78)	1.48 (3.33)				
Neutral	2.04 (4.60)					

boredom vs. sadness although its proportion has even decreased. In exchange, the confusion of boredom and neutral has lost its importance dramatically.

In the remainder of this section we will go into detail about our approaches to achieve a better recognition performance. Before finally analyzing the performance of the two-step speech–emotion recognizer setup, we will now have a quick glance at the impact of reducing the emotion set as already done for plain emotion recognition. Applying the distance measures among $E(U)$ as described above is of course not motivating to change anything regarding the emotions as there will always be confusions between anger, fear and happiness. However, applying the usual hard measures, one might expect to see better recognition rates.

Here, we distinguish the omission of disgust, the merging of neutral and boredom and a combination of both leading to a reduced emotion set of four emotions plus neutral. The omission of disgust does not lead to any significant improvement of the recognition rates: using plain MFCCs as features, the maximum word–emotion recognition rates are 35.7% (word–emotion loop), 74.8% (sentences) and 57.4% (bi-gram). Compared to the numbers in Table 6.5, these numbers are to some extent even lower. The best word–emotion recognition rates among all feature sets are achieved with the MFCPAC-40 features – here we obtain 35.8%, 75.8% and 60.1% which is also just slightly better than the equivalent performance of the regular MFCC-based recognizer.

Merging boredom and neutral, i.e., including boredom in the neutral models, we achieve slightly better results when using MFCC features. Here, the maximum word–emotion recognition rates are 35.8% (word–emotion loop), 78.2% (sentences) and 61.9% (bi-gram). Among the other feature sets, the best performance is obtained with the MFCPAC-44 and MFCPAC-40 features, namely 35.8%, 80.0% and 61.5%. Comparing these numbers with the ones in Tables 6.5 and 6.6, it is noticeable that the most significant improvement is achieved with the sentences grammar. I.e., a-priori knowledge about the text helps to improve emotion recognition.

Similar results are achieved with the combination of both approaches. For MFCCs, we obtain 37.3%, 78.2% and 62.1%, among the other feature sets, we obtain, e.g., 36.3%, 79.3% and 61.9% when using MFCPAC-44 features. An overview on the MFCC results is given in Table 6.9. Whereas the word recognition rates do not improve significantly (as expected), the emotion recognition rates increase when merging boredom and neutral. Although the recognition rates in Table 6.9

Table 6.9 Evaluation of the speech–emotion recognizer using Mel-frequency cepstral coefficients and their first and second order regression coefficients (MFCCDA-39) by means of word–emotion (**a**), word (**b**) and emotion (**c**) recognition rates with a reduced emotion set

(**a**) Word–emotion recognition rates in %

	W.-e. loop	Sentence	Bi-gram
Stage-1	37.3	77.6	58.5
Stage-2	35.0	78.2	62.2
Stage-3	30.1	74.6	60.1

(**b**) Word recognition rates in %

	W.-e. loop	Sentence	Bi-gram
Stage-1	53.6	>98	82.1
Stage-2	53.3	>98	88.4
Stage-3	48.9	>98	87.1

(**c**) Emotion recognition rates in %

	W.-e. loop	Sentence	Bi-gram
Stage-1	67.4	72.6	66.3
Stage-2	67.4	72.6	65.3
Stage-3	70.5	69.5	69.5

are determined for both omitting disgust and merging boredom and neutral, we can achieve similar numbers when only merging boredom and neutral. Using other features than MFCCs, we achieve even higher emotion recognition rates of up to 75.8% (MFCPAC-41 and MFCPAC-44) still without distinguishing female and male speakers.

The what we refer to as two-step approach, described in Section 5.1.2, is mainly motivated by the consideration how much the emotion recognizer part of the speech–emotion recognizer can benefit when information about the speech recognizer output is already present. Simulation results for the two-step recognition approach with all seven emotions are listed in Table 6.10. Here, we distinguish two scenarios/assumptions:

- The speech recognizer output is not 100% reliable, i.e., not all words are necessarily correctly recognized. Here, we use the output of the MFCC recognizer with a bi-gram recognizer which features a word accuracy of 94.7%. The results are shown in Table 6.10(a).
- The speech–emotion recognizer in the second step can avail itself of the full knowledge about the recognized text (fictive word accuracy of 100%). The respective results are shown in Table 6.10(b).

Looking at the recognition rates, it can be seen that the word–emotion recognition rates increase when the reliability of the speech recognizer increases. All in all, however, the performances in both cases are not exhilarating, the values for 100% speech recognizer accuracy more or less concur with the recognition rates when using the sentences grammar as shown for selected features in Tables 6.5 and 6.6.

Keeping these recognition rates in mind, we now consider the reduction of the emotion set. With respect to our previous considerations on the different approaches

Table 6.10 Evaluation of two-step speech–emotion recognizers (seven emotions) by word–emotion recognition rates (in %)

(**a**) Word accuracy 94.7%

	Stage-1	Stage-2	Stage-3	Stage-4	Stage-5
MFC	71.5	71.0	67.0		
PAC-24	49.2	46.5			
MFCPAC-40	67.5	67.5	64.2	69.6	69.7
MFCPAC-41	69.3	70.4	67.1	69.4	68.0
MFCPAC-44	69.0	70.4	69.4	70.3	70.4
MFCPAC-46	69.3	69.7	69.7	71.6	67.5
MFCPAC-56	69.5	69.5	69.0	71.8	71.2

(**b**) Word accuracy 100%

	Stage-1	Stage-2	Stage-3	Stage-4	Stage-5
MFC	74.2	75.8	70.7		
PAC-24	50.6	49.4			
MFCPAC-40	72.4	70.7	67.5	70.8	70.6
MFCPAC-41	72.1	74.1	69.1	71.2	70.7
MFCPAC-44	72.7	74.1	69.8	74.0	72.7
MFCPAC-46	71.9	70.8	72.2	74.3	68.0
MFCPAC-56	72.6	72.2	72.3	73.2	73.5

to reducing the emotion set for one-step speech–emotion recognition, we only merge boredom and neutral as the omission of disgust does not lead to significant improvements. In contrary, experiments show that the combination of both approaches deteriorate the recognition rates compared to only merging boredom and neutral. The respective results are listed in Table 6.11. The word–emotion recognition rates are shown in Table 6.11(a). Comparing these numbers to the respective ones in Table 6.10 a significant increase is noticeable. I.e., by merging boredom and neutral word–emotion recognition rates of up to 80% (with MFCPAC-44 features) are achieved. Accordingly, also the plain emotion recognition rates increase. Now, the maximum emotion recognition rate, also achieved using MFCPAC-44 features, is 79.3% (six emotions, female and male speakers are not distinguished) given the speech recognizer output with an accuracy of 94.7%.

Summarizing, despite its noticeably higher complexity as opposed to plain speech or emotion recognizers, the combined speech–emotion recognition approach features a comparably good and also a large potential for optimizations. For seven emotions, the "standard" speech–emotion recognizer achieves emotion recognition rates of 67.9% when using MFCC features and up to 71.7% with mixed features such as the MFCPAC-48 feature set. A reduction of the emotion set increases the emotion recognition rates, especially when boredom and neutral are merged. For five emotions, the speech–emotion recognizer achieves emotion recognition rates of 72.6% with MFCC features and up to 75.8%, e.g., with the MFCPAC-41 or -44 feature sets. At a first glance, for all seven emotions the two-step recognizer does not lead to (expected) significant improvements. With (partial) knowledge about the textual content of the utterances, the word–emotion recognition rates resemble those

Table 6.11 Evaluation of two-step speech–emotion recognizers (six emotions – boredom and neutral are merged) assuming a speech recognizer accuracy of 94.7%

(**a**) Word–emotion recognition rates in %

	Stage-1	Stage-2	Stage-3	Stage-4	Stage-5
MFC	78.2	77.7	75.7		
PAC-24	57.8	52.2			
MFCPAC-40	76.3	76.7	74.7	76.9	77.3
MFCPAC-41	76.8	78.3	76.2	79.3	79.2
MFCPAC-44	77.0	77.9	75.1	79.6	80.0
MFCPAC-46	75.7	74.9	77.8	79.3	79.1
MFCPAC-56	75.8	76.4	76.1	76.5	76.3

(**b**) Emotion recognition rates in %

	Stage-1	Stage-2	Stage-3	Stage-4	Stage-5
MFC	74.5	73.6	70.8		
PAC-24	49.1	42.5			
MFCPAC-40	70.8	71.7	70.8	74.5	71.7
MFCPAC-41	71.7	73.6	70.8	73.6	72.6
MFCPAC-44	71.7	73.6	71.7	79.3	79.3
MFCPAC-46	70.8	72.6	75.5	72.6	73.6
MFCPAC-56	71.7	72.6	71.7	77.4	76.4

of the regular speech–emotion recognizers using the (rather unfair) sentences grammar. However, when reducing the emotion set by merging boredom and neutral, the two-step approach unfolds its potential. For six emotions emotion recognition rates of 74.5% with MFCC features and up to 79.3% with MFCPAC-44 features are achieved, outperforming the plain emotion recognizer as described in Section 6.3.1.

6.3.3 Combining Multiple Speech–Emotion Recognizers

When we compare the output of different speech recognizers, it can be noticed that the errors occur at different positions even though the overall word accuracies are similar. These differences are exploited by the ROVER approach which is targeted on improving the recognition performance by combining the output of different speech recognizers as described by Fiscus (1997). The extension of this approach to speech–emotion recognition is described in Section 5.2.

In our experiments, we distinguish two different approaches to the word–emotion transition network alignment: timestamps alignment (with respect to the beginning and end times of word–emotions) and Levenshtein alignment (considering the minimum edit distance between two word–emotion sequences). Furthermore, we consider five values for the weighting factor α in the scoring module: 0.0 (score is exclusively dependent on the frequency of occurrence of a word, emotion or word–emotion), 0.25, 0.5, 0.75 and 1.0 (score is only dependent on the recognizer confidence measures).

We choose the potential input systems among our standard one-step speech–emotion recognizers distinguishing seven emotions without any optimizations. These are, in descending order of their overall word–emotion recognition rates:

- WETN-1: speech–emotion recognizer based on MFCPAC-40 features, using a bi-gram language model.
  - 59.5% word–emotion accuracy
  - 89.1% word accuracy
  - 68.9% emotion recognition rate
- WETN-2: speech–emotion recognizer based on MFCPAC-48 features, using a bi-gram language model.
  - 59.0% word–emotion accuracy
  - 82.5% word accuracy
  - 68.9% emotion recognition rate
- WETN-3: speech–emotion recognizer based on MFCPAC-44 features, using a bi-gram language model.
  - 58.4% word–emotion accuracy
  - 84.7% word accuracy
  - 67.0% emotion recognition rate
- WETN-4: speech–emotion recognizer based on MFC-39 features, using a bi-gram language model.
  - 58.1% word–emotion accuracy
  - 87.7% word accuracy
  - 63.2% emotion recognition rate
- WETN-5: speech–emotion recognizer based on MFCPAC-41 features, using a bi-gram language model.
  - 57.1% word–emotion accuracy
  - 87.0% word accuracy
  - 66.0% emotion recognition rate

On the basis of these input WETNs we consider four ROVER systems: ROVER-2 combining WETN-1 and WETN-2, ROVER-3 combining WETN-1, WETN-2 and WETN-3, ROVER-4 combining WETN-1, WETN-2, WETN-3 and WETN-4 and ROVER-5 combining all five of the above WETNs.

The word–emotion accuracies when applying the standard ROVER approach to these WETNs are listed in Table 6.12. Looking at these numbers, it can be observed that the difference between both alignment methods is not very noteworthy. This can be reproduced by means of the output of speech or speech–emotion recognizers where the words or word–emotions typically co-occur to some extent by what

Table 6.12 Word–emotion recognition rates (in %) with a standard ROVER approach using timestamps (*left numbers*) and Levenshtein (*right numbers*) alignment applied to up to five WETNs

α	ROVER-2	ROVER-3	ROVER-4	ROVER-5
0.00	58.8/58.7	59.3/59.1	59.6/59.7	60.1/59.8
0.25	58.8/58.7	59.7/59.1	59.7/59.9	59.8/59.9
0.50	58.8/58.7	59.9/59.2	60.0/60.2	59.9/60.2
0.75	58.8/58.7	59.8/59.4	60.1/60.2	59.9/60.1
1.00	59.5/59.5	59.7/59.2	60.5/59.9	59.9/60.5

both alignment methods provide similar results. Regarding the number of included recognizers, it can be seen that the ROVER-2 system, although combining the two best recognizers, leads to worse results than the single recognizers. The combination of the three best recognizers already leads to some slight improvement over the best single recognizer and the ROVER-4 and ROVER-5 system lead to (although still not flabbergasting) improvements regardless of the value of α. The maximum word–emotion recognition rate of 60.5% are achieved with four or five recognizers which is 1% better than the best single word–emotion recognition rate of 59.5%.

In the modified approach to using the ROVER idea for word–emotions, words and emotions are separated into word transition networks and emotion transition networks which are evaluated individually as described in Section 5.2.2. The recognition results for this approach are presented in Table 6.13. Comparing the word–emotion recognition rates in Table 6.13(a) with the numbers in Table 6.12, a strong improvement (up to 72.4% from 60.5% in the standard ROVER and 59.5% of the best single recognizer) can be noticed. These rates are determined after words and emotions pass through separated ROVER processes the output WTNs and ETNs of which are recombined to the output WETN. Again, the best results are obtained with four or five recognizers, although the combination of recognizers also leads to some remarkable improvements in the word–emotion recognition rates. A breakdown of the reasons leading to the improvement of this approach can be found when considering the word recognition rates (b) and emotion recognition rates (c) separately. The word recognition rate ranges from 87.8% to 89.8% constituting rather a decrease than an increase from the best single recognizer at 89.1%. I.e., here, on the word level, the ROVER idea does not contribute to any significant improvements justifying the enormous effort of multiple recognizers.

As described in Section 5.2.2, no alignment is applied to the emotion transition networks so that the voting module directly operates on all emotions, calculating the scores not for each discrete time slot but for all emotions in all ETNs together at once. Integrating this approach into the ROVER system for speech–emotion recognition, we achieve emotion recognition rates of up to 76.4% which is significantly higher than the emotion recognition of 68.9% for the best single speech–emotion recognizer. Further examining these numbers, it is remarkable that for any $\alpha \geq 0.5$ and for any number of speech–emotion recognizers in the system the emotion recognition rates are above those of the best single recognizer.

Table 6.13 Recognition rates with the modified speech–emotion ROVER approach using timestamps (*left numbers*) and Levenshtein (*right numbers*) alignment applied to up to five WETNs

(**a**) Word–emotion recognition rates in %

α	ROVER-2	ROVER-3	ROVER-4	ROVER-5
0.00	64.3/64.6	62.1/62.3	63.8/63.9	65.0/65.0
0.25	65.8/66.1	65.4/65.1	68.1/68.5	69.7/69.6
0.50	67.1/67.3	69.7/69.4	71.5/71.6	72.0/71.9
0.75	68.4/68.6	68.6/68.4	71.8/71.7	72.4/72.3
1.00	69.8/69.8	68.1/68.0	69.8/69.7	69.8/69.8

(**b**) Word recognition rates in %

α	ROVER-2	ROVER-3	ROVER-4	ROVER-5
0.00	88.4/88.7	87.8/87.8	89.2/89.6	89.1/89.2
0.25	88.4/88.7	88.3/87.8	89.3/89.7	88.7/88.8
0.50	88.4/88.7	88.3/88.0	89.4/89.7	88.9/89.1
0.75	88.4/88.7	88.3/87.9	89.5/89.7	89.0/89.2
1.00	89.1/89.1	88.3/88.3	89.8/89.7	89.3/89.2

(**c**) Emotion recognition rates in %

α	ROVER-2	ROVER-3	ROVER-4	ROVER-5
0.00	67.0	64.2	65.1	66.0
0.25	69.8	68.9	71.7	72.6
0.50	71.7	74.5	75.5	76.4
0.75	71.7	71.7	74.5	75.5
1.00	71.7	70.8	70.8	71.7

Hypothetically, given a large number of input systems tending to infinity, the output recognition rates (at least their upper bounds) of a ROVER system are supposed to approach 100%. Apart from the problem that this is quite difficult to realize, we also face the problem that the errors are not as ideally distributed as the whole idea presumes. In addition to the system including the five best recognizers as described above we also examine a system including the six best recognizers where, needless to say, the sixth recognizer's performance is lower than the other five recognizers' performance. Here, a turnaround is recognizable: the recognition rates (word–emotion as well as words and emotions separately) decrease visibly, in some cases even below the recognition rates of the individual recognizers. The problem with such an approach is that, choosing the n best recognizers, the error rate of each additional recognizer increases. By that, we have an increasing number of recognizers (which should increase the robustness of the combined output) which unfortunately leads to the inclusion of more errors (which has a negative influence on the robustness of the combined output). In our case, the trade-off is achieved with five recognizers. Further experiments with reduced emotion sets show that the combination of multiple recognizers may also lead to improvements of the overall recognizer performance in these cases. E.g., considering five emotions, the recognition rates is slightly increased to up to 77.9% (from 75.8% and less).

With respect to the rather unusual output of our speech–emotion recognizers which can contain multiple emotions (one per word), the results of the ROVER idea motivate also to apply the emotional scoring as described in Eq. 5.15 on single WETNs, i.e., the output of a single speech–emotion recognizer in order to determine the predominant emotion of an utterance. In practice, this means, for an utterance we calculate the score $S'(e)$ of each emotion e as

$$S'(e) = \alpha \cdot N'(e) + (1 - \alpha) \cdot C'(e), \tag{6.12}$$

where $N'(e)$ and $C'(e)$ are the normalized overall durations and confidence measures of e in the utterance.

Altogether, the idea to combine the information provided by multiple recognizers carries a large potential to achieve lower error rates. The opportunities and risks of using such a complex approach can be estimated in advance with the aid of upper and lower bounds on the recognition rates on the basis of existing data as described in Section 5.2. In our simulations, even the standard ROVER approach, originally designed for plain speech recognition, leads to, although not significant, improvements of the word–emotion recognition rate of up to 60.5% compared to the rates of the single recognizers which are at 59.5% or lower. Applying a modified ROVER approach processing words and emotions separately, significant improvements can be achieved, especially among the emotion recognition rate. The word–emotion recognition rates reach values up to 72.4% still compared to 59.5% and lower of the single recognizers, the emotion recognition rates increase from 68.9% to 76.4%.

6.3.4 Emotion Recognition by Linguistic Analysis

For the evaluation of our speech signal-based emotion recognizers we follow, with slight variations, the standard evaluation procedure of speech recognizers: a set of utterances is labeled by human experts and then processed by the recognizer under test. The respective recognition rates are determined by comparing reference labels and recognizer output. By contrast, to estimate the capabilities of our approach to emotion recognition by linguistic analysis, we put the cart before the horse. On the basis of the employed affective grammar in combination with the application-specific grammar, we generate a large number of possible sentences which are in accordance to these grammars. An arbitrary number of these sentences (e.g., 100) are randomly selected from this set and human experts (test persons) are then asked to assess the emotional content of these sentences in a questionnaire.

This assessment can be done in different ways: either the test persons assign each word of the sentence an emotional state, e.g., *"Great, I want to go to London."* could be labeled as "happiness, neutral neutral neutral neutral neutral neutral", or the sentence is assigned one emotional state, either by name ("anger") or valence representations ("++", "+", "o", ...). Then, the performance, i.e., the credibility of

`Great, please give me a ticket to Seattle.`	*happiness*	+
`Shut up, idiot, from JFK.`	*anger*	−
`That's frustrating.`	*sadness*	−
`From L A X to Miami.`	*neutral*	*o*
`That's unacceptable.`	*anger*	−
`Such a pleasure, to Dallas you stupid thing.`	*anger*	+−
`Oh, how humiliating from Denver, great.`	*happiness*	−+
`...`		

Fig. 6.6 Excerpt of a questionnaire to assess the performance of emotion recognition by texts and linguistic analysis

the linguistic analysis is also determined as a recognition rate by comparing the recognizer output (which we certainly know) and the labels of the human "experts" (which may naturally differ).

An excerpt of one of the questionnaires used in our survey is shown in Fig. 6.6. Here, the test persons are asked to choose among seven emotions (anger, boredom, disgust, fear, happiness, sadness and neutral) and among three valences (positive "+", neutral "o" and negative "−") for each utterance. In addition to the affective grammar, an exemplary grammar for an air travel information system on flights within the United States is considered here to define the application scenario. The sentences are generated on a random basis with the aid of HTK (HSGen, see Young et al., 2006). By that, depending on the flexibility of the grammar, not all of these sentences actually make sense, are grammatically correct or contain any utilizable information. Accordingly, also the emotional keywords are arbitrarily blended leading to confusion and disagreement among the test persons. Whereas, e.g., *"Great, please give me a ticket to Seattle."* (punctuation manually added after the text generation) is unequivocally classifiable as happy, sentences like *"Such a pleasure, to Dallas you stupid thing."* rather lead to varying opinions. At first glance, such a sentence may seem to be very unlikely to occur in dialogues. However, the linguistic analysis in an interactive voice response system (IVR) also needs to be able to cope with error-prone speech recognizer output, where a lot of errors occur in the recognition of emotional keywords (Pittermann et al. 2008b). This requires an adequate error handling, i.e., as commonly done in information-theoretical problems like channel (de)coding, a majority decision is applied: Considering the valences, we have happiness ("+") at the beginning and we have anger ("−") at the end of the sentence adding (or averaging) up to "o", i.e., neutral. Here, each utterance is considered individually and not in the context with previous or following utterances.

On the other hand, such a sentence can not be labeled as a typical neutral sentence. I.e., depending on the test person's gut feeling and creativity, the sentence could be labeled as, e.g., anger, happiness, or happiness-anger. To obtain a better overview, in the questionnaire, we leave it up to the test persons how to assign the labels.

The evaluation is performed on 500 sample sentences. Depending on how strictly we compare the system-generated sentences and the labels in the questionnaires, we achieve different recognition rates:

- Demanding a strict congruence of the labels, i.e., the experts totally agree with the system about each word, 34% are correctly interpreted.
- Demanding a congruence of the predominant emotion in each utterance, the recognition rate increases to 70%.
- Comparing the valence tendencies by sign ("–"/ "o"/"+"), 90% of the sentences are correctly interpreted.

Regarding these numbers and taking into account the problem that most of the words in any language can not be unambiguously assigned a certain emotional state, it can be concluded that the strength of linguistic approaches to emotion recognition is not the classification of distinct emotions but rather the detection of emotional tendencies, e.g., whether the user is in a positive, neutral or negative state. This would be sufficient, e.g., in dialogue systems which need to determine whether the human–computer dialogue is likely to end successfully or whether it is reasonable to put the (mostly angry) user through to a human call center agent (see also Burkhardt et al. 2006a). Alternatively, the linguistic analysis is a suitable supplement to further emotion recognizers contributing to a more robust overall recognition performance assuming an expedient fusion of recognition results as partly described in Section 5.2.2. Furthermore, a combination of speech signal-based and linguistic analysis open up new vistas such as irony detection presuming the both recognizers separately provide reliable recognition results.

6.4 Evaluation of Our Dialogue Manager

Constituting the central component in an SLDS, the dialogue manager is not only expected to manage the application access and provide the appropriate system reaction but also to do its job robustly without being tripped up by (eventually unexpected) user input. If we assess the performance of, e.g., an automatic speech recognizer implemented on the basis of existing (purchased) tools, we typically trust in the fact that these tools serve the purpose as described in their specifications and we limit our evaluation considerations to the concepts and data applied to these tools unless some strange behavior leads to suspiciousness about the tools' functionality. The same applies to dialogue systems, the dialogue manager of which is typically implemented in existing frameworks like, e.g., the CSLU Toolkit. Here, with respect to the fact that most parts of the dialogue manager are implemented independent of the underlying VoiceXML framework, the evaluation of this dialogue manager puts forth at two main levels: the robust functionality of the underlying system and the performance and usability of the concept implementation based on this system.

A rough overview on whether the system actually functions correctly can be obtained by implementing a short dialogue, here in the air travel information domain, and testing this dialogue:

1 System: *Welcome to the flight information system. Where do you want to travel to?*
2 User: *To Colorado.*
3 System: *Which of the following airports in Colorado would you prefer – Aspen (ASE), Colorado Springs (COS), Denver International Airport (DEN), Grand Junction (GJT) or Pueblo (PUB)?*
4 User: *Denver.*
5 System: *Sorry, I didn't understand you. Which of the following airports in Colorado would you prefer – Aspen (ASE), Colorado Springs (COS), Denver International Airport (DEN), Grand Junction (GJT) or Pueblo (PUB)?*
6 User: *D E N.*
7 System: *From where would you like to depart?*
8 User: *Trenton, New Jersey.*
9 System: *There are no flights available from Trenton, New Jersey (TTN) to Denver International Airport, Colorado (DEN). Alternatively, you could go from Newark International Airport, New Jersey (EWR) to Denver International Airport, Colorado (DEN). Thank you for using this flight system.*

Apart from the problem that the system is not able to interpret *"Denver"* correctly (here, the user is supposed the say/enter *"Denver International Airport"*), which is rather a grammar problem than a fault of the underlying system, the system seems to be functional. With respect to the fact that one single test like this dialogue example is not representative at all, a larger number of further tests need to be accomplished. As these are quite time-consuming and tedious to accomplish, we use an automated approach bombarding the system (similar to the approach proposed in Ito et al. (2006)) with a huge number of generated sentences like those in Fig. 6.6 plus other sentences which are not covered by this example system's grammar. Storing the generated input and the respective system reaction, most problems concerning the system's functionality can be addressed. Whereas the functionality of a rule-based dialogue approach can be reconstructed with relatively low effort, a large amount of data is required for the involved statistical processes to see, e.g., if the selection of prompts is actually in accordance with the probabilities established by the dialogue model. Here, in the ideal case, it is necessary to determine reliable probability density functions for all occurring transitions, which would, however, exceed the scopes. Thus, typically, a few of these cases are covered by the in-depth tests.

Assuming the underlying system is working reliably, in the second step, we assess the usability and user acceptance of the adaptive dialogue concept. This is typically accomplished on the basis of user surveys about the implementation of the respective dialogue concept. Here, we put forth the example of the air travel information system as described above and we extend it with the ability to react on the user's emotional states as determined by the linguistic analysis. The experiments are not conducted in a WOZ setting but in a fully automated environment. I.e., the users are not only made believe that they are interacting with a computer, they are actually on their own without any supervision. By that, on the one hand, we are able to achieve realistic dialogues, on the other hand, the user is facing a higher number of understanding problems due to speech recognizer inaccuracies.

The dialogue manager functionality of such an example system is implemented in VoiceXML and then integrated into an end-to-end framework by compilation to ECMAScript as described by (Bühler and Hamerich 2005). This Java-based framework includes a Sphinx-4 speech recognizer (see Lamere et al. 2003) including diversified models for universal use in test environments and a FreeTTS text-to-speech synthesis (Walker et al. 2002). The system accepts both typed and spoken input: For functionality tests, we prefer typed input as this also allows input from large text files by the "<" operator on the console. Here, however, the keyboard is disabled and for the users there is nothing else to it but to interact with the system in natural language. Concerning the output modalities, the system also shows the spoken output as text on the screen. As for standard VoiceXML interfaces, it is also possible to include prerecorded files (jingles, sounds, music or text) in the audio output or to replace the text-to-speech synthesis totally by prerecorded system prompts and replies.

An excerpt of the dialogue description is shown in Fig. 6.7. Each of the field includes its own texts for standard prompts and for extra events like help texts or replies to no input or no match. E.g., a bell sounds if the user's input is not utilizable. The main processing is launched after the user's reply to the prompts. Here, the user's emotional state is determined on the basis of the current input and the previous emotional states as stored in the dialogue history. Accordingly, an acknowledgment to the user's input is realized for the determined emotional state. It should be noted that such a chronological order is technically equivalent to the implementation as described in Section 5.3 where the actual prompts are realized according to the previously determined emotional state. Here, there is one acknowledgment output per emotion for each field. These prompts differ from field to field. Depending on the application it is also possible to pool all associated prompts centrally and access these, e.g., randomly, from the individual fields. If no emotional state is determined, the system assumes that the user is in a neutral state.

In the same manner, also fields for departure state, destination state and destination airport as well as an explicit confirmation at the end are implemented. The test users are given an instruction sheet describing the tasks they are required to accomplish. Such an instruction sheet is shown in Fig. 6.8. Here, the test users are asked to find out about flight connections between American airports. They are also encouraged to express their emotions by certain keywords in order to assess how well the system reacts to these clues.

```
<!--Departure airport-->
<field name="dep_airport">
  <prompt count="1">
    The Airports available in <value expr="dep_state"/> are:
    <value expr="getAirports(dep_state)"/>
    Please choose one of the available airports!
  </prompt>
  <prompt count="2">
    <audio src="bell.wav"/>
    Please tell me the airport's name or its three-letter code.
  </prompt>
  <help>
    Choose an airport from the list, say its name or code.
  </help>
  <noinput  count="1">
    <audio src="bell.wav"/>
    I'm sorry, I didn't hear anything, please repeat.
  </noinput>
  ...

  <filled>
    <if cond="emotion_state==undefined">
      <assign name="emotion_state" expr="'neutral'"/>
    </if>
    <!-- Calculate the cumulative emotional state -->
    <assign name="emo_state"
      expr="calcEmotion(emotionList,emotion_state)"/>
     <prompt cond="emo_state=='angry'">
      Please don't be frustrated!
      I'll offer you the best possible service!
    </prompt>
    <prompt cond="emo_state=='happy'">
      Great, good choice!
    </prompt>
    <prompt cond="emotion_state2=='sad'">
      Hey, come on, cheer up!
    </prompt>
    <prompt cond="emotion_state2=='fear'">
      Don't worry! It'll be great.
    </prompt>
    <prompt cond="emotion_state2=='neutral'">
      Thank you!
    </prompt>
  </filled>
</field>
```

Fig. 6.7 VoiceXML excerpt of the adaptive dialogue description for the dialogue manager evaluation

Having accomplished all tasks, the users are asked to complete a questionnaire parts of which are shown in Fig. 6.9. Typically, users are surveyed about further details about themselves (age, gender, experience with computers, experience with (spoken) human–computer interfaces, etc.) and the questions are also layed out such that further information about the user's psychological background can be extracted.

Find out about the flight connections from and to the following locations:

Task 1
From: Tucson International Airport, Arizona
To: Seattle Tacoma International Airport, Washington

Task 2
From: Phoenix Sky Harbor International Airport, Arizona
To: San Jose, California

Task 3
From: Yuma International Airport, Arizona
To: Los Angeles International Airport, California

...

Please follow the system's instructions. On success, the system will list all available flights. You may say "help" or "exit" anytime to obtain further information or to exit the dialogue.

You are able to express your emotional state using the following keywords:
- happiness: wow, great, excellent
- ...

Fig. 6.8 Instruction sheet containing the tasks to be completed while interacting with the dialogue system under test

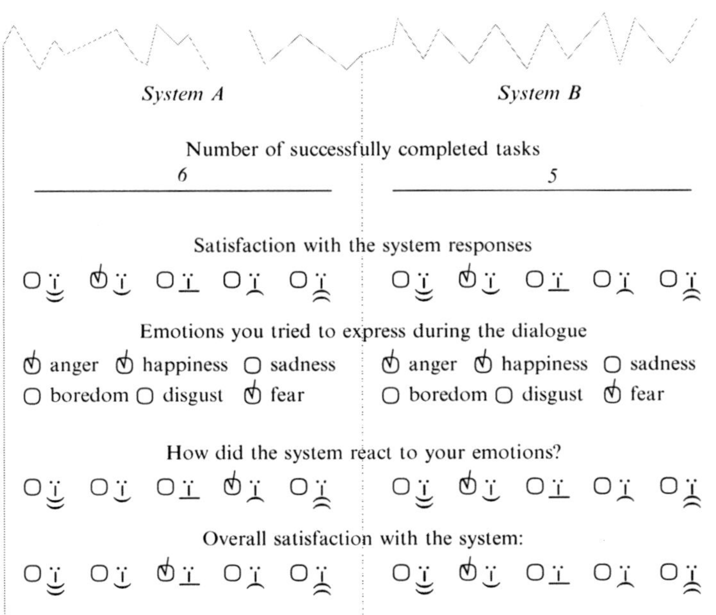

Fig. 6.9 Questionnaire about the users' experience with the dialogue system

Here, knowing that most of the test users are international students around the age of 20, experienced with computers and electronics but not experienced with spoken dialogue systems, at this point, we set aside these details. The questionnaire is subdivided into two parts corresponding to two systems the users are asked to interact with, without being told whether and how these systems differ. If System A and System B are identical, there is more elbowroom for the interpretation of the results – either a user has the same opinion on both systems (which one would actually expect) or a user has different opinions on these systems (which might indicate that this user probably is not as assiduous as other users). Here, both systems actually differ: they are indeed based on the same implementation, but only System B is responsive to the emotional cues as described in Fig. 6.7, whereas System A does not consider the emotional states and, across the board, presumes that the user is in a neutral state. This direct comparison is required to obtain the relative user satisfaction which is more meaningful than conducting two user studies (one for each system) with different test users. Accordingly, the evaluation results for both systems are slightly different.

Preliminary experiments with the system show that the performance of the speech recognizer is directly depending on the size and complexity of the task grammar. I.e., it makes a significant difference whether the grammar contains 184 airports in 50 states or just the 12 airports in 7 states covered by the task description. Similarly, the inclusion of the affective grammar with a total of 376 keywords leads to a strong decrease of the speech recognizer performance. Even with a reduced affective grammar, the regular system (System A) yields better word recognition rates than the adaptive system (System B). Keeping this in mind, it is not surprising that the task success rate of the adaptive system is lower than the task success rate of the regular system. In situations where the system repeatedly misunderstands the input, users tend to exit the dialogue and start over hoping the system's performance improves after the restart. Although not mentioned in Fig. 6.9, the users are also asked what they like and dislike most about the systems. Here, a strong agreement about System B's low speech recognition performance is identifiable. Nevertheless, most of the test persons feel that System B is more responsive to the user's emotional state than System A. Accordingly, their judgment about the system responses is more positive for System B. The most commonly used emotions are anger and happiness, sadness plays a rather minor role whereas boredom, disgust and fear are almost not involved at all. For this particular scenario, we can observe a sort of trade-off between user-friendliness and usability resulting in a relatively constant overall satisfaction for both systems.

Leaving aside interpretation errors, which could be reduced by using more accurate and specialized models, the integration of emotions as done in the adaptive system evokes a different behavior among the users. As a computer is typically not expected to react on the users' emotional state, the users' curiosity is aroused and they consider it entertaining or even funny to see how the system reacts in different cases. This, not uncommonly, even ends with users laughing (a behavior which has been observed quite often during the recordings of the spontaneous speech database) or showing their friends *"Look at that! Isn't that funny?"*. Furthermore, the evaluation results still leave open some questions, especially when and to what extent it is useful or sensible to integrate emotions into spoken dialogue and when users actually expect a dialogue system to be responsive to their emotions. Depending on the application domain and the anticipation or expectations of the users, the use of emotional cues like in the described scenario can either be appropriate or breaking a fly on the wheel. Nevertheless, we argue that there is a large potential for the integration of emotions into spoken dialogue.

It remains to be seen whether an emotion-sensitive information kiosk, e.g., a train timetable system in stations, is actually preferred to hitherto existing systems or whether such a system is rather smiled at by its users. One promising application in call centers, as already mentioned in the previous section, is the assessment whether an ongoing dialogue is likely to finish successfully or not, based on the current state. Here, in addition to multiple factors such as dialogue measures like the number of repair requests or no-input/no-match user utterances also emotions play an important role in this assessment. If the system detects a certain risk that the caller hangs up before the dialogue finishes successfully, the caller is typically put through to a human call center agent. In the majority of the cases, people call service hotlines when they require help with a product or when something went completely wrong and not when they just feel like conversing with someone. Accordingly, voice portal systems like the one described in Burkhardt et al. (2006a) are targeted on detecting when the caller becomes angry or, in situations where the callers are already angry before they decide to call, when a caller's anger level exceeds a certain critical point. Due to their modularity, our approaches to emotion recognition as described in Chapter 4 and in Section 6.3 above are in principle insertable without difficulty in such a system as, especially when reducing the emotion set to "angry" and "not angry" or to "+" and "–", these recognizers feature a sufficient recognition accuracy. Also, the dialogue models in the dialogue manager may be extended by an "abort" state in which the dialogue manager hands over the caller to a human contact person.

6.5 Discussion

As opposed to plain speech recognition, the evaluation of emotion recognizers faces the problem of inconsistent performance evaluation criteria making it difficult to actually compare different approaches and classification methods. In our work, to maintain a certain degree of comparability, we limit our considerations on matching predominant emotions on the utterance level without regarding "soft" distance measures, e.g., in the valence-arousal space. The huge number of experiments which we conduct with different feature sets, models, and recognizer setups typically entails also an accordingly huge number of results. For the sake of a better overview, we collect the most important results in this chapter.

Summarizing, we propose four different approaches and modifications to speech-based emotion recognition: plain emotion recognition, combined speech–emotion recognition, two-step speech–emotion recognition and the adapted

ROVER approach combining the outputs of multiple speech–emotion recognizers. In the following, we briefly discuss the underlying ideas and summarize their performance.

- The plain emotion recognizer is designed as an independent stand-alone module including its own feature extraction and classification capabilities. The acoustic model either includes one HMM per emotion or separate HMMs for female and male speakers leading to a significant improvement of the recognition performance. Our emotion recognizer implementation achieves an accuracy of approximately 62% for seven emotions and 72% for five emotions.
- Our approach to combined speech–emotion recognition reduces the overall system complexity by consolidating feature extraction and classification of speech and emotion recognizers. With respect to the multiplication of HMMs in the acoustic model, the word recognition performance decreases (compared to a plain speech recognizer trained on the same data) whereas the emotion recognition accuracy remains on a satisfactory level. Modifications on the feature set, however, allow optimizations of either word or emotion recognition performance. Using standard MFCC features, the system achieves emotion recognition accuracies of 63.2% (seven emotions) and 69% (six emotions) with a word accuracy of 87.3%. Enriching the feature set with pitch, intensity and formants (MFCPAC-44), the emotion accuracies increase to 67% (seven emotions) and 75.8% (six emotions) with a word accuracy of 84.7%.
- Aiming at an improvement of the overall word and emotion recognition performance we surrender (to some extent) the idea of a compact speech–emotion recognizer when introducing a two-step approach. This approach involves a common feature extraction but inserts a plain speech recognition using MFCC features before the combined speech–emotion recognizer. Based on the output of the preceding speech recognizer, a minimized language model for the speech–emotion recognizer is derived reducing its decoding complexity and increasing the overall performance. With this approach, we achieve a word accuracy of 94.7% and emotion accuracies of 73% (seven emotions) and 79.% (six emotions) also with the MFCPAC-44 features.
- Independent of the speech–emotion recognizer implementation (regular or two-step), we propose an adapted ROVER method (see Fiscus, 1997) which combines the output of multiple recognizers into one sequence of word–emotions with a lower number of errors. Our approach includes the standard alignment and voting of the words like in the original ROVER plus an extra voting module for the emotions. Considering five different speech–emotion recognizers with word accuracies between 82.5% and 89.1% and emotion accuracies between 63.2% and 68.9% our implementation achieves an overall emotion accuracy of up to 76.4% and a word accuracy of 89.7%. Whereas the improvement of the word accuracy is negligible, we observe a significant increase in the emotion recognition performance.

A general recommendation which technique may be preferred in any environment is difficult to be given since the actual advantages of a certain approach are directly affected by individual scenarios and preferences. On the one hand, the plain emotion recognizer is absolutely suitable as an add-on to existing dialogue frameworks requiring a satisfactory recognition performance without an excessive processing overhead. On the other hand, the two-step speech–emotion recognizer constitutes an all-in-one solution featuring better emotion recognition capabilities in conjunction with speech recognition, e.g., enabling a better integration of further emotion recognition concepts like linguistic cues into an end-to-end system. As for the selection of useful features we suggest to combination of MFCCs with few prosodic and acoustic features such as pitch, intensity and formants portending that plain MFCCs may also lead to satisfactory results.

The evaluation of the adapted ROVER method to combine multiple recognizers shows visible and promising improvements of the emotion recognition rate. Its actual application in SLDS frameworks seems rather like breaking a butterfly on a wheel due to the enormous complexity introduced by the multiple recognizers and the alignment and voting procedures. Thus, the system developer needs to trade off well whether potential improvements legitimate the high complexity. Our experiments show that the best results can be achieved when combining four or five recognizers – fewer recognizers typically do exhibit the required differences in the nature of errors whereas a larger number of recognizers introduce more errors in general. It should be noted that we choose the five best recognizers from a given set by what each additional recognizer itself exhibits a lower performance which has an unfavorable effect on the overall performance.

Our proposed approach to integrating emotional cues into dialogue management is established on the adaptation of the stylistic realization of system prompts following Brown and Levinson (1987) and Walker et al. (1997a) as proposed by André et al. (2004). In our straightforward rule-based approach, the appropriate prompt is selected with respect to the detected emotional state according to predefined rules. In excess thereof, we introduce a semi-stochastic dialogue model consisting of predefined states the transitions between which are determined according to bi- and tri-turn transition probabilities derived from preprocessed dialogue data. This does not only allow the adaptation of the prompt style but also an alignment of the dialogue flow according to an arbitrary number of dialogue control parameters such as emotions, speech recognizer confidence measures, etc. User studies of an emotion-sensitive implementation show a higher user-friendliness and user acceptance compared to a standard dialogue implementation.

Chapter 7
Conclusion and Future Directions

With an increasing performance of their components such as automatic speech recognition, linguistic analysis as well as naturally sounding text-to-speech synthesis, spoken language dialogue systems take on an important place in everyday life. SLDSs have evolved from the first commercially available systems which did not feature much more than isolated word or command recognition with a very limited vocabulary to natural language systems including continuous word recognition, covering multiple domains and even involving a certain degree of "intelligence". Nevertheless, surveys conducted on different occasions give the impression that the acceptance of SLDSs, especially in interactive voice response systems such as voice portals or call center applications which are considered as annoying, is relatively low despite their actual capabilities.

There exist a large variety of approaches, in current research and available products, to rendering spoken human–computer interfaces more efficient, more natural and more user-friendly. In this book, we have addressed the question of how SLDSs can and shall react to the user's emotional state. Consequently, we have concentrated, for one, on the recognition of emotions from the speech signal in cooperation with the speech recognizer and, for another, on the integration of emotions into the dialogue flow via an extended user state and dialogue manager.

For the integration of emotional cues into dialogue management, in Section 3.5, we have described a rule-based approach which adapts the stylistic realization of system prompts following Brown and Levinson (1987) and Walker et al. (1997a) as described by André et al. (2004). To accomplish that, the output of the emotion recognizer is transformed into a value between 0 and 2 according to the emotion's position in the valence-arousal space and the appropriate prompt is selected depending on that value. With respect to the complexity introduced by the additional control parameter rendering the dialogue design process rather inflexible, we have proposed a semi-stochastic dialogue model to adapt the dialogue flow and style to the user's input and emotional state (Pittermann and Pittermann, 2007). Compared to other stochastic approaches like (Partially Observable) Markov Decision Processes, our dialogue model features a lower complexity in terms of the involved (internal) parameters. Analogously, the dialogue flow is modeled by transitions between dialogue states in a network. As opposed to existing work considering bi-turn transitions (from the previous state to the current), we have also included tri-turn

J. Pittermann et al., *Handling Emotions in Human-Computer Dialogues*,
DOI 10.1007/978-90-481-3129-7_7,

transitions (from the penultimate state to the previous state to the current state) leading to a more consistent model. The respective probabilities are trained on preprocessed dialogue data. Using appropriate training data, our model can also allow for an error-prone emotion recognizer output (see Sections 3.6–3.9). User studies on an exemplary implementation of the dialogue manager as described in Section 5.3 corroborate the presumed user-friendliness and naturalness of our approach.

Our approaches to speech-based emotion recognition are largely geared to common approaches to automatic speech recognition. I.e., we concentrate on the combination of MFCCs with prosodic and acoustic features which are not calculated for the utterance as a whole but every 10 ms to capture the temporal characteristics of these features during the utterance. Concerning the classification we do not explicitly distinguish between speech, emotion or speech–emotion recognition as we persistently use three-state left-to-right HMMs with Gaussian mixtures operating on the phoneme, emotioneme or emophoneme level.

In our approach to plain emotion recognition as described in Section 4.5, we have overstrained these HMM-based acoustic models by considering an emotion as one "emotioneme" (in analogy to a phoneme) which, however, may span a longer period than a typical phoneme. Our choice of features includes the most common prosodic and acoustic features like pitch, intensity, formants, jitter, harmonicity plus their computational statistics such as (global) minimum, mean, maximum and variance or local characteristics applied to voiced and unvoiced parts of the utterances. This choice has been motivated by the dominant features in human perception of emotions – e.g., angry persons tend to speak faster, more loudly and with a broader pitch range than sad persons.

With a total of 24 prosodic and acoustic features, recognition rates of up to 72% have been achieved on the acted emotional speech data provided by the Berlin Database of Emotional Speech (Burkhardt et al. 2005) with a reduced set of five emotions (see Section 5.1.1). This performance is comparable to the average performance of existing plain emotion recognizers. However, our recognizer's complexity is considerably reduced with particular respect to our approach to combining speech and emotion recognition. Our associated experiments have been conducted on randomly selected utterances of the corpus regardless of how convincing they were perceived by the annotators. Analyzing the results of different test series, we have found that a better performance is yielded when training different models for female and male speakers and when considering boredom and neutral as one "mainly neutral" superclass. The differentiation of female and male speakers, on the one hand, doubles the number of acoustic models, but, on the other hand, leads to more precise models finally providing a more reliable recognizer output. We have combined neutral and boredom with respect to their acoustic similarity and due to the fact that their influence in an adaptive dialogue system is almost identical. This combination occurs in accordance with the emotional labels but leads to a more vague and imprecise boredom-neutral model. I.e., other emotions are more likely to be confused with this superclass, which, however, is bearable compared to the overall performance gain. Considering that disgust is not relevant for the application of adaptive dialogue management, we have also left out disgust utterances and models in our plain emotion recognition experiments, although this has not led to a performance improvement or degradation. With respect to the HMM properties, also the labeling of the emotions, especially their duration, has a significant influence on the recognizer performance. However, despite the fact that the HMMs are better suited for phoneme-level recognition, they also perform well when labeling the emotions on a word level.

In excess of plain emotion recognition we have considered a combined approach to speech and emotion recognition which we refer to as "speech–emotion recognition" as described in Section 4.6. Here, emotions are considered on the phoneme level, i.e., each (speech recognizer) phoneme is assigned different emotional states. Accordingly, each word is extended to a word–emotion, e.g., "PLEASE" (p l iy z) becomes "PLEASE-ANGER" (pa la iya za), "PLEASE-BOREDOM" (pb lb iyb zb), etc., and we assume that the speaker's emotional state does not change within one word. Considering seven emotional states, the number of phonemes (which we refer to as "emophonemes" in this context) and, thus, the number of HMMs in the acoustic model is multiplied by seven. To counter the loss of performance due to this multiplication of models, robust features for recognizing both text and emotional states from speech are required. Preliminary experiments have shown that the use of plain prosodic and acoustic features does not lead to satisfactory results for the recognition of text, whereas plain MFCC features already provide reasonable recognition rates for both text and emotions.

Using plain MFCC features and distinguishing seven emotions, we have obtained an overall word–emotion recognition accuracy of 58.1% which includes a word accuracy of 87.3% and an emotion accuracy of 63.2%. Having performed an exhaustive search on promising feature combinations, we have proposed the combination of MFCCs with a comparatively small number of prosodic and acoustic features. Using MFCCs with pitch, intensity and three formants, our system has yielded a word–emotion accuracy of 58.4% with 84.7% word accuracy and 67% emotion accuracy. Reducing the emotion set we have achieved accordingly better results, especially when merging neutral and boredom, whereas, as expected, the omission of disgust has not influenced the recognition performance. Distinguishing anger, disgust, fear, happiness, sadness and the boredom-neutral superclass, the recognizer has yielded a word–emotion accuracy of 60.1% with the same word accuracy of 87.1% and an emotion accuracy of 69.5% also using MFCC features (Meng et al., 2007). Compared to existing (plain) speech or emotion recognizers, these numbers seem rather low but prove the feasibility of our straightforward approach.

Further developing the idea of combined speech–emotion recognition, we have also examined how the knowledge about the textual content of an utterance helps to improve the emotion recognition performance. To accomplish that, we have proposed a two-step approach to speech–emotion recognition as described in Section 5.1.2 (see Fig. 5.5). In this approach, all features for speech–emotion recognition are extracted and in the first step, only the MFCC features are used to perform plain speech recognition on the speech signal using a flexible stochastic language model. Based on this recognizer output, an optimized sentence-based language model is created for the speech–emotion recognizer in the second step.

Here, we have distinguished perfect knowledge (100% word accuracy) and realistic knowledge (approx. 94% word accuracy) of the textual content and we have been able to achieve up to 73% (seven emotions) and 79.3% (six emotions) emotion accuracy with this approach. Particularly, for six emotions, the emotion recognition performance of our speech–emotion recognizer is above most existing (plain) emotion recognizers, while simultaneously also accomplishing speech recognition at a reasonable performance.

The exploitation of differences in the nature of errors occurring in the output of multiple (different) speech recognizers has been addressed by Fiscus (1997). In Section 5.2, we have picked up the ROVER idea and adapted it to our speech–emotion recognizer. With respect to the different evaluation criteria for speech and emotion recognition, the proposed ROVER algorithm for speech–emotion recognition includes a specific emotion scoring and voting module in addition to the original speech recognizer output alignment and voting modules. For our experiments we have chosen the five "best" recognizers for one-step speech emotion recognition using different feature sets. Their single word–emotion accuracies are between 57.1% and 59.5% and the respective emotion accuracies are ranging from 63.2% to 68.9%. Combining the outputs of all five recognizers, our approach has yielded an overall word–emotion accuracy of 72.4% and an emotion accuracy of 76.4% constituting an absolute increase of 7.5% compared to the emotion accuracy of the best single recognizer (Pittermann et al. 2007b). Comparable work on the use of ROVER-like methods for emotion recognition is not reported in literature. In the field of speech recognition, however, lower improvements have been reported with the ROVER algorithm (5% increase at word accuracies below 55%, Fiscus 1997).

Our speech–signal-based emotion recognizers have been complemented by a linguistic approach to emotion recognition based on a what we refer to as "affective grammar". We have evaluated the performance of this affective grammar on the basis of user studies who were asked to assess the emotional content of sample sentences. Depending on the evaluation criteria, recognition rates between 60% and 80% have been yielded.

In this work, we have proposed four different approaches and modifications to speech-based emotion recognition each of which has its individual advantages and weak points and we have demonstrated the feasibility and functionality of the ROVER approach on the basis of five speech–emotion recognizers. Considerations on the theoretical limits have shown that higher recognition rates may be yielded when combining multiple recognizers. The developer, however, should trade off whether such a gain is justifiable by the enormous effort involved in this procedure. The use of an affective grammar as an independent emotion recognizer seems promising but we expect a higher recognition reliability when combining the linguistic analysis with our proposed acoustic emotion recognizers.

A summarizing overview on the performance of our proposed systems is illustrated in Fig. 7.1. The average recognition rate of the linguistic analysis (strict evaluation, i.e., the predominant emotions must be recognized) of 60% as well as the human emotion recognition accuracy of 85.2% (see also Section 6.3.1) are represented by light grey bars for comparison. Determined from the corpus annotators'

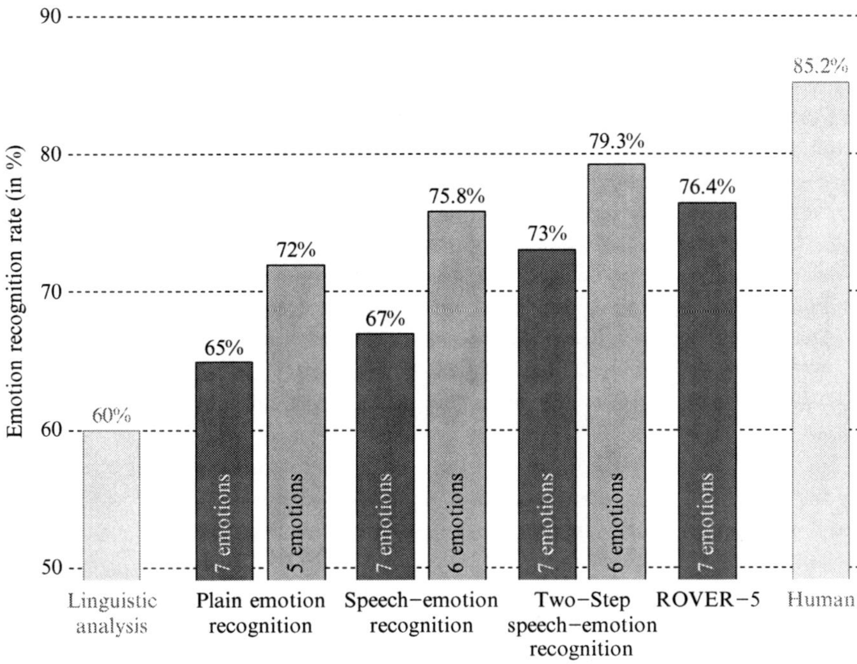

Fig. 7.1 Emotion recognition performance comparison of the proposed speech-based emotion recognizers

labels (cf. Burkhardt et al. 2005) and forming the basis for the training of our recognizers, the human recognition accuracy constitutes the upper bound for the performance of the recognizers. Our plain emotion recognizer yields 65% accuracy for all seven emotions and 72% for five emotions and our one-step speech–emotion recognizer achieves 67% and 75.8% for seven and six emotions, respectively. A visible performance gain is obtained with the two-step speech–emotion recognizer, which yields 73% and 79.3% accuracy for the classification of seven and six emotions. The highest performance has been achieved when combining multiple speech–emotion recognizers – a barely acceptable trade-off between performance and complexity is the combination of five speech–emotion recognizers which yields an overall recognition rate of 76.4% for all seven emotions.

Nevertheless, the proposed approaches still exhibit some weaknesses which establish a wide range of possibilities for improvements. The performance of speech and emotion recognizers in general stands or falls with the quality and the size of the corpora used for the training of the statistical models. Despite the fact that the Berlin Database of Emotional Speech provides a sound solid basis for our experiments, its actual (available) size is too small for more detailed considerations. Using a larger amount of training and test data, including a larger variety of speakers, we expect to be able to perform a better fine-tuning of the parameters. Particularly, in our combined speech–emotion recognizer we accept a compromise of complexity and performance with respect to the limited amount of data, leading to the two-step version of a system which is originally meant to be a one-step system. Experiments with our two-step system have shown that the knowledge about the textual content of an utterance contributes to a better emotion recognition performance. In this approach, up to now, we have only considered the best speech recognizer output string as the basis for the language model of the speech–emotion recognizer in the second step. With a larger number of data available, we expect an even better performance when also integrating the second best, third best, etc. recognizer hypothesis into this adapted language model. Focusing on the further development of our combined speech–emotion recognizer, possible improvements of the plain emotion recognizer will be directly integrated into the speech–emotion recognizer. Experiments have shown that MFCC features also contribute to a better emotion recognition performance, virtually turning the plain emotion recognizer into a speech–emotion recognizer.

The ROVER method combining multiple speech–emotion recognizers is, as demonstrated in our experiments, also applicable to our speech–emotion recognizers, where particularly the emotion recognition rate improves. In our according experiments, the nature of errors in the output of the individual recognizers has been sufficiently different as respects the emotion recognition portion, whereas for the word recognition portion, the output of the recognizers have resembled more strongly. The effect is clearly observable – the overall emotion recognition performance improves whereas the word recognition performance does not outperform the best single word recognition performance significantly. Experiments with plain speech recognizers have shown that such a strong degree of dissimilarity of multiple recognizers is difficult to achieve virtually rendering the ROVER algorithm useless in certain scenarios. To avoid frustration in such cases, we envisage more meaningful bounds as indicators whether the use of the ROVER method is useful in certain situations. Upper and lower bounds assuming the best and worst case of a ROVER implementation have been already described in Section 5.2 and in Pittermann and Pittermann (2006b). The calculation of these performance bounds is planned to be extended by probabilistic considerations so that these measures allow a more realistic assessment about the usefulness of the ROVER method without having to implement a complete system beforehand. In order to improve the performance of a system combining multiple recognizers, Hillard et al. (2007) propose the application of a stochastic classifier instead of the voting module showing promising results. Accordingly, we also plan to include aspects of this "*i*ROVER" idea in our adapted ROVER method for speech–emotion recognition.

Further attention for our future work on emotion recognition is paid to the combination of linguistic and paralinguistic cues to achieve a better and more reliable hybrid recognizer performance. This involves an optimization of the phoneme-level models tying these to a certain emotional state over a longer period than the duration of one phoneme while still being able to recognize (correct) phoneme sequences. One predominant problem in the development of speech-based emotion recognition is the availability of useful emotional speech data. Whereas it is comprehensible that acted emotional speech differs significantly from spontaneous emotional speech, the quality of spontaneous speech corpora is rather unsatisfactory as most of the included utterances are typically neutral plus a few angry or happy utterances. Especially our combined speech–emotion recognizer suffers from the fact that emotional speech corpora do not include a wide range of text which makes these rather unsuitable for the training of large vocabulary continuous speech recognizers. We plan to address this problem by spicing up neutral speech corpora for speech recognizer training with emotional characteristics on the feature level by generating differential models between neutral and other emotional states which shall then be "added" to the plain speech data. To improve the performance of the emotional linguistic analysis we plan to relocate the classification from the rule-based grammar approach to a stochastic language modeling approach as described by Minker et al. (1999) or to combine the advantages of both methods. For the combination of linguistic and paralinguistic emotion recognition, we also plan to pick up and extend the three approaches described in Section 4.7. These include an extended linguistic analysis taking into account the emotions from the speech–emotions or an extra module performing the fusion of both information sources (Pittermann et al. 2008b).

Our semi-stochastic dialogue model described in Chapter 3 has proved to be straightforward to implement and suitable for selected applications (Pittermann et al. 2007b). In excess of its application to only one dialogue control parameter (here: emotion), the model is also capable of including a virtually unlimited number of further parameters as discussed in Section 3.9. By nature, the model complexity increases with every additional state per parameter and, even more significant, with every additional dimension (parameter). Simultaneously, a (significantly) larger amount of dialogue data is required for the training of an accurate model. This entails the same problem which we are also facing with our speech–emotion recognizer: the (unfortunately limited) availability of adequate data. In the current structure of the model, we compensate for the lack of training data by a simplified backoff strategy adding a comparatively small fixed number to all state transitions regardless of whether these occur in the training data or not. We plan to enhance this fixed number by an adaptive factor which is determined on the basis of similar transitions in the model. E.g., assuming that the dialogue training data does not provide any information about transitions from the "departure_city:happy", but includes reliable data for the transitions from the "departure_city:angry" and the "destination:happy" states, the backoff summands for the "departure_city:happy" are adapted according to the transition probabilities of the other two states. Going beyond this simple example, we envisage a more sophisticated calculation of the summands involving all cross-relations between states and transitions. Furthermore, we plan to apply methods to reduce the number of states by estimating the significance of the individual states and replacing the least significant states by one dummy state. Accordingly, the selection process also needs to be adapted to these strategies. Such an extended system shall then be adapted to two different application scenarios – in IVR systems on the one hand, and in ambient intelligent systems on the other hand.

We are currently witnessing an increasing use of SLDS technology in today's IVR systems such as call centers to unburden the call center employees from routine requests, and to lower the number of calls on hold. These systems enable a quicker and more efficient handling of the calls and, last but not least, lower the costs for customer service significantly. Typically, SLDSs are deployed in front-ends for call-routing (e.g., *"If you experience trouble with our products please press '1' or say 'Support', if you want to purchase one of our products, please press '3' or say 'Sales'…"*), for plain information retrieval like train schedule information systems or as automated agents providing fully automated (technical) support to customers. However, speech recognition errors, badly designed dialogues or the customers' inability to interact with SLDSs may lead to decreasing customer satisfaction and in the worst case to a hang up on the part of the customer. Particularly with regard to the assessment whether a dialogue is likely to end successfully or not, emotion recognition can contribute to the detection of such problems emerging during the call and may, in combination with other indicators, lead to a change in the dialogue strategy or in an escalation to a human operator. To achieve a robust behavior of the dialogue estimation we focus our consideration on the detection of anger as proposed by Burkhardt et al. (2006a) by limiting the emotion set to "positive", "neutral" and "negative". For the classification of these three states we expect a visibly better performance than for the above described classification of five, six or seven emotional states. Moreover, the linguistic analysis shall concentrate on negative keywords, especially swearwords which contribute to a very robust detection of negative emotional states. An appropriate system shall be implemented and tested on existing dialogue data as well as on human subjects in live tests (Pittermann et al. 2008b).

For the integration of intelligence and emotion awareness into ambient SLDSs, we are currently working on a system that enables the user to interact with one or multiple applications in a natural and user-friendly way. Such a system is able to detect conflicts arising from the cooperation of different applications and communicates these to the user. In this way, the system and the user can negotiate solutions whereupon the user is not burdened too much with brooding over possible solutions. Apart from its problem solving capabilities, the system also needs to be able to adapt to the user's emotional state and/or (cognitive) stress level to communicate the required information appropriately. A typical application scenario can be found in the automotive sector. Here the driver's attentiveness and ability to react is significantly impaired by having to handle different communication and (route) planning devices while driving. Despite the fact that speech technology is already successfully utilized in this area, the construction and management of complex tasks and constraints for isolated applications still constitute a high cognitive burden on the driver. Another scenario relates to the care of the elderly or to situations in hospitals where employees and physicians as well as patients and residents interact with a variety of applications provided by a central computer system. These may be reminders for medication, emergency calls, patients' requests for service, a patients database or an electronic working plan and task scheduler for employees. Here problems and conflicts may arise ranging from minor issues like the precise timing of medication intake up to critical emergencies. In either case, the people involved may suffer from high stress or even fear, so that the communication provided by the dialogue system should be adapted accordingly. In order to achieve these goals we envisage the development of a generalized concept for the reconciliation of the user's state, situation and issues arising from problem solving. Based on the components developed to date, a prototype system is currently being implemented. It includes the described dialogue manager, problem solving assistant and speech-based emotion recognizer. With the aid of WOZ experiments conducted with the prototype system, application-dependent dialogue flow definitions will be developed. Moreover, the system will be equipped with additional functionality including user preferences and personalization features. I.e., apart from regular user-dependent settings, a set of parameters will be stored user-independently in the system. These, however, can be negotiated during the dialogue, when certain situation- or user-specific priorities are explicitly requested from the user. Furthermore, typical user preferences will be gathered by observation and automatic learning of the user's behavior, also including the current user's emotional state. Here, the appropriate parameters, settings and preferences need to be identified by the system and then used in the adaptation process (Pittermann et al. 2007a).

Appendix A
Emotional Speech Databases

Basque: Audiovisual Database of Emotional Speech in Basque by Navas et al. (2004a).
Emotions: *Anger, disgust, fear, happiness, sadness, surprise, neutral*
Elicitation: Audio–visual recordings of a professional actress uttering isolated words and digits as well as sentences of different length, both with emotional content and emotion-independent content
Size: 450 utterances with emotional content, 665 utterances with emotion-independent content, 1 female speaker.

Basque: Emotional Speech Database for Corpus Based Synthesis in Basque by Saratxaga et al. (2006).
Emotions: *Anger, disgust, fear, happiness, sadness, surprise, neutral*
Elicitation: Recordings of two speakers reading texts
Size: 702 sentences per emotion (20 h of recordings in total), two speakers (one female, one male).

Burmese: Emotional speech corpus by Nwe et al. (2001).
Emotions: *Anger, dislike, fear, happiness, sadness, surprise*
Elicitation: Recordings of two speakers uttering sentences experienced in daily life in different emotional states including rehearsals
Size: 144 sentences (54 sentences of the first speaker, 90 sentences of the second speaker, 0.6 to 1.6 s per sentence), two speakers.

Chinese: Emotional speech corpus by Yang and Campbell (2001).
Emotions: *Wide range of emotions*
Elicitation: Recordings of Mandarin Chinese speakers, combination of acoustic data from spontaneous conversation and experimental data from perceptual tests
Size: 6 h of recorded speech.

Chinese: Emotional speech database by Yuan et al. (2002).
Emotions: *Anger, fear, joy, sadness*
Elicitation: Speakers were asked to read the first paragraphs of a story evoking certain emotions. Recordings of the last paragraph plus two target sentences uttered emotionally
Size: 288 target sentences, nine female speakers.

Chinese: Drama corpus (Chuang and Wu 2004).
Emotions: *Anger, disgust, fear, happiness, sadness, surprise, neutral*

Elicitation: Recordings of professional impersonators in different emotional states
Size: 2,100 sentences in 440 dialogues by two speakers (1,085 sentences in 227 dialogues from the leading man, 1,015 sentences in 213 dialogues from the leading woman).

Chinese: Emotional speech database by Dan-Ning Jiang (2004).
Emotions: *Anger, fear, happiness, sadness, surprise, neutral*
Elicitation: Recordings of an amateur actress uttering different sentence types (statements, questions)
Size: Approx. 200 utterances per emotion, neutral database with approx. 300 utterances, one female speaker.

Chinese: Mandarin Emotional Speech Corpus I. (Pao et al. 2004).
Emotions: *Anger, boredom, happiness, sadness, neutral*
Elicitation: Recordings of short utterances expressed by native Mandarin speakers in different emotional states
Size: 558 utterances, 12 speakers (seven females, five males).

Chinese: Mandarin Emotional Speech Corpus II used by Pao et al. (2004).
Emotions: *Anger, boredom, happiness, sadness, neutral*
Elicitation: Recordings of utterances expressed by professional Mandarin speakers in different emotional states
Size: 503 utterances, two speakers (one female, one male).

Chinese: Spontaneous speech corpus used by Tao (2004).
Emotions: *Anger, fear, hate, joy, sadness, surprise, neutral*
Elicitation: No further information available
Size: 835 sentences, 5,000 words used by Tao (2004).

Chinese: Acted speech corpus by Tao et al. (2006).
Emotions: *Anger, fear, happiness, sadness, neutral*
Elicitation: A professional actress reading texts from a *Reader's Digest* collection
Size: 1,500 utterances, 3,649 phrases, one speaker.

Chinese: Emotional speech corpus (Wu et al. 2006).
Emotions: *Anger, fear, happiness, sadness, neutral*
Elicitation: Recordings of 25 actors uttering short passages with emotional content and command phrases
Size: 150 short passages (30–50 s), 5,000 command phrases (2–10 s), 50 speakers (25 females, 25 males).

Chinese: Speech database by Zhang et al. (2006).
Emotions: *Anger, fear, joy, sadness, neutral*
Elicitation: Recordings of eight speakers (acoustically isolated room)
Size: 2,400 sentences (20 sentences, uttered three times each, for every emotion), eight speakers (four females, four males).

Chinese, English, Japanese: JST/CREST database by Campbell (2002).
Emotions: *Wide range of emotional states and emotion-related attitudes*
Elicitation: Natural emotions of volunteers recording their domestic and social spoken interactions for extended periods throughout the day
Size: Recordings still ongoing, target: 1,000 h over 5 years.

Chinese, English, Japanese: Speech corpus by Jiang et al. (2005).
Emotions: *Angry, calm, happy, sad, surprise*
Elicitation: Recordings of a speaker uttering a sentence in three languages and in different emotional states
Size: 750 utterances (50 utterances per language and emotion), one speaker.

Danish: Danish Emotional Speech Database (Engberg et al. 1997).
Emotions: *Anger, happiness, sadness, surprise, neutral*
Elicitation: Recordings of actors familiar with radio theater uttering single words, sentences and passages of fluent speech in different emotional states
Size: Approx. 10 min of speech in total (30 s per emotion per speaker) plus neutral recordings (2 speakers) plus extra recordings (three speakers), four speakers in total (two females, two males).

Dutch: Emotional database by Van Bezooijen (1984).
Emotions: *Anger, contempt, disgust, fear, interest, joy, sadness, shame, surprise, neutrality*
Elicitation: Recordings of speakers reading semantically neutral phrases
Size: four phrases, eight speakers (four females, four males).

Dutch: Groningen corpus S0020 ELRA (1996) (see `http://www.elra.info/`).
Emotions: *Mostly neutral, few emotions*
Elicitation: Recordings of speakers reading short texts, sentences, numbers, monosyllabic words, long vowels and an extended word list
Size: 20 h of speech, 238 speakers.

Dutch: Emotional database by Mozziconacci and Hermes (1999).
Emotions: *Anger, boredom, fear, indignation, joy, sadness, neutrality*
Elicitation: Recordings of actors expressing semantically neutral sentences in different emotional states after the respective emotion has been elicited by reading a semantically emotional sentence
Size: three speakers (one female, two males), 315 utterances, five sentences.

Dutch: Experiment at Tilburg University (Wilting et al. 2006).
Emotions: *Acted negative, acted positive, negative, positive, neutral*
Elicitation: Recordings of participants reading sentences in different (partly acted) emotional states according to the mood induction procedure proposed by Velten (1968)
Size: 50 participants (31 females, 19 males), each reading 40 sentences of 20 s length.

English: Database produced by Cowie and Douglas-Cowie (1996).
Emotions: *Anger, fear, happiness, sadness, neutral*
Elicitation: Recorded passages of speakers from the Belfast area
Size: 40 speakers (20 females, 20 males), passages of 25–30 s length.

English: SUSAS database (Hansen et al. 1998).
Emotions: *Talking styles (angry, clear, fast, loud, question, slow, soft), single tracking task (high stress, Lombard effect, moderate), dual tracking task (high stress, moderate), actual speech under stress (anxiety, fear, G-force, Lombard effect, noise), psychiatric analysis (angry, anxiety, depression, fear)*

Elicitation: Recordings of isolated-word utterances under simulated or actual stress in several scenarios, e.g., amusement park roller-coaster, helicopter cockpit, patient interviews
Size: Approx. 16,000 utterances of 36 speakers (13 females, 23 males) in total.

English: Emotional speech database by Li and Zhao (1998).
Emotions: *Anger, fear, happy, sad, surprised, neutral*
Elicitation: Recordings of actors uttering 20 sentences with emotional content and three sentences without emotional content in different emotional states
Size: 5 untrained speakers (two females, three males), 23 sentences per speaker.

English: Emotional speech database by Whiteside (1998).
Emotions: *Cold Anger, elation, happiness, hot anger, interest, sadness, neutral*
Elicitation: Recordings of actors uttering sentences in different emotional states
Size: 70 utterances, two speakers (one female, one male), five different short sentences.

English: Emotional corpus by Cowie et al. (1999b).
Emotions: *Wide range of emotions as defined in the FEELTRACE tool*
Elicitation: Video tape recordings of groups of three friends each discussing about issues they strongly felt about
Size: Recordings of 1 h per group, nine speakers (three groups of three friends).

English: Emotional speech database by Robson and Mackenzie-Beck (1999).
Emotions: *Smiling, neutral*
Elicitation: Recordings of speakers uttering sentences in a neutral state and while smiling
Size: 66 utterances, 11 speakers, three sentences.

English: Reading/Leeds Emotional Speech Corpus (Greasley et al. 2000).
Emotions: *Anger, disgust, fear, happiness, sadness, neutral*
Elicitation: Recordings of interviews on radio/television, speakers asked by interviewers to relive emotionally intense experiences
Size: Approx. 5 h of samples of emotional speech.

English: Emotional speech database by McGilloway et al. (2000).
Emotions: *Anger, fear, happiness, sadness, neutral*
Elicitation: Recordings of speakers reading emotional texts in appropriate style
Size: 40 speakers, five texts (100 syllables each).

English: Emotional speech database by Pereira (2000).
Emotions: *Cold anger, happiness, hot anger, sadness, neutral*
Elicitation: Recordings of actors uttering two sentences in different emotional states
Size: 80 utterances (two repetitions of 40 utterances), two speakers.

English: Database produced by Polzin and Waibel (2000).
Emotions: *Anger, sadness, neutrality (other emotions as well, but in insufficient numbers to be used)*
Elicitation: Audio–visual data, i.e., sentence-length segments taken from acted movies
Size: 1,586 angry segments, 1,076 sad segments, 2,991 neutral segments.

English: Belfast Naturalistic Database (Douglas-Cowie et al. 2000).
Emotions: *Wide range of emotions*

Elicitation: Audio–visual recordings of people discussing emotive subjects with each other/the research team plus recordings of extracts from television programs, i.e., members of the public interacting in a way that appears essentially spontaneous
Size: 239 clips (209 from TV Recordings, 30 from interview recordings, clip durations: 10–60 s), 125 speakers (94 females, 31 males).

English: Database by France et al. (2000).
Emotions: *Depression, suicidal state, neutrality*
Elicitation: Recordings of spontaneous dialogues between patients and therapists in therapy sessions, phone conversations and post therapy evaluation sessions
Size: 115 speakers (48 females, 67 males).

English: DARPA Communicator corpus (Walker et al. 2001).
Emotions: *Annoyance, frustration*
Elicitation: Users making air travel arrangements over the phone
Size: Recordings of simulated interactions with a call center, 13,187 utterances in total (1,750 emotional utterances).

English: Capital Bank Service and Stock Exchange Customer Service (Devillers et al. 2002).
Emotions: *Anger, excuse, fear, satisfaction, neutral*
Elicitation: Human–human interaction in a stock exchange customer service (call) center
Size: 100 dialogues, 5,229 speaker turns.

English: Emotional speech database by Fernandez and Picard (2003).
Emotions: *Stress*
Elicitation: Recordings of speakers solving mathematical problems while driving in a car simulator
Size: four speakers, four situations, 598 utterances, length varying from 0.5 to 6 s.

English: Speech database by Lee and Narayanan (2003).
Emotions: *Negative, non-negative*
Elicitation: Users interacting with a machine agent in a call center
Size: 1,367 utterances (776 utterances of female speakers, 591 utterances of male speakers).

English: LDC Emotional Prosody Speech and Transcription used by Liscombe et al. (2003) and Yacoub et al. (2003).
Emotions: *Anxiety, boredom, cold anger, contempt, despair, disgust, elation, happy, hot anger, interest, panic, pride, sadness, shame, neutral*
Elicitation: Professional actors reading short (4-syllables each) dates and numbers
Size: eight actors (five females, thre males), 44 utterances used by Liscombe et al. (2003), 2,433 utterances used by Yacoub et al. (2003) .

English: ORESTEIA database by McMahon et al. (2003).
Emotions: *Irritation, shock, stress*

Elicitation: Audio-physiological (and partly visual) recordings of driving persons encountering various problems (deliberately positioned obstructions, dangers, annoyances "on the road")
Size: 90 min per session and speaker, 29 speakers.

English: Sensitive Artificial Listener (SAL) database by Cowie et al. (2004).
Emotions: *Wide range of emotions or emotion related states*
Elicitation: Audio–visual recordings of speakers interacting with an artificial listener with different personalities
Size: Recordings of approx. 10 h, 20 speakers.

English: Speech database by Lee et al. (2004) and Yildirim et al. (2004).
Emotions: *Angry, happy, sad, neutral*
Elicitation: Recordings of a semi-professional actress uttering 112 unique sentences in four emotions
Size: 880 utterances, one female speaker.

English: Emotional speech synthesis database (Tsuzuki et al., 2004).
Emotions: *Anger, happiness, sadness, neutral*
Elicitation: Recordings of a non-professional male speaker uttering short declarative sentences with emotional content
Size: 363 utterances, one male speaker.

English: Modified LDC CallFriend corpus prepared by Yu et al. (2004).
Emotions: *Boredom, happy, hot anger, interest, panic, sadness, no emotion plus numerical values (on a discretized scale from 1 to 5) for each of arousal, valence and engagement*
Elicitation: Recordings of social telephone conversations between friends
Size: 1,888 utterances (1,011 utterances from female speakers, 877 utterances from male speakers, eight speakers (four females, four males).

English: WOZ data corpus (Zhang et al., 2004).
Emotions: *Confidence, puzzle, hesitation*
Elicitation: Audio–visual recordings of children interacting with an intelligent tutoring system for learning basic concepts of Mathematics and Physics
Size: 714 students' utterances (approx. 50 min of clean speech), 4.2 s of speech and 8.1 words per utterance, 17 speakers.

English: Speech database by Lee et al. (2005).
Emotions: *Angry, happy, sad, neutral*
Elicitation: Recordings of a male speaker producing sentences with non-emotional content in the respective emotional states (including biosensor data)
Size: 280 utterances, 14 sentences, one male speaker.

English: HMIHY speech database (Liscombe et al. 2005).
Emotions: *Positive/neutral, somewhat angry, somewhat frustrated, somewhat other negative, very angry, very frustrated, very other negative*
Elicitation: Recordings of callers interacting with an automated agent concerning account balance, explanation of bill charges, AT&T rates and calling plans, etc.
Size: 5,690 dialogues, 20,013 user turns.

English: Expressive spoken corpus of children's stories modified by Alm and Sproat (2005).
Emotions: *Angry, disgusted, fearful, happy, sad, surprised, neutral*
Elicitation: Recordings of a semi-professional female speaker reading two children's stories in an extremely expressive mode
Size: Approx. 10 min of speech, 128 sentences, one female speaker.

English: ITSPOKE corpus (Ai et al. 2006).
Emotions: *Positive, negative, neutral*
Elicitation: Students interacting with an interactive tutoring system
Size: 100 dialogues, 20 students, 2,252 student turns, 2,854 tutor turns plus a set of human–human corpus.

English: Situation Analysis in a Fictional and Emotional (SAFE) corpus (Clavel et al., 2006).
Emotions: *Fear, other negative emotions, positive emotions, neutral*
Elicitation: Audio–visual excerpts taken from movie DVDs, abnormal contexts
Size: 400 sequences, total length 7 h, 4,724 segments of speech (up to 80 s).

English: Castaway database (Devillers et al. 2006).
Emotions: *Wide range of emotions*
Elicitation: Audio–visual recordings of a reality TV show
Size: 10 recordings (30 min each), 10 speakers.

English: Speech database by Lee et al. (2006).
Emotions: *Angry, happy, sad, neutral*
Elicitation: Recordings and magnetic resonance images of a male speaker uttering a set of four sentences
Size: 80 utterances (four sentences, five repetitions, four emotions), one male speaker.

English: Emotional speech corpus by Kumar et al. (2006).
Emotions: *Inappropriateness, lack of clarity, uncertainty, neutral*
Elicitation: Recordings of participants interacting with an SLDS in terms of a customer survey about grocery stores plus answering of a questionnaire
Size: 257 utterances, 17 participants (10 females, 7 males).

English: ISL Meeting corpus (Neiberg et al. 2006).
Emotions: *Negative, positive, neutral*
Elicitation: Recordings of 18 meetings with a total of 92 persons and an average duration of 35 min accompanied by orthographic transcription
Size: 12,068 utterances, thereof 424 negative, 2,073 positive and 9,571 neutral.

English: "Yeah right" corpus by Tepperman et al. (2006).
Emotions: *Sarcastic, neutral*
Elicitation: "Yeah right" utterances taken from the Switchboard and Fisher corpora of spontaneous telephone dialogues
Size: 131 utterances.

English, French, German: Speech database by Klasmeyer et al. (2000).
Emotions: *Emotional, neutral*
Elicitation: Recordings of English, French and German speakers reading sentences and uttering passages of spontaneous speech

Size: Approx. 13,000 utterances, 120 speakers (25 English, 65 French, 30 German).

English, French, German, Italian: Geneva Airport Lost Luggage Speech Database by Scherer and Ceschi (1997).
Emotions: *Anger, good humor, indifference, resignation, worry*
Elicitation: Audio–visual recordings of passengers at the lost luggage desk at Geneva Airport plus interviews
Size: 112 passengers, 12 airline employees, 10–20 min of interaction per passenger at the desk.

English, French, Slovenian, Spanish: INTERFACE Emotional Speech Synthesis Database (IESSDB, Nogueiras et al. 2001).
Emotions: *Anger, disgust, fear, joy, sadness, surprise, neutral*
Elicitation: Six different kinds of sentences (affirmative, exclamatory, interrogative, paragraphs of approx. five sentences, isolated words and digits) spoken by professional actors in each language and each emotion
Size: Two actors (one female, one male), 150–190 utterances for each of the six emotional styles in four languages.

English, German: AIBO (Erlangen database, Batliner et al. 2004a).
Emotions: *Angry, bored, emphatic, helpless, joyful, motherese, reprimanding, surprised, touchy (irritated), neutral and rest*
Elicitation: Children interacting with a WOZ robot
Size: 51 German children (30 females, 21 males, 51,393 words, 9.2 h of speech), 30 English children (5,822 words, 1.5 h of speech).

English, German: Corpus in the framework of the FERMUS-III project (Rigoll et al. 2005).
Emotions: *Anger, disgust, fear, sadness, surprise, neutral*
Elicitation: First set: Actors uttering sentences in different emotional states, second set: utterances of automotive infotainment speech interaction dialogues
Size: 3,529 utterances (first set: 2,829 utterances, second set: 700 utterances), 13 speakers (one female, 12 males).

English, German, Japanese: Material taken from the TV series *Ally McBeal* (Braun and Katerbow 2005).
Emotions: *Cold anger, fear, hot anger, joy, sadness, neutral*
Elicitation: Audio–visual data taken from a DVD
Size: six speakers (three females, three males), 135 utterances in total (45 utterances per language).

English, Japanese: Emotional database by Shigeno (1998).
Emotions: *Anger, disgust, fear, happiness, sadness, surprise*
Elicitation: Audio–visual recordings of actors uttering short sentences and words in English and Japanese
Size: two speakers (one American, one male), 36 audio–visual stimuli.

English, Korean: Emotional database by Chung (2000).
Emotions: *Joy, sadness, neutral*
Elicitation: Audio–visual recordings of female speakers in Korean and American television shows talking about problems in their lives, expressing sadness and joy

Size: one Korean speaker (eight utterances, each lasting 1 min), five American speakers (one neutral and one emotional utterance, each lasting 1 min, per speaker).

English, Malay: Emotional speech database by Razak et al. (2005).
Emotions: *Anger, disgust, fear, happy, sad, surprise*
Elicitation: Recordings of learning actors uttering sentences frequently used in everyday communication
Size: 1,200 utterances, four different sentences, female and male speakers.

English, Swedish: Speech database by Laukka (2004).
Emotions: *Anger, disgust, fear, happiness, sadness, neutral*
Elicitation: Recordings of British and Swedish actors uttering short sentences
Size: 176 utterances, eight speakers (four females, four males, four British, four Swedish).

Farsi/Persian: Farsi emotional speech corpus (Gharavian and Ahadi 2005).
Emotions: *Angry, sad, neutral*
Elicitation: Sentences of a non-emotional corpus reuttered angrily, sadly and neutrally
Size: 1,518 utterances, one male speaker.

Finnish: MediaTeam emotional speech corpus (Väyrynen et al. 2003).
Emotions: *Anger, happiness/joy, sadness, neutral*
Elicitation: Recordings of professional actors reading out a phonetically rich Finnish passage with non-emotional content
Size: 56 monologues, 14 speakers (six females, eight males).

French: Emotional database by Johnstone and Scherer (1999).
Emotions: *Anxious, bored, depressed, happy, irritated, tense, neutral*
Elicitation: Recordings of students playing a manipulated computer space game and making statements to their emotions, furthermore recordings of biosignals
Size: 36 males speakers.

French: Messages corpus by Chateau et al. (2004).
Emotions: *Positive emotion, negative emotion, no particular emotion (neutral)*
Elicitation: Recordings of France Telecom customers describing their opinions about the customer care service
Size: 103 messages split into 478 emotional utterances, 103 speakers.

French: EmoTV corpus (Abrilian et al. 2005).
Emotions: *Anger, despair, disgust, doubt, exaltation, fear, irritation, joy, pain, sadness, serenity, surprise, worry, neutral plus 176 fine-grain categories*
Elicitation: Audio–visual recordings of TV news interviews
Size: 51 recordings, 48 speakers, 14 s per recording, 2,500 words.

French: Talkapillar corpus by Beller and Marty (2006).
Emotions: *Anger, boredom, disgust, happiness, indignation, sadness, surprise (negative and positive), neutral and neutral question*
Elicitation: Recordings of one actor reading semantically neutral sentences in different emotional states
Size: 539 utterances, one speaker, 26 sentences, 2 h of speech.

French: CEMO corpus (Devillers et al. 2006).
Emotions: *8 coarse-grained classes: anger, fear, hurt, other positive, relief, sadness, surprise, neutral, additionally 21 fine-grained classes*
Elicitation: Real agent-client recordings obtained from a medical emergency call center offering medical advice
Size: 688 agent-client dialogues of around 20 h, seven agents (four females, three males), 784 clients (513 females, 271 males), 48 turns per dialogue in average, 262,000 words thereof 92,000 distinct words.

German: Emotional speech database by Tolkmitt and Scherer (1986).
Emotions: *Cognitive and emotional stress*
Elicitation: Recordings of speakers who were shown slides either containing logical problems which they had to solve verbally or photos of injured people which they were asked to comment on
Size: 60 speakers (27 females, 33 males), 20 slides, max. 40 s speech per slide.

German: Geneva Vocal Emotion Expression Stimulus Set (GVEESS, Banse and Scherer, 1996).
Emotions: *Anxiety, boredom, cold anger, contempt, disgust, elation, happiness, hot anger, interest, panic fear, pride, sadness, shame*
Elicitation: Audio–visual recordings of actors acting scripted emotion-eliciting scenarios for each emotion
Size: 12 actors (six females, 6 males), 224 recordings.

German: Speech database by Dellaert et al. (1996).
Emotions: *Anger, fear, happiness, sadness, neutral*
Elicitation: Recordings of speakers reading a variety of sentences in different emotional states
Size: five actors, 50 sentences, 1,000 utterances.

German: Speech database by Klasmeyer (1996).
Emotions: *Anger, boredom, fear, happiness, sadness, neutral*
Elicitation: Recordings of actors uttering short sentences with non-emotional content
Size: 10 sentences per emotion, three actors.

German: Emotional speech material used by Alter et al. (1999) and Alter et al. (2000).
Emotions: *Anger, happiness, neutral*
Elicitation: Recordings of sentences with emotional content spoken by a trained female speaker in a sound proof room at the University of Leipzig, ratings by 30 subjects on a 5-point scale indicating the three emotions
Size: 148 sentences, one speaker.

German: Database used and produced by Batliner et al. (2000).
Emotions: *Anger, neutral*
Elicitation: Recordings of an acting person produced within the VERBMOBIL scenario, recordings of naive subjects reading emotional sentences, recordings of angry and neutral persons in a WOZ scenario
Size: Acted data: 1,240 neutral turns and 96 angry turns, Read data: 50 neutral and 50 emotional sentences, WOZ data: 2,395 turns (20 dialogues) planned to be extended.

German: Database of affect bursts (Schröder 2000).
Emotions: *Admiration, anger, boredom, contempt, disgust, elation, relief, startle, threat, worry*
Elicitation: Speakers reading silently a frame story, recordings of an produced affect burst of their choice plus two produced affect bursts from a list
Size: 180 vocalizations (30 vocalization per speaker), six speakers (three females, three males, thereof four amateur actors).

German: Lego corpus (Kehrein 2001).
Emotions: *Wide range of emotions*
Elicitation: Recordings of dialogues between two persons interactively trying to build a Lego kit without seeing each other
Size: 180 min of speech thereof 372 emotional turns.

German: SmartKom database (Schiel et al., 2002; Wahlster, 2006).
Emotions: *Anger, gratification, helplessness, irritation, joy, pondering, reflecting, surprise, neutral, unidentifiable episodes*
Elicitation: Audio–visual recordings of human–computer information system dialogues in a WOZ scenario
Size: 224 speakers, 448 recordings, 4-5 min sessions.

German: Speech database used by Tato et al. (2002).
Emotions: *Angry, bored, happy, sad, neutral*
Elicitation: Recordings of speakers put in an emotional state and reading commands to the Sony entertainment robot AIBO
Size: 2,800 utterances, 40 commands, 14 speakers (seven females, seven males).

German: SYMPAFLY database (Batliner et al. 2004b).
Emotions: *Angry, compassionate, emphatic, helpless, ironic, joyful, panic, surprised, touchy, neutral*
Elicitation: Naive users book flights using a machine dialogue system
Size: 270 dialogues and 29,200 words in total, three parts with increasing system performance and 62–110 speakers per part.

German: Berlin Database of Emotional Speech (Burkhardt et al. 2005).
Emotions: *Anger, boredom, disgust, fear, joy, sadness, neutral*
Elicitation: Recordings of non-professional actors uttering sentences with non-emotional content in each emotion
Size: More than 800 utterances, 10 speakers (five females, five males).

German: EMO-SI database (Schuller et al. 2005).
Emotions: *Anger, disgust, fear, joy, sadness, surprise, neutrality*
Elicitation: Spontaneous and acted emotions in short phrases of car interaction dialogues
Size: 39 speakers (three females, 36 males), 2,730 samples (70 samples per speaker).

German: Emotional database by Kim and André (2006).
Emotions: *High arousal (negative valence, positive valence), low arousal (negative valence, positive valence)*
Elicitation: Recordings of users playing a quiz (including biosensor data)
Size: 45 min per speaker, three speakers.

Greek: Greek Emotional Database (Fakotakis 2004).
Emotions: *Anger, fear, joy, sadness, neutral*
Elicitation: Recordings of a professional actress reading semantically neutral words and sentences in different emotions
Size: 10 single words, 20 short sentences, 25 long sentences, 12 passages of fluent speech, one female speaker.

Hebrew: Emotional speech corpus by Amir et al. (2000).
Emotions: *Anger, disgust, fear, joy, sadness, neutral*
Elicitation: Recordings of students telling of personal experiences evoking certain emotions (including biosensor data)
Size: 31 speakers (15 females, 16 males).

Japanese: Nicholson et al. (1999).
Emotions: *Anger, disgust, fear, joy, sadness, surprise, teasing, neutral*
Elicitation: Speakers were asked to read a word list in eight emotions trying to imitate emotional recordings produced by radio actors
Size: 50 females and 50 males native Japanese speakers uttering a list of 100 context-free Japanese words eight times (once per emotion), each of the Japanese phonemes equally represented within the list.

Japanese: Speech database by Oudeyer (2003).
Emotions: *Anger, joy/pleasure, sorrow/sadness/grief, normal/neutral*
Elicitation: Recordings of professional speakers uttering short sentences or phrases and imagining themselves uttering these sentences to a pet robot
Size: 4,800 utterances (200 per speaker and emotion), six speakers (female and male).

Japanese: Prosodic corpus by Hirose et al. (2004).
Emotions: *Anger, calm, joy, sadness*
Elicitation: Recordings of a female narrator reading sentences with emotional content
Size: Approx. 1,600 utterances (around 400 sentences per emotion), one female speaker.

Japanese: Emotional speech database by Iwai et al. (2004).
Emotions: *Anger, joy, sadness, neutral*
Elicitation: Recordings of students uttering the word "okaasan" (Japanese: "mother") in four emotions
Size: 766 utterances, three male speakers.

Japanese: Emotional speech database by Takahash et al. (2005).
Emotions: *Angry, bright, excited, raging, neutral*
Elicitation: Recordings of expressive speech sounds narrated by professional actors
Size: 1,500 expressive speech sounds, eight speakers.

Japanese: Emotional speech database by Nisimura et al. (2006).
Emotions: *Anger, contempt, contentment, depression, excitement, fear, joy, mirth, pleasure, pressure, sadness, surprise, tension, tiredness, displeasure, neutral*
Elicitation: Recordings of children's utterances extracted from a public spoken

dialogue system
Size: 2699 utterances.

Korean: Emotional speech database by Kim et al. (2004b).
Emotions: *Anger, joy, sadness, neutral*
Elicitation: Recordings of speakers uttering short sentences in different emotional states
Size: 400 utterances, five different sentences (less than 1.5 sec. duration), four male speakers.

Korean: Database produced by Media and Communication Signal Processing Laboratory, Prof. C.Y. Lee of Yonsei University (Kim et al. 2005).
Emotions: *Angry, joyful, sad, neutral*
Elicitation: Ten speakers uttering dialogic sentences expressing natural emotions with easy pronunciation; afterwards subjective emotion recognition by human listeners for verification
Size: 5,400 sentences: 45 dialogic sentences, three repetitions, four emotions, 10 speakers (five females, five males).

Russian: RUSLANA database (Makarova et al. 2002).
Emotions: *Anger, fear, happiness, sadness, surprise, neutral*
Elicitation: Recordings of actors expressing emotional sentences
Size: 61 actors (49 females, 12 males), 610 utterances.

Spanish: SES Spanish Emotional Speech database (Montero et al. 1999).
Emotions: *Anger, happiness, sadness, surprise, neutral*
Elicitation: Recordings of an actor reading neutral texts in different emotional states
Size: three passages and 15 sentences acted by one speaker in four emotions plus neutral style.

Spanish: Emotional speech database by Iriondo et al. (2000).
Emotions: *Desire, disgust, fear, fury, joy, sadness, surprise*
Elicitation: Recordings of actors reading texts in different emotional states and intensities
Size: eight actors (four females, four males), three intensities, 336 utterances.

Spanish: Emotional speech database by Álvarez Martínez and Barrientos Cruz (2005).
Emotions: *Anger, fear, happiness, sadness, neutral*
Elicitation: Recordings of actors and actresses uttering sentences in different emotional states plus extracted utterances from DVD movies
Size: 380 utterances (300 utterances with four different sentences as synthetic data set (actors), 80 utterances as real data set (DVD movies)), 15 non-professional speakers (female and male) in the synthetic data set.

Swedish: Emotional speech database by Abelin and Allwood (2000).
Emotions: *Anger, disgust, dominance, fear, joy, sadness, shyness, surprise*
Elicitation: Recordings of a speaker uttering a non-emotional phrase in different emotional states
Size: 1 male speaker.

Swedish: VeriVox database by Karlsson (1999).
Emotions: *Stress ranked from zero to nine identifying 5 as normal stress level*
Elicitation: Recordings of male speakers reading texts in different tasks and stress levels
Size: 50 males speakers, 30 min per speaker.

Swedish: The Voice Provider Material (VP, Neiberg et al. 2006).
Emotions: *Emphatic, negative, neutral*
Elicitation: Recordings of voice-controlled telephone services (traffic information, postal assistance, etc.)
Size: 7619 utterances, thereof 160 emphatic, 335 negative and 7,124 neutral.

No specific language: Corpus of infants' cries (Matsunaga et al. 2006).
Emotions: *Anger, hunger, pampered, sadness, sleepiness*
Elicitation: Infants' cries recorded by their mothers at home using a digital recorder, emotional judgment by the mothers taking into consideration facial expressions, behavior, etc., emotional intensity ranked from zero (emotion not contained at all) to four (emotion fully contained)
Size: 402 cries, 23 infants (12 females, 11 males, age: 8–13 months).

(see also "The HUMAINE Portal" website at `http://emotion-research.net/wiki/Databases`)

Appendix B
Used Abbreviations

ANN	Artificial Neural Network
ANS	Autonomic Nervous System
ASR	Automatic Speech Recognition
ATIS	Air Travel Information System
BDI	Beliefs-Desires-Intentions
BEEP	British English Example Pronunciations dictionary
BEEV	British English Emotional Vocabulary
CSLU	Center for Spoken Language Understanding (Oregon Health & Science University)
DBN	Dynamic Belief Network
DETT	Disposition–Emotion–Trigger–Tendency
DVD	Digital Versatile Disk
EBNF	Extended Backus-Naur Form
ELRA	European Language Resources Association
ETN	Emotion Transition Network
FIA	Form Interpretation Algorithm
GMM	Gaussian Mixture Model
GSM	Global System for Mobile communications
HMIHY	How May I Help You?
HMM	Hidden Markov Model
HTK	Hidden Markov Model Toolkit
IPA	International Phonetic Alphabet
ITU-T	Telecommunication Standardization Sector of the International Telecommunication Union
IVR	Interactive Voice Response
JPEG	Joint Photographic Experts Group
JSGF	Java Speech Grammar Format
KNN	k-Nearest Neighbor
LDC	Linguistic Data Consortium (University of Pennsylvania)
LP	Linear Prediction
MDP	Markov Decision Process
MFCC	Mel-Frequency Cepstral Coefficient
MIMO	Multiple Input Multiple Output

MP3	MPEG-1 Audio Layer 3
MPEG	Moving Picture Experts Group
OCC	Ortony, Clore, Collins
PARADISE	PARAdigm for DIalogue System Evaluation
PDA	Personal Digital Assistant
PLP	Perceptual Linear Prediction
POMDP	Partially Observable Markov Decision Process
PROMISE	PROcedure for Multimodal Interactive System Evaluation
RIFF	Resource Interchange File Format
ROVER	Recognizer Output Voting Error Reduction
SAMPA	Speech Assessment Methods Phonetic Alphabet
SLDS	Spoken Language Dialogue System
SRGS	Speech Recognition Grammar Specification
SVM	Support Vector Machine
TRINDI	Task oRiented INstructional DIalogue
W3C	World Wide Web Consortium
WETN	Word-Emotion Transition Network
WOZ	Wizard-of-OZ
WTN	Word Transition Network
XML	eXtensible Markup Language

References

Abdennaher S, Aly M, Bühler D, Minker W, Pittermann J (2007) BECAM tool – A semi-automatic tool for bootstrapping emotion corpus annotation and management. In: European conference on speech and language processing (EUROSPEECH), Antwerp, Belgium

Abdi H, Valentin D, Edelman B (1999) Neural networks. Sage, Newbury Park, USA

Abelin Å, Allwood J (2000) Cross linguistic interpretation of emotional prosody. In Proceedings of ISCA workshop on speech and emotion, Belfast, United Kingdom

Abrilian S, Devillers L, Buisine S, Martin J-C (2005) EmoTV1: Annotation of real-life emotions for the specification of multimodal affective interfaces. In: 11th international conference on human–computer interaction (HCII'2005), Las Vegas, USA

Ai H, Litman DJ, Forbes-Riley K, Rotaru M, Tetreault J, Purandare A (2006) Using system and user performance features to improve emotion detection in spoken tutoring dialogues. In: International conference on speech and language processing (ICSLP), Pittsburgh, USA, pp 797–800

Alexandris C, Fotinea S-E (2004) Reusing language resources for speech applications involving emotion In: International conference on language resources and evaluation (LREC), Lisbon, Portugal, pp 1383–1386

Allen J (1995) Natural Language Understanding. Benjamin Cummings, Menlo Park, USA

Alm CO, Sproat R (2005) Perceptions of emotions in expressive storytelling. In: International conference on speech and language processing (ICSLP), Lisbon, Portugal, pp 533–536

Alter K, Rank E, Kotz SA, Pfeifer E, Besson M, Friederici AD, Matiasek J (1999) On the relations of semantic and acoustic properties of emotions. In: International congress of phonetic sciences (ICPhS), San Francisco, USA

Alter K, Rank E, Kotz SA, Toepel U, Besson M, Schirmer A, Friederici AD (2000) Accentuation and emotions – Two different systems? In: Proceedings of ISCA workshop on speech and emotion, Belfast, United Kingdom

Álvarez Martínez C, Barrientos Cruz A (2005) Emotion recognition in non-structured utterances for human-robot interaction. In: IEEE International Workshop on Robot and Human Interactive Communication, Nashville, USA, pp 19–23

Amir N, Ron S, Laor N (2000) Analysis of an emotional speech corpus in Hebrew based on objective criteria. In: Proceedings of ISCA workshop on speech and emotion, Belfast, United Kingdom

Andersson G, Paiva A, Arafa Y, Piedade M, Botelho L, Ramos P, Mourão D, Sengers P (2002) Shell for affective user modelling. SAFIRA Deliverable D3.2

André E (2000) The generation of multimedia presentations. In: Dale R, Moisl H, Somers H (eds) Handbook of natural language processing. Marcel Dekker, New York, USA, pp 305–327

André E, Rehm M, Minker W, Bühler D (2004) Endowing spoken language dialogue systems with emotional intelligence. In: Tutorial and Research Workshop Affective Dialogue Systems, Irsee, Germany, pp 178–187

Androutsopoulos I, Aretoulaki M (2003) Natural language interaction. In: Mitkov R (ed) Handbook of computational linguistics, chapter 35. Oxford University Press, New York, USA, pp 629–649

Ang J, Dhillon R, Krupski A, Shriberg E, Stolcke A (2002) Prosody-based automatic detection of annoyance and frustration in human–computer dialog. In: International conference on speech and language processing (ICSLP), Denver, USA, pp 2037–2040

Anscombe E, Geach P (1970) Descartes philosophical writings. The Open University, Nelson

Arnold M (1960) Emotion and personality. Columbia University Press, New York, USA

Aubergé V, Audibert N, Rilliard A (2003) Why and how to control the authentic emotional speech corpora. In: European conference on speech and language processing (EUROSPEECH), Geneva, Switzerland, pp 185–188

Averill JR (1980) A constructivist view of emotion. In: Plutchik R, Kellerman H (eds) Emotion: Theory, research and experience, vol 1. Academic, New York, USA, pp 305–339

Azzini I, Falavignat D, Gretter R, Lanzola G, Orlandi M (2001) First Steps Toward an Adaptive Spoken Dialogue System in Medical Domain. In: European conference on speech and language processing (EUROSPEECH), Aalborg, Denmark, pp 1327–1330

Baken RJ, Orlikoff RF (2000) Clinical measurements of speech and voice. Singular Thomson learning, San Diego, USA

Bakis R (1976) Continuous speech word recognition via centisecond acoustic states. In: Proceedings of the 91st annual meeting of the acoustical society America, Washington DC, USA

Banse R, Scherer KR (1996) Acoustic profiles in vocal emotion expression. J Person Soc Psychol 70(3):614–636

Bard P (1934) On emotional expression after decortication with some remarks on certain theoretical views. Part II. Psychol Rev 41:424–449

Bartels A, Sendlmeier WF, Rolfes M, Burkhardt F (2006) Emo-DB. http://pascal.kgw.tu-berlin.de/emodb/

Batliner A, Fischer K, Huber R, Spilker J, Nöth E (2000) Desperately seeking emotions or: Actors, wizards, and human beings. In: Proceedings of ISCA workshop on speech and emotion, Belfast, United Kingdom

Batliner A, Hacker C, Steidl S, Nöth E, D'Arcy S, Russell M, Wong LP (2004a) "You stupid tin box" – Children interacting with the AIBO robot: A cross-linguistic emotional speech corpus. In: International conference on language resources and evaluation (LREC), Lisbon, Portugal, pp 171–174

Batliner A, Hacker C, Steidl S, Nöth E, Haas J (2004b) User states, user strategies, and system performance: How to match the one with the other. In: International conference on language resources and evaluation (LREC), Lisbon, Portugal, pp 171–174

Batliner A, Steidl S, Hacker C, Nöth E, Niemann H (2005) Private emotions vs. social interaction – towards new dimensions in research on emotion. In: Carberry S, de Rosis F (eds) Proceedings of the workshop on adapting the interaction style to affective factors at UM'05, Edinburgh, United Kingdom

Baum LE, Eagon JA (1967) An inequality with applications to statistical estimation for probalistic functions of markov processes and to a model for ecology. Am Math Soc Bull 73:360–363

Baum LE, Petrie T, Soules G, Weiss N (1970) A maximization technique occurring in the statistical analysis of probabilistic functions of Markov chains. Ann Math Stat 41(1):164–171

Baum LE, Sell GR (1968) Growth functions for transformations on manifolds. Pac J Math 27:211–227

Beller G, Marty A (2006) Talkapillar: Outil d'analyse de corpus oraux. In: Rencontres Jeunes Rechercheurs RJC-ED268, Paris, France

Benyon D, Turner P, Turner S (2005) Designing interactive systems: People, activities, contexts, technologies. Addison-Wesley, Harlow, United Kingdom

Beringer N, Kartal U, Louka K, Schiel F, Türk U (2002) PROMISE – A procedure for multimodal interactive system evaluation. In: Proceedings of the LREC workshop on multimodal resources and multimodal systems evaluation, Las Palmas, Spain, pp 77–80

Bernsen NO, Dybkjær H, Dybkjær L (1994) Wizard of Oz prototyping: How and when? In: CCI working papers in cognitive science and HCI, WPCS-94-1. Centre for Cognitive Science, Roskilde University, Roskilde, Denmark

Boersma P (2001) Praat, a system for doing phonetics by computer. Glot Int 5(9/10):341–345

Boersma P (2002) Praat Website. http://www.praat.org/

Bogert BP, Healy MJR, Tukey JW (1963) The quefrency alanysis of time series for echoes: Cepstrum, pseudo-autocovariance, cross-cepstrum, and Saphe cracking. In: Rosenblatt M (ed) Proceedings of the symposium on time series analysis, Providence, USA, pp 209–243

Bolinger D (1989) Intonation and its uses. Stanford University Press, Stanford, USA

Bosma W, André E (2004) Exploiting emotions to disambiguate dialogue acts. In: International conference on intelligent user interfaces (IUI), ACM Press, Funchal, Portugal, pp 85–92

Bourgeois J (2005) An LMS viewpoint on the local stability of second order blind source separation. In: IEEE workshop on statistical signal processing, Bordeaux, France, pp 1096–1101

Bourgeois J, Freudenberger J, Lathoud G (2005) Implicit control of noise canceller for speech enhancement. In: European conference on speech and language processing (EUROSPEECH), Lisbon, Portugal, pp 2065–2068

Braun A, Katerbow M (2005) Emotions in dubbed speech: An intercultural approach with respect to F0. In: International conference on speech and language processing (ICSLP), Lisbon, Portugal, pp 521–524

Brennan SE, Hulteen EA (1995) Interaction and feedback in a spoken language system: A theoretical framework. Knowl Syst 9:143–151

Brown P, Levinson SC (1987) Politeness – Some universals in language Use. Cambridge University Press, Cambridge, United Kingdom

Bühler D, Hamerich SW (2005) Towards VoiceXML compilation for portable embedded applications in ubiquitous environments. In: European conference on speech and language processing (EUROSPEECH), Lisbon, Portugal, pp 3397–4000

Bühler D, Riegler M (2005) Integrating dialogue management and domain reasoning. In: Proceedings of SPECOM, Patras, Greece, pp 409–412

Bui TH, Zwiers J, Poel M, Nijholt A (2006) Toward affective dialogue modeling using partially observable Markov decision processes. In: Proceedings of workshop emotion and computing, 29th annual German conference on artificial intelligence, Bremen, Germany

Burkhardt F, Ajmera J, Englert R, Stegmann J, Burleson W (2006a) Detecting anger in automated voice portal dialogs. In: International conference on speech and language processing (ICSLP), Pittsburgh, USA, pp 1053–1056

Burkhardt F, Audibert N, Malatesta L, Türk O, Arslan L, Aubergé V (2006b) Emotional prosody – Does culture make a difference? In: 3rd international conference on speech prosody, Dresden, Germany, pp 245–248

Burkhardt F, Paeschke A, Rolfes M, Sendlmeier WF, Weiss B (2005) A database of german emotional speech. In: European conference on speech and language processing (EUROSPEECH), Lisbon, Portugal, pp 1517–1520

Burton RR (1976) Semantic grammar: An engineering technique for constructing natural language understanding systems. Technical report 3353, Bolt Beranek & Newman, Inc, Cambridge, USA

Burzio L (1993) English stress, vowel length, and modularity. J Linguist 29(2):359–418

Busso C, Narayanan S (2006) Interplay between linguistic and affective goals in facial expression during emotional utterances. In: Proceedings of 7th international seminar on speech production (ISSP), Ubatuba, Brazil, pp 549–556

Campbell JP (1997) Speaker recognition: A tutorial. Proc IEEE 85(9):1437–1462

Campbell N (2000) Databases of emotional speech. In: Proceedings of ISCA workshop on speech and emotion, Belfast, United Kingdom

Campbell N (2002) The Recording of emotional speech – JST/CREST database research. In: International conference on language resources and evaluation (LREC), vol 6. Las Palmas, Spain, pp 2029–2032

Campbell N, Devillers L, Douglas-Cowie E, Aubergé V, Batliner A, Tao J (2006) Resources for the processing of affect in interactions. In: ELRA (ed) International conference on language resources and evaluation (LREC) Genova, Italy, pp 25–28

Cannon WB (1927) The James-Lange theory of emotion: A critical examination and an alternative theory. Am J Psychol 39:106–124

Carletta JC (1996) Assessing the reliability of subjective codings. Computat Ling 22(2):249–254

Carlson R, Granström B (1977) Speech synthesis. In: Hardcastle WJ, Laver J (eds) The handbook of phonetic sciences. Blackwell, Oxford, United Kingdom, pp 768–788

Carofiglio V, de Rosis F (2005) In favour of cognitive models of emotions. In: Workshop on mind-minding agents at AISB, Hatfield, United Kingdom

Carofiglio V, de Rosis F, Grassano R (2002) Mixed emotion modeling. In: Proceedings of the AISB02 symposium on animating expressive characters for social interactions, London, United Kingdom, pp 5–10. The Society for the Study of Artificial Intelligence and Simulation of Behaviour

Carofiglio V, de Rosis F, Grassano R (2003) Dynamic models of mixed emotion activation. In: Cañamero L, Aylett R (eds) Animating expressive characters for social interactions. John Benjamins, Amsterdam, The Netherlands

Cauldwell RT (2000) Where did the anger go? The role of context in interpreting emotion in speech. In: Proceedings of ISCA workshop on speech and emotion, Belfast, United Kingdom

Chateau N, Maffiolo V, Blouin C (2004) Analysis of emotional speech in voice mail messages: The influence of speakers' gender. In: International conference on speech and language processing (ICSLP), Jeju, Korea

Chu S-W, O'Neill I, Hanna P, McTear M (2005) An approach to multi-strategy dialogue management. In: International conference on speech and language processing (ICSLP), Lisbon, Portugal, pp 865–868

Chuang Z-J, Wu C-H (2004) Emotion recognition using acoustic features and textual content. In: Proceedings of the IEEE international conference on multimedia and expo (ICME), vol 1. Taipei, Taiwan, pp 53–56

Chung S (2000) L'expression et la perception de l'émotion extraite de la parole spontanée: Évidences du coréen et de l'anglais. PhD thesis, Université, de la Sorbonne Nouvelle, Paris III

Clavel C, Vasilescu I, Devillers L, Ehrette T (2004) Fiction Database for Emotion Detection in Abnormal Situations. In: International conference on speech and language processing (ICSLP), Jeju, Korea

Clavel C, Vasilescu I, Devillers L, Ehrette T, Richard G (2006) Fear-type emotions of the SAFE corpus: Annotation issues. In: International conference on language resources and evaluation (LREC), Genova, Italy, pp 1099–1104

Cohen J (1960) A coefficient of agreement for nominal scales. Educ Psychol Measure 20:37–40

Cohen MH, Giangola JP, Balogh J (2004) Voice user interface design. Addison-Wesley, Harlow, United Kingdom

Cohen PR (1995) Empirical methods for artificial intelligence. MIT Press, Boston, USA

Conati C, Gertner A, VanLehn K (2002) Using Bayesian networks to manage uncertainty in student modeling. J User Model User-Adapted Interact 12(4):371–417

Cornelius RR (2000) Theoretical approaches to emotion. In: Proceedings of ISCA workshop on speech and emotion, Belfast, United Kingdom

Cowie R (2000) Describing the emotional states expressed in speech. In: Proceedings of ISCA workshop on speech and emotion, Belfast, United Kingdom

Cowie R, Douglas-Cowie E (1996) Automatic statistical analysis of the signal and prosodic signs of emotion in speech. In: International conference on speech and language processing (ICSLP), vol 3. Philadelphia, USA, pp 1989–1992

Cowie R, Douglas-Cowie E, Apolloni B, Taylor J, Romano A, Fellenz W (1999a) What a neural net needs to know about emotion words. In: Proceedings of the 3rd IMACS international multiconference on circuits, systems, communications and computers (CSCC '99), Athens, Greece, pp 5311–5316

Cowie R, Douglas-Cowie E, Cox C, Cemgil AT (2004) D09: Final version of non-verbal speech parameter extraction module. ERMIS project IST-2000-29319

Cowie R, Douglas-Cowie E, Romano A (1999b) Changing emotional tone in dialogue and its prosodic correlates. In: ESCA tutorial and research workshop on dialogue and prosody, Veldoven, the Netherlands, pp 41–46

Cowie R, Douglas-Cowie E, Savvidou S, McMahon E, Sawey M, Schröder M (2000) "Feeltrace": An instrument for recording perceived emotion in real time. In: Proceedings of ISCA workshop on speech and emotion, Belfast, United Kingdom

Cowie R, Douglas-Cowie E, Tsapatsoulis N, Votsis G, Kollias SD, Fellenz WA, Taylor JG (2001) Emotion recognition in human-computer interaction. IEEE Signal Proc Mag 12:32–80

Cowie R, Schröder M (2004) Piecing together the emotion jigsaw. In: Workshop on multimodal interaction and related machine learning algorithms (MLMI04), Martigny, Switzerland, pp 305–317

Craggs R (2004) Annotating emotion in dialogue – Issues and approaches. In: 7th annual CLUK research colloquium, Birmingham, United Kingdom

Craggs R, McGee Wood M (2003) Annotating emotion in dialogue. In: Proceedings of the 4th SIGdial workshop on discourse and dialogue, Sapporo, Japan

Cristiani N, Shawe-Taylor J (2000) An introduction to support vector machines and other kernel-based learning methods. Cambridge University Press, Cambridge, United Kingdom

Crystal D (1969) Prosodic systems and intonation in English. Cambridge University Press, Cambridge, United Kingdom

d'Alessandro C, Mertens P (1995) Automatic pitch contour stylization using a model of tonal perception. Comput Speech Lang 9(3):257–288

Dan-Ning Jiang L-HC (2004) Classifying emotion in Chinese speech by decomposing prosodic features. In: International conference on speech and language processin (ICSLP), Jeju, Korea

Darwin C (1872) The expression of the emotions in man and animals. John Murray, London, United Kingdom

Dasarathy BV (1991) Nearest neighbor (NN) norms: NN pattern classification techniques. IEEE Computer Society Press, Los Alamitos, USA

De Mori R (1998) Spoken dialogue with computers. Academic Press, Orlando, USA

Dellaert F, Polzin T, Waibel A (1996) Recognizing emotion in speech. In: International conference on speech and language processing (ICSLP), Philadelphia, USA, pp 1970–1973

Deller Jr. JR, Proakis JG, Hansen JH (1993) Discrete-time processing of speech signals. Prentice Hall PTR, Upper Saddle River, USA

Delorme F, Lehuen J (2003) Dialog planning and domain knowledge modeled in terms of tasks and methods – A flexible framework for dialog managing. In: Proceedings of the international symposium on methodologies for intelligent systems, Maebashi City, Japan, pp 689–693

Descartes (1649) Les Passions de l'âme. Paris, France

Devillers L, Cowie R, Martin J-C, Douglas-Cowie E, Abrilian S, McRorie M (2006) Real life emotions in French and English TV video clips: An integrated annotation protocol combining continous and discrete approaches. In: International conference on language resources and evaluation (LREC), Genova, Italy, pp 1105–1110

Devillers L, Lamel L, Vasilescu I (2003) Emotion detection in task-oriented spoken dialogues. In: proceedings of the IEEE international conference on multimedia and expo (ICME), vol 3. Baltimore, USA, pp 549–552

Devillers L, Vasilescu I, Lamel L (2002) Annotation and detection of emotion in a task-oriented human-human dialog corpus. In: ISLE Workshop on dialogue tagging, Edinburgh, United Kingdom

Douglas-Cowie E, Cowie R, Schröder M (2000) A new emotion database: Considerations, sources and scope. In: proceedings of ISCA workshop on speech and emotion, Belfast, United Kingdom

Duff D, Gates B, LuperFoy S (1996) An architecture for spoken dialogue management. In: International conference on speech and language processing (ICSLP), vol 2. Philadelphia, USA, pp 1025–1028

Durbin R, Eddy SR, Krogh A, Mitchison G (1999) Biological sequence analysis: Probabilistic models of proteins and nucleic acids. Cambridge University Press, Cambridge, United Kingdom

Dybkjær L, Bernsen NO (2001) Usability evaluation in spoken language dialogue systems. In: Proceedings of the ACL workshop on evaluation methodologies for language and dialogue systems, Toulouse, France, pp 9–18

Ecma International (1999) ECMAScript language specification. ECMA Standard 262, 3rd Edition
Ecma International (2005) ECMAScript for XML (E4X) specification. ECMA Standard 357, 2nd Edition
Ekman P (1977) Biological and cultural contributions to body and facial movement. In: Blacking J (ed) The anthropology of the body. Academic, London, United Kingdom, pp 34–84
Ekman P (1992) An argument for basic emotions. Cogn Emot 6(3–4):169–200
Ekman P (1999) Basic emotions. In: Dalgleish T, Power M (eds) Handbook of cognition and emotion. John Wiley & Sons, Ltd., New York, USA, pp 301–320
Ekman P, Friesen WV, Ellsworth P (1982) What emotion categories or dimensions can observers judge from facial behaviour? In: Ekman P (ed) Emotion in the human face. Cambridge University Press, New York, USA, pp 39–55
Engberg IS, Hansen AV, Andersen O, Dalsgaard P (1997) Design, recoding and verification of a danish emotional speech database. In: European conference on speech and language processing (EUROSPEECH), Rhodes, Greece, pp 1695–1698
Fakotakis N (2004) Corpus design, recording and phonetic analysis of Greek emotional database. In: International conference on language resources and evaluation (LREC), Lisbon, Portugal, pp 1391–1394
Fallside F, Woods WA (1985) Computer speech processing. Prentice-Hall, Englewood Cliffs, USA
Fék M, Németh G, Olaszy G, Gordos G (2004) Design of a hungarian emotional database for speech analysis and synthesis. In: Proceedings of affective dialogue systems, tutorial and research workshop, ADS 2004, Kloster Irsee, Germany, pp 113–116
Feldman JA, Ballard DH (1982) Connectionist models and their properties. Cogn Sci 6(3):205–254
Fernandez R, Picard RW (2003) Modeling drivers' speech under stress. Speech Commun 40: 145–149
Fillmore CJ (1968) The case for case. In: Bach E, Harms R (eds) Universals in linguistic theory. Holt, Rhinehart and Winston, New York, USA, pp 1–88
Fiscus JG (1997) A post-processing system to yield reduced word error rates: Recogniser output voting error reduction (ROVER). In: Proceedings of the IEEE workshop on automatic speech recognition and understanding, Santa Barbara, USA, pp 347–352
Fleiss JL (1971) Measuring nominal scale agreement among many raters. Psychol Bull 76(5): 378–382
Forbes-Riley KM, Litman DJ (2004) Predicting emotion in spoken dialogue from multiple knowledge sources. In: Proceedings of the human language technology conference of the north american chapter of the association for computational linguistics (NAACL-HLT), pp 201–208
Fotinea S-E, Bakamidis S, Athanaselis T, Dologlou I, Carayannis G, Cowie R, Douglas-Cowie E, Fragopanagos NF, Taylor JG (2003) Emotion in speech: Towards an integration of linguistic, paralinguistic, psychological analysis. In: Joint international conference ICANN/ICONIP, Istanbul, Turkey, pp 1125–1132
Fox NA (1992) If it's not left it's right. Am Psychol 46:863–872
France DJ, Shiavi RG, Silverman S, Silverman M, Wilkes DM (2000) Acoustical properties of speech as indicators of depression and suicidal risk. IEEE Trans Biomed Eng 47(7):829–837
Frederking RE (1996) Grice's maxims: Do the right thing. In: Working notes of the AAAI'96 spring symposium on computational implicature: Computational approaches to interpreting and generating conversational implicature, Palo Alto, USA, pp 21–26
Freund Y, Schapire RE (1997) A decision-theoretic generalization of on-line learning and an application to boosting. J Comput Syst Sci 55(1):119–139
Frijda NH (1970) Emotion and recognition of emotion. In: Arnold M (ed) Feelings and emotions. Academic, New York, USA
Frijda NH (1986) The emotions. Cambridge University Press, Cambridge, United Kingdom
Fujisawa T, Cook ND (2004) Identifying emotion in speech prosody using acoustical cues of harmony. In: International conference on speech and language processing (ICSLP), Jeju, Korea
Gharavian D, Ahadi SM (2005) The effect of emotion on Farsi speech parameters: A statistical evaluation. In: Proceedings of SPECOM, Patras, Greece, pp 463–466

Gibbon D, Lüngen H (1999) Consistent vocabularies for spoken language machine translation systems. In: Multilinguale Corpora. Codierung, Strukturierung, Analyse, Frankfurt, Germany, pp 169–178

Glass J, Goddeau D, Hetherington L, McCandless M, Pao C, Phillips M, Polifroni J, Seneff S, Zue V (1995) The MIT ATIS system: December 1994 progress report. In: Proceedings of the ARPA spoken language systems technology workshop, Austin, USA

Glass J, Seneff S (2003) Flexible and personalizable mixed-initiative dialogue systems. In: Proceedings of the HLT-NAACL 2003 workshop on research directions in dialogue processing, Edmonton, Canada

Gmytrasiewicz PJ, Lisetti CL (2000) Using decision theory to formalize emotions for multi-agent system applications: Preliminary report. In: 4th international conference on multi-agent systems, Boston, USA, pp 391–392

González-Bernal JA, López-López A, Muñoz-Arteaga J, Montes y Gómez M, Reyes-García CA, Villaseñor-Pineda L (2002) Natural language dialogue system for information retrieval. In: Proceedings of the international workshop on research and development of human communication technologies for conversational interaction and learning, Puebla, Mexico

Gorin AL, Riccardi G, Wright JH (1997) How may I help you? Speech Commun 23(1/2):113–127

Gratch J, Marcella S (2004) A domain-independent framework for modeling emotion. J Cogn Syst Res 5(4):269–306

Gray JA (1982) The neuropsychology of anxiety. Oxford University Press, Oxford, United Kingdom

Greasley P, Sherrard C, Waterman M (2000) Emotion in language and speech: Methodological issues in naturalistic approaches. Lang Speech 43:355–375

Grice HP (1975) Logic and conversation. In: Cole P, Morgan JL (eds) Syntax and semantics, Vol. 3: Speech Acts. Academic, New York, USA, pp 41–58

Gussenhoven C (2004) The phonology of tone and intonation. Cambridge University Press, Cambridge, United Kingdom

Gutiérrez-Arriola JM, Montero JM, Vallejo JA, Córdoba R, San-Segundo R, Pardo JM (2001) A new multi-speaker formant synthesizer that applies voice conversion techniques. In: European conference on speech and language processing (EUROSPEECH), Aalborg, Denmark, pp 357–360

Hagen E, Popowich F (2000) Flexible speech act based dialogue management. In: Dybkjær L, Hasida K Traum D (eds) Proceedings of the first SIGdial workshop on discourse and dialogue. Association for computational linguistics, Somerset, NJ, pp 131–140

Hajdinjak M, Mihelič F (2006) The Paradise evaluation framework: Issues and findings. Computat Linguist 32(2):263–272

Hansen JH, Bou-Ghazale SE, Sarikaya R, Pellom B (1998) Getting started with the SUSAS: Speech under simulated and actual stress database. Technical report RSPL-98-10, Robust Speech Processing Laboratory, Duke University, Durham, USA

Hassel L, Hagen E (2005) Adaptation of an automotive dialogue system to users' expertise. In: 6th SIGdial workshop on discourse and dialogue, Lisbon, Portugal, pp 222–226

Henton C (2005) Bitter pills to swallow. ASR and TTS have drug problems. Int J Speech Technol 8(3):247–257

Hermansky H (1990) Perceptual linear predictive (PLP) analysis of speech. J Acoust Soc Am 87(4):1738–1752

Hermansky H, Morgan N (1994) RASTA processing of speech. IEEE Trans Speech Audio Process 2(4):578–589

Hillard D, Hoffmeister B, Ostendorf M, Schlüter R, Ney H (2007) iROVER: Improving system combination with classification. In: Proceedings of the human language technology conference of the north American chapter of the association for computational linguistics (NAACL-HLT), Rochester, USA, pp 44–51

Hirose K, Sato K, Minematsu N (2004) Improvement in corpus-based generation of F_0 contours using generation process model for emotional speech synthesis. In: International conference on speech and language processing (ICSLP), Jeju, Korea

Holzapfel H, Fügen C, Denecke M, Waibel A (2002) Integrating emotional cues into a framework for dialogue management. In: proceedings of the international conference on multimodal interfaces, Pittsburgh, USA, pp 141–148

Howard RA (1971) Dynamic probabilistic systems, Vol 1: Markov chains. John Wiley & Sons, Ltd, New York, USA

Hunt A, McGlashan S (2004) Speech recognition grammar specification, Version 1.0. `http://www.w3.org/TR/speech-grammar/`

Hurtado LF, Griol D, Segarra E, Sanchis E (2006) A stochastic approach for dialog management based on neural networks. In: International conference on speech and language processing (ICSLP), Pittsburgh, USA, pp 49–52

Iida A, Campbell N, Higuchi F, Yasumura M (2003) A corpus-based speech synthesis system with emotion. Speech Commun 40:161–187

Ikehara S, Miyazaki M, Shirai S, Yokoo A, Nakaiwa H, Ogura K, et al (1999) A japanese lexicon. Iwanami Shoten, Tokyo, Japan

Ioannou SV, Raouzaiou AT, Tzouvaras VA, Mailis TP, Karpouzis KC, Kollias SD (2005) Emotion recognition through facial expression analysis based on a neurofuzzy network. Neural Networks 18:423–435

Iriondo I, Guaus R, Rodríguez A, Lázaro P, Montoya N, Blanco JM, Bernandas D, Oliver JM, Tena D, Longhi L (2000) Validation of an acoustical modelling of emotional expression in spanish using speech synthesis techniques. In: Proceedings of ISCA workshop on speech and emotion, Belfast, United Kingdom

Isard SD, Miller DA (1986) Diphone synthesis techniques. In: Proceedings of IEE international conference on speech input/output, London, United Kingdom, pp 77–82

Ito A, Shimada K, Suzuki M, Makino S (2006) A user simulator based on VoiceXML for evaluation of spoken dialog systems. In: International conference on speech and language processing (ICSLP), Pittsburgh, USA, pp 1045–1048

ITU-T P.supp24 (2005) P-series recommendation – Supplement 24. Parameters describing the interaction with spoken dialogue systems

Iwai A, Yano Y, Okuma S (2004) Complex emotion recognition system for a specific user using SOM based on prosodic features. In: International conference on speech and language processing (ICSLP), Jeju, Korea

Izard CE (1971) The face of emotion. Appleton century crofts, New York, USA

James W (1884) What is an emotion? Mind, 9:188–205

Jelinek F, Lafferty J, Magerman DM, Ratnaparkhi A, Roukos S. (1994) Decision tree parsing using a hidden derivation model. In: Proceedings of the ARPA workshop on human language technology, pp 260–265

Jelinek F, Lafferty J, Mercer R (1992) Basic methods of probabilistic context free grammars. Speech recognition and understanding. Recent advances, 75:345–360

Jiang X, Tian L, Han M (2005) Separability and recognition of emotion states in multilingual speech. In: Proceedings of the international conference on communications, circuits and systems, vol 2. Hong Kong, China, pp 861–864

Johannesson R (1988) Informationsteori – Grundvalen för telekommunikation. Studenlitteratur AB, Lund, Sweden

Johnston M, Bangalole S, Vasireddy G, Stent A, Eblen P, Walker MA, Whittaker S, Maloor P (2002) MATCH: An architecture for multimodal dialogue systems. In: Proceedings of the 40th annual meeting of the association for computational linguistics (ACL2002), Philadelphia, USA, pp 376–383

Johnstone T, Scherer KR (1999) The effects of emotions on voice quality. In: International congress of phonetic sciences (ICPhS), San Francisco, USA, pp 2029–2032

Jokinen K, Kerminen A, Kaipainen M, Jauhiainen T, Wilcock G, Turunen M, Hakulinen J, Kuusisto J, Lagus K (2002) Adaptive dialogue systems – interaction with interact. In: Proceedings of the 3rd SIGdial workshop on discourse and dialogue, Philadelphia, USA, pp 64–73

Jönsson A, Dahlbäck N (2000) Distilling dialogues – A method using natural dialogue corpora for dialogue systems development. In: 6th applied natural language processing conference (ANLP), Seattle, USA, pp 44–51

Juang B-H, Rabiner LR (1991) Hidden markov models for speech recognition. Technometrics 33(3):251–272

Jun S-A (2005) Prosodic typology. The phonology of intonation and phrasing. Oxford University Press, New York, USA

Kalman RE (1960) A new approach to linear filtering and prediction problems. Trans ASME J Basic Eng 82:35–45

Kamm C, Narayanan S, Dutton D, Ritenour R (1997) Evaluating spoken dialog systems for telecommunication systems. In: European conference on speech and language processing (EUROSPEECH), Rhodes, Greece, pp 22–25

Karlsson I (1999) Within-speaker variability in the VeriVox database. In: Proceedings of the twelfth swedish phonetics conference (Fonetik 99), number 81 in Gothenburg Papers in Theoretical Linguistics, Gothenburg, Sweden, pp 93–96

Kehrein R (2001) Linguistische und psychologische Aspekte der Erforschung des prosodischen Emotionsausdrucks. In: Germanische Linguistik (GL), 157–158:91–123

Kim EH, Hyun KH, Kwak YK (2005) Robust emotion recognition feature, frequency range of meaningful signal. In: IEEE international workshop on robots and human interactive communication (RO-MAN), Nashville, USA, pp 667–671

Kim I-S (2006) Automatic speech recognition: Reliability and pedagogical implications for teaching pronunciation. Educ Technol Soc 9(1):322–334

Kim J, André E (2006) Emotion recognition using physiological and speech signal in short-term observation. In: Tutorial and research workshop on perception and interactive technologies (PIT06), Irsee, Germany, pp 53–64

Kim KH, Bang SH, Kim SR (2004a) Emotion recognition system using short-term monitoring of physiological signals. Med Biol Eng Comput 42:419–427

Kim S-J, Kim K-K, Hahn M (2004b) Study on emotional speech features in Korean with its application to voice color conversion. In: International conference on speech and language processing (ICSLP), Jeju, Korea

Kipp M (2001) ANVIL – A generic annotation tool for multimodal dialogue. In: European conference on speech and language processing (EUROSPEECH), Aalborg, Denmark, pp 1367–1370

Klasmeyer G (1996) Perceptual cues for emotional speech. In: Workshop on the auditory basis of speech perception, Keele, United Kingdom, pp 154–157

Klasmeyer G, Johnstone T, Bänzinger T, Sappok C, Scherer KR (2000) Emotional voice variability in speaker verification. In: Proceedings of ISCA workshop on speech and emotion, Belfast, United Kingdom

Kleinginna PR, Kleinginna AM (1981) A categorized list of emotion definitions, with suggestions for a consensual definition. Motivat Emot 5(4):345–359

Kochanski G, Grabe E, Coleman J, Rosner B (2005) Loudness predicts prominence: Fundamental frequency lends little. J Acoust Soc Am 118(2):1038–1054

Koike K, Suzuki H, Saito H (1998) Prosodic parameters in emotional speech. In: International conference on speech and language processing (ICSLP), Sydney, Australia, pp 679–682

Komatani K, Kawahara T (2000) Flexible mixed-initiative dialogue management using concept-level confidence measures of speech recognizer output. In: Proceedings of the 18th conference on computational linguistics, Saarbrücken, Germany, pp 467–473

Krahmer E, Landsbergen J, Pouteau X (1997) How to obey the 7 commandments for spoken dialogue systems. In: Proceedings of the (E)ACL workshop on interactive spoken dialog systems, Madrid, Spain, pp 82–89

Kshirsagar S, Magnenat-Thalmann N (2002) A Multilayer Personality Model. In: Proceedings of the 2nd international symposium on smart graphics, Hawthorne, USA, pp 107–115

Kumar R, Rosé CP, Litman DJ (2006) Identification of confusion and surprise in spoken dialog using prosodic features. In: International conference on speech and language processing (ICSLP), Pittsburgh, USA, pp 1842–1845

Kwon O-W, Chan K, Hao J, Lee T-W (2003) Emotion recognition by speech signals. In: European conference on speech and language processing (EUROSPEECH), Geneva, Switzerland, pp 125–128

Ladd DR (1996) Intonational phonology. Cambridge University Press, Cambridge, United Kingdom

Lamere P, Kwok P, Walker W, Gouvea E, Singh R, Raj B, Wolf P (2003) Design of the CMU Sphinx-4 decoder. In: European conference on speech and language processing (EUROSPEECH), Geneva, Switzerland, pp 1191–1184

Lange CG (1885) Om Sindsbevægelser. Et psyko-fysiologisk Studie. Rasmussen, Copenhagen, Denmark

Larson JA (2001) VoiceXML 2.0 and the W3C Speech Interface Framework. In: IEEE Workshop on automatic speech recognition and understanding (ASRU), Madonna di Campiglio, Italy, pp 5–8

Larson JA (2002) VoiceXML: Introduction to developing speech applications. Prentice Hall, Upper Saddle River, USA

Larsson S (2000) Godis Demo. `http://www.ling.gu.se/leifg/WebDemos/godis2.0a/`

Larsson S, Berman A, Grönqvist L, Kronlid F (2002) TRINDIKIT 3.0 Manual, D6.4, Siridus Project

Larsson S, Traum D (2000) Information state and dialogue management in the TRINDI dialogue move engine toolkit. Nat Lang Eng 6:323–340

Laskowski K, Burger S (2006) Annotation and analysis of emotionally relevant behavior in the ISL meeting corpus. In: International conference on language resources and evaluation (LREC), Genova, Italy, pp 1111–1116

Laukka P (2004) Vocal expression of emotion. PhD thesis, Uppsala University

Lazarus RS (1982) Thoughts on the relations between emotion and cognition. Am Psychol 37:1019–1024

Lazarus RS (1991) Emotion and adaptation. Oxford University Press, New York, USA

Lazarus RS, Smith CA (1988) Knowledge and appraisal in the cognition-emotion relationship. Cogn Emot 2:281–300

Lecœuche R (2001) Learning optimal dialogue management rules by using reinforcement learning and inductive logic programming. In: Proceedings of the 2nd meeting of the NAACL, Pittsburgh, USA

Lee CM, Narayanan S (2003) Emotion recognition using a data-driven fuzzy interference system. In: European conference on speech and language processing (EUROSPEECH), Geneva, Switzerland, pp 157–160

Lee CM, Yildirim S, Bulut M, Kazemzadeh A, Busso C, Deng Z, Lee S, Narayanan S (2004) Emotion recognition based on phoneme classes. In: International conference on speech and language processing (ICSLP), Jeju, Korea

Lee S, Bresch E, Adams J, Kazemzadeh A, Narayanan S (2006) A study of emotional speech articulation using a fast magnetic resonance imaging technique. In: International conference on speech and language processing (ICSLP), Pittsburgh, USA, pp 2234–2237

Lee S, Yildirim S, Kazemzadeh A, Narayanan S (2005) An articulatory study of emotional speech production. In: International conference on speech and language processing (ICSLP), Lisbon, Portugal, pp 497–500

Lefèvre F Gauvain J-L Lamel L (2001) Improving genericity for task-independent speech recognition. In: European conference on speech and language processing (EUROSPEECH), Aalborg, Denmark, pp 1241–1244

Lesher GW, Moulton BJ, Higginbotham DJ (1999) Effects of ngram order and training text size on word prediction. In: Proceedings of the RESNA 1999 annual conference, Long Beach, USA, pp 52–54

Levenshtein VI (1966) Binary codes capable of correcting deletions, insertions, reversals. Sov Phys Doklady 10(8):707–710

Levin E, Narayanan S, Pieraccini R, Biatov K, Bocchieri E, Di Fabbrizio G, Eckert W, Lee S, Pokrovsky A, Rahim M, Ruscitti P, Walker M (2000a) The AT&T DARPA communicator mixed-initiative spoken dialog system. In: International conference on speech and language processing (ICSLP), Beijing, China, pp 122–125

Levin E, Pieraccini R (1995) "CHRONUS", the next generation. In: Proceedings of the DARPA Speech and Natural Language Workshop, Austin, USA, pp 269–271

Levin E, Pieraccini R, Eckert W (2000b) A Stochastic Model of Human Machine Interaction for Learning Dialog Strategies. IEEE Trans Speech Audio Process 8(1):11–23

Li J, Najmi A, Gray RM (2000) Image classification by a two-dimensional hidden Markov model. IEEE Trans Signal Process 48(2):517–533

Li Y, Zhao Y (1998) Recognizing Emotions in Speech Using Short-term and Long-term Features. In: International conference on speech and language processing (ICSLP), Sydney, Australia, pp 2255–2258

Linhard K, Haulick T (1999) Noise subtraction with parametric recursive gain curves. In: European conference on speech and language processing (EUROSPEECH), Budapest, Hungary, pp 2611–2614

Lippmann RP (1997) Speech recognition by machines and humans. Speech Commun 22(1):1–15

Liscombe J, Riccardi G, Hakkani-Tür D (2005) Using context to improve emotion detection in spoken dialog systems. In: International conference on speech and language processing (ICSLP), Lisbon, Portugal, pp 1845–1848

Liscombe J, Venditti J, Hirschberg J (2003) Classifying subject ratings of emotional speech using acoustic features. In: Proceedings of the 8th European conference on speech communication and technology, Geneva, Switzerland, pp 725–728

Litman DJ, Kearns MS, Singh SB, Walker MA (2000) Automatic optimization of dialogue management. In: Proceedings of the 17th conference on computational linguistics, Saarbrücken, Germany, pp 502–508

Litman DJ, Pan S (2002) Designing and evaluating an adaptive spoken dialogue system. User Model User-Adapted Interact 12:111–137

Lopes Rodrigues LM, Carvalho M (2004) Emotional and motivational ITS architecture. In: Proceedings of the IEEE international conference on advanced learning techniques (ICALT), Joensuu, Finland, pp 276–280

López-Jaquero V, Montero F, Fernández-Caballero A, Lozano MD (2005) Towards adaptive user interfaces generation. In: Camp O, Filipe JBL, Hammoudi S, Piattini M, (eds) Enterprise systems V. Springer, the Netherlands, pp 226–232

Luengo I, Navas E, Hernáez I, Sánchez J (2005) Automatic emotion recognition using prosodic parameters. In: International conference on speech and language processing (ICSLP), Lisbon, Portugal, pp 493–496

LuperFoy S, Loehr D, Duff D, Miller K, Reeder F, Harper L (1998) An architecture for dialogue management, context tracking, pragmatic adaptation in spoken dialogue systems. In: Proceedings of the 17th international conference on computational linguistics, Montreal, Canada, pp 794–801. Association for Computational Linguistics

Ma X, Lee H, Bird S, Maeda K (2002) Models and tools for collaborative annotation. In: International conference on language resources and evaluation (LREC), Las Palmas, Spain, pp 2066–2073

Macherey K, Ney H (2003) Features for tree based dialogue management. In: European conference on speech and language processing (EUROSPEECH), Geneva, Switzerland, pp 601–604

Makarova V, Petrushkin VA, Ruslana VA (2002) A database of russian emotional utterances. In: International conference on speech and language processing (ICSLP), Denver, USA, pp 2041–2044

Makhoul J (1975) Linear prediction: A tutorial review. Proc IEEE 63(4):561–580

Marin J-M, Mengersen K, Robert CP (2005) Bayesian modelling and inference on mixtures of distributions. In: Dey D, Dao CR (eds) Handbook of statistics, vol 25. Elsevier-Sciences, Amsterdam, the Netherlands

Markov AA (1907) Extension of the limit theorems of probability theory to a sum of variables connected in a chain (in Russian). Bulletin of the imperial academy of sciences, XXII. reprinted and translated to English in Appendix B of Howard (1971)

Martin J-C, Caridakis G, Devillers L, Karpouzis K, Abrilian S (2006) Manual annotation and automatic image processing of multimodal emotional behaviours: Validating the annotation of TV interviews. In: International conference on language resources and evaluation (LREC), Genova, Italy, pp 1127–1132

Martinovski B, Traum D (2003) Breakdown in human-machine interaction: The error is the clue. In: Proceedings of the ISCA tutorial and research workshop on Error handling in dialogue systems, pp 11–16

Masters T (1994) Signal and image processing with neural networks. John Wiley & Sons, Ltd, New York, USA

Matsumoto K, Minato J, Ren F, Kuroiwa S (2005) Estimating human emotions using wording and sentence patterns. In: Proceedings of the IEEE international conference on information acquisition, Hong Kong and Macau, China, pp 421–426

Matsunaga S, Sakaguchi S, Yamashita M, Miyahara S, Nishitani S, Shinohara K (2006) Emotion detection in infants' cries based on a maximum likelihood approach. In: International conference on speech and language processing (ICSLP), Pittsburgh, USA, pp 1834–1837

McCulloch W, Pitts W (1943) A logical calculus of the ideas immanent in nervous activity. Bull Math Biophys 7:115–133

McDougall W (1926) An introduction to social psychology. Luce, Boston, USA

McGilloway S, Cowie R, Douglas-Cowie E (1995) Prosodic signs of emotion in speech: Preliminary results from a new technique for automatic statistical analysis. In: International congress of phonetic sciences (ICPhS), vol 1. Stockholm, Sweden, pp 250–253

McGilloway S, Cowie R, Douglas-Cowie E, Gielen S, Westerdijk M, and Stroeve S (2000) Approaching automatic recognition of emotion from voice: A rough benchmark. In: Proceedings of ISCA workshop on speech and emotion, Belfast, United Kingdom

McMahon E, Cowie R, Kasderidis S, Taylor JG, Kollias S (2003) What chance that a DC could recognise hazardous mental states from sensor outputs? In: DC Tales, Santorini, Greece

McTear MF (2004) Spoken dialogue technology – Toward the conversational user interface. Springer, London, United Kingdom

Meng H, Pittermann J, Pittermann A, Minker W (2007) Combined speech-emotion recognition for spoken human-computer interfaces. In: IEEE international conference on signal processing and communications (ICSPC), Dubai, United Arab Emirates

Miceli M, de Rosis F, Poggi I (2006) Emotional and non-emotional persuasion. Appl Artif Intell Int J 20(10):849–879

Miller S, Bobrow R, Ingria R, Schwartz R (1994) Hidden understanding models of natural language. In: Proceedings of the 32nd annual meeting of the association for computational linguistics (ACL1994), Las Cruces, USA, pp 25–32

Minker W, Albalate A, Bühler D, Pittermann A, Pittermann J, Strauss P-M, Zaykovskiy D (2006a) Recent trends in spoken language dialogue systems. In: ITI 4th international conference on information and communications technology (ICICT 2006), Cairo, Egypt

Minker W, Pittermann J, Pittermann A, Strauss P-M, Bühler D (2006b) Next-generation human-computer interfaces – Towards intelligent, adaptive and proactive spoken language dialogue systems. In:2nd IEE international conference on intelligent environments, Athens, Greece

Minker W, Waibel A, Mariani J (1999) Stochastically-based semantic analysis, vol 514 of The Kluwer international series in engineering and computer science. Kluwer, Boston, USA

Miwa H, Umetsu T, Takanishi A, Takanobu H (2000) Robot personalization based on the mental dynamics. In: IEEE/RSJ conference on intelligent robots and systems, vol 1, pp 8–14

Montero JM, Gutiérrez-Arriola J, Colás J, Macías-Guarasa J, Enríquez E, Pardo JM (1999) Development of an emotional speech synthesiser in spanish. In: European conference on speech and language processing (EUROSPEECH), Budapest, Hungary, pp 2099–2102

Mowrer OH (1960) Learning theory and behaviour. John Wiley & Sons, Ltd, New York, USA

Mozziconacci S. JL, Hermes DJ (1999) Role of intonation patterns in conveying emotion in speech. In: International congress of phonetic sciences (ICPhS), San Francisco, USA, pp 2001–2004

Nasoz F, Alvarez K, Lisetti CL, Finkelstein N (2003) Emotion recognition from physiological signals for presence technologies. Int J Cogn Technol Work 6(1):4–14

Nass C, Lee KM (2001) Does computer-synthesized speech manifest personality? Experimental tests of recognition, similarity-attraction, consistency-attraction. J Exper Psychol Appl 7(3):171–181

Navas E, Castelruiz A, Luengo I, Sánchez J, Hernáez I. (2004a) Designing and recording an audiovisual database of emotional speech in Basque. In: International conference on language resources and evaluation (LREC), Lisbon, Portugal, pp 1387–1390

Navas E, Hernaéz I, Castelruiz A, Luengo I (2004b) Obtaining and evaluating an emotional database for prosody modelling in standard Basque. In: Sojka P Kopeček I Pala K (eds) Proceedings of the 7th international conference on text, speech and dialogue (TSD). Springer, Brno, Czech Republic, pp 393–400

Navas E, Hernáez I, Luengo I (2006) An objective and subjective study of the role of semantics and prosodic features in buidling corpora for emotional TTS. IEEE Trans Audio Speech Lang Proces 14(4):1117–1127

Neiberg D, Elenius K, Laskowski K (2006) Emotion recognition in spontaneous speech using GMMs. In: International conference on speech and language processing (ICSLP), Pittsburgh, USA, pp 809–812

Nicholson J, Takahashi K, Nakatsu R (1999) Emotion recognition in speech using neural networks. In: Proceedings of the 6th international conference on neural information processing (ICONIP), vol 2. Perth, Australia, pp 495–501

Nisimura R, Omae S, Kawahara H, Irino T (2006) Analyzing dialogue data for real-world emotional speech classification. In: International conference on speech and language processing (ICSLP), Pittsburgh, USA, pp 1822–1825

Nogueiras A, Moreno A, Bonafonte A, no J. BM (2001) Speech emotion recognition using hidden Markov models. In: European conference on speech and language processing (EUROSPEECH), Aalborg, Denmark, pp 2679–2682

Nuttall AH (1981) Some windows with very good sidelobe behavior. IEEE Trans Acoust Speech Signal Proces ASSP-29(1):84–91

Nwe TL, Wei FS, De Silva LC (2001) Speech based emotion classification. In: Proceedings of the IEEE region 10 international conference on electrical and electronic technology (TENCON), vol 1. Phuket Island, Singapore, pp 297–301

Oatley K, Jenkins JM (1996) Understanding emotions. Blackwell, Oxford, United Kingdom

Oatley K, Johnson-Laird PN (1987) Towards a cognitive theory of emotions. Cogn Emot 1:29–50

Oatley K, Johnson-Laird PN (1995) Communicative theory of emotions: Empirical tests, mental models & implications for social interaction. In: Martin LL, Tesser A (eds) Goals and affect. Erlbaum, Hillsdale, USA

O'Brien D, Monaghan A. IC (2001) Concatenative synthesis based on a harmonic model. IEEE Trans Speech Audio Process, 9(1):11–20

Okada N, Inui K, Tokuhisa M (1999) Towards affective integration of vision, behavior, speech processing. In: Proceedings of integration of speech and image understanding, Corfu, Greece, pp 49–77

O'Malley MH (1990) Text-to-speech conversion technology. Computer 23(8):17–23

O'Neill I, Hanna P, Liu X, McTear M (2003) The queen's communicator: An object-oriented dialogue manager. In: European conference on speech and language processing (EUROSPEECH), Geneva, Switzerland, pp 593–596

Oppenheim AV, Schafer RW (2004) From frequency to quefrency: A history of the cepstrum. IEEE Signal Proces Mag 21(5):95–106

Ortony A, Clore G, Foss M (1987) The referential structure of the affective lexicon. Cogn Sci 11:341–346

Ortony A, Clore GL, Collins A (1988) The cognitive structure of emotions. Cambridge University Press, Cambridge, United Kingdom

Ortony A, Turner TJ (1990) What's basic about basic emotions. Psychol Rev 97(3):315–331

Osgood CE, Suci GJ, Tanenbaum PH (1957) The measurement of meaning. University of Illinois Press, Urbana, USA

Oudeyer P-Y (2003) The production and recognition of emotions in speech: Features and algorithms. Int J Hum-Comput St 59(1-2):157–183

Oviatt SL (1997) Multimodal interactive maps: Designing for human performance. Hum-Comput Int (Special Issue on Multimodal Interfaces) 12:93–129

Oviatt SL (2000) Taming recognition errors with a multimodal interface. Commun ACM 43(9): 45–51

Page L, Brin S, Motwani R, Winograd T (1998) The PageRank citation ranking: Bringing order to the Web. Technical report, Stanford digital library technologies project

Panksepp J (1982) Toward a general psychobiological theory of emotions. Behav Brain Sci 5:407–467

Pao T-L, Chen Y-T, Yeh J-H, Lu J-J (2004) Detecting emotions in mandarin speech. In: Proceedings of the 16th conference on computational linguistics and speech processing ROCLING, Taipei, Taiwan

Park C-H, Si K-B (2003) Emotion recognition and acoustic analysis from speech signal. In: Proceedings of the international joint conference on neural networks (IJCNN), vol 4. Portland, USA, pp 2594–2598

Parunak H. VD, Bisson R, Brueckner S, Matthews R, Sauter J (2006) A model of emotions for situated agents. In: Proceedings of the fifth international joint conference on autonomous agents and multiagent systems, Hakodate, Japan

Pereira C (2000) Dimensions of emotional meaning in speech. In: Proceedings of ISCA workshop on speech and emotion, Belfast, United Kingdom

Peter C, Beale R (eds) (2008) Affect and emotion in human-computer interaction: From theory to applications. (Lecture notes in computer science) Springer, Berlin, Germany

Peter C, Herbon A (2006) Emotion representation and physiology assignments in digital systems. Interact Comput 18(2):139–170

Peter G, Rösner D (1994) User-model-driven generation of instructions. User Model User-Adapted Interact 3(4):289–319

Picard RW (2000a) Affective computing. The MIT Press, Cambridge, USA

Picard RW (2000b) Toward computers that recognize and respond to user emotion. IBM Syst J 39(3,4):705–718

Pierrehumbert J, Hirschberg J (1990) The meaning of intonational contours in the interpretation of discourse. In: Cohen PR, Morgan J, Pollack ME (eds) Intentions in communication. MIT Press, Cambridge, USA, pp 271–311

Pittermann A, Pittermann J (2006a) Getting bored with HTK? Using HMMs for emotion recognition. In: 8th international conference on signal processing (ICSP), vol 1. Guilin, China, pp 704–707

Pittermann J, Minker W, Pittermann A, Bühler D (2007a) ProblEmo – Problem solving and emotion awareness in spoken dialogue systems. In: 3rd IET international conference on intelligent environments, Ulm, Germany

Pittermann J, Pittermann A (2006b) A post-processing approach to improve emotion recognition rates. In: 8th international conference on signal processing (ICSP), vol 1. Guilin, China, pp 708–711

Pittermann J, Pittermann A (2006c) An 'emo-statistical' model for flexible dialogue management. In: 8th international conference on signal processing (ICSP), vol 2. Guilin, China, pp 1599–1602

Pittermann J, Pittermann A (2006d) Integrating emotion recognition into an adaptive spoken language dialogue system. In: 2nd IET international conference on intelligent environments, vol 1. Athens, Greece, pp 197–202

Pittermann J, Pittermann A (2007) A data-oriented approach to integrate emotions in adaptive dialogue management. In: International conference on intelligent user interfaces (IUI), Honolulu, USA, pp 270–273

Pittermann J, Pittermann A, Meng H, Minker W (2007b) Towards an emotion-sensitive spoken dialogue system – Classification and dialogue modeling. In: 3rd IET international conference on intelligent environments, Ulm, Germany

Pittermann J, Pittermann A, Minker W (2007c) Design and implementation of adaptive dialogue strategies for speech-based interfaces. J Ubiquitous Comput Intell 1(2):145–152

Pittermann J, Pittermann A, Schmitt A, Minker W (2008a) Comparing evaluation criteria for (automatic) emotion recognition. In: 4th IET international conference on intelligent environments, Seattle, USA

Pittermann J, Rittinger A, Minker W (2005) Flexible dialogue management in intelligent human-machine interfaces. In: The IEE international workshop on intelligent environments, University of Essex, Colchester, United Kingdom

Pittermann J, Schmitt A, Fawzy El Sayed N (2008b) Integrating linguistic cues into speech-based emotion recognition. In: 4th IET international conference on intelligent environments, Seattle, USA

Plutchik R (1980a) A generalpsychorevolutionary theory of emotion. In: Plutchik R, Kellerman H (eds) Emotion: Theory, research, experience: Vol. 1. Theories of emotion. Academic, New York, USA, pp 3–31

Plutchik R (1980b) Emotion: A psychorevolutionary synthesis. Harper & Row, New York, USA

Plutchik R (1994) The psychology and biology of emotion. Harper Collins College Publishers, New York, USA

Polifroni J, Chung G (2002) Promoting portability in dialogue management. In: International conference on speech and language processing (ICSLP), Denver, USA, pp 2721–2724

Polzin TS, Waibel A (1998) Detecting emotions in speech. In: Proceedings of the CMC, Tilburg, The Netherlands

Polzin TS, Waibel A (2000) Emotion-sensitive human-computer interfaces. In: ITRW on speech and emotion, ISCA, pp 201–206

Potapova R, Potapov V (2005) Identification of prosodic features of emotional state of a speaker. In: Proceedings of SPECOM, Patras, Greece, pp 25–32

Power M, Dalgleish T (1997) Cognition and emotion: From order to disorder. Pschology Press, Hove, United Kingdom

Rabiner LR (1989) A tutorial on hidden Markov models and selected applications in speech recognition. Proc IEEE 77(2):257–286

Rank E, Pirker H (1998) Generating emotional speech with a concatenative synthesizer. In: International conference on speech and language processing (ICSLP), Sydney, Australia

Rao AS, Georgeff MP (1991) Modeling agents within a BDI architecture. In: Proceedings of the 2nd international conference on principles of knowledge representation and reasoning (KR-91), pp 473–484

Razak AA, Komiya R, Abidin M. IZ (2005) Comparison between fuzzy and NN method for speech emotion recognition. In: Proceedings of the 3rd IEEE international symposium on information technology and applications (ICITA), vol 1. Sydney, Australia, pp 297–302

Razak AA, Yusof M. HM, Komiya R (2003) Towards automatic recognition of emotion in speech. In: Proceedings of the 3rd IEEE international symposium on signal processing and information technology (ISSPIT), Darmstadt, Germany, pp 548–551

Reeves B, Nass C (1996) The media equation: How people treat computers, television, new media like real people and places. Cambridge University Press, Cambridge, United Kingdom

Reidsma D, Heylen D, Ordelman R (2006) Annotating emotions in meetings. In: International conference on language resources and evaluation (LREC), Genova, Italy, pp 1117–1122

Reiter E, Dale R (2000) Building natural language generation systems. Cambridge University Press, Cambridge, United Kingdom

Reza FM (1961) An introduction to information theory. McGraw-Hill, New York, USA

Rigoll G, Müller R, Schuller B (2005) Speech emotion recognition exploiting acoustic and linguistic information sources. In: Proceedings of SPECOM, Patras, Greece, pp 61–67

Ripley BD (1996) Pattern recognition and neural networks. Cambridge University Press, Cambridge, United Kingdom

Roark B, Saraclar M, Collins M (2007) Discriminative n-gram language modeling. Comput Speech Lang 21(2):373–392

Robinson T (1997a) Speech analysis – Lecture notes, lent term. Cambridge University, Cambridge, (Online tutorial)
Robinson T (1997b) The British English example pronunciation dictionary
Robson J, Mackenzie-Beck J (1999) Hearing smiles – Perceptual, acoustic and production aspects of labial spreading. In: International congress of phonetic sciences (ICPhS), San Francisco, USA, pp 219–222
Roseman IJ (1979) Cognitive aspects of emotion and emotional behaviour. In: 87th annual convention of the American Psychological Association, New York, USA
Roseman IJ, Spindel MS, Jose PE (1990) Appraisals of emotion-eliciting events: Testing a theory of discrete emotions. J Personality Soc Psychol 59:899–915
Rossi G, Schwabe D, Guimarães R (2001) Designing personalized Web applications. In: Proceedings of the 10th international World Wide Web conference, Hong Kong, China, pp 275–284
Rotaru M, Litman DJ (2005) Using word-level pitch features to better predict student emotions during spoken tutoring dialogues. In: International conference on speech and language processing (ICSLP), Lisbon, Portugal, pp 1845–1848
Rotaru M, Litman DJ, Forbes-Riley K (2005) Interactions between speech recognition problems and user emotions. In: International conference on speech and language processing (ICSLP), Lisbon, Portugal, pp 2481–2484
Roy N, Pineau J, Thrun S (2000) Spoken dialogue management using probabilistic reasoning. In: Proceedings of the 38th annual meeting of the association for computational linguistics (ACL2000), Hong Kong, China
Russell JA (1980) A circumplex model of affect. J Personality Soc Psychol 39:1161–1178
Russell JA, Mehrabian A (1977) Evidence for a three-factor theory of emotions. J Res Personality 11:273–294
Sagisaka Y (2001) Speech synthesis. J ASJ 57(1):11–20
Sakoe H, Chiba S (1978) Dynamic programming algorithm optimization for spoken word recognition. IEEE Trans Acoust Speech Signal Proces (ASSP) 26(1):43–49
Saratxaga I, Navas E, Hernáez I, Luengo I (2006) Designing and recording an emotional speech database for corpus based synthesis in Basque. In: International conference on language resources and evaluation (LREC), Genova, Italy, pp 2126–2129
Schachter S, Singer JE (1962) Cognitive, social, physiological determinants of emotional state. Psychol Rev 69:379–399
Scherer KR (1988) Criteria for emotion-antecedent appraisal: A review. In: Hamilton V, Frijda NH (eds) Cognitive perspectives on emotion and motivation. Kluwer, Dordrecht, the Netherlands, pp 89–126
Scherer KR (2000) Psychological models of emotion. In: Borod JC (ed) The neuropsychology of emotion. Oxford University Press, New York, USA, pp 137–162
Scherer KR, Bänziger T (2004) Emotional expression in prosody: A review and an agenda for future research. In: Proceedings of speech prosody 2004, Nara, Japan, pp 359–366
Scherer KR, Ceschi G (1997) Lost luggage: A field study of emotion-antecedent appraisal. Motivat Emot 21(3):211–235
Schiel F, Steininger S, Türk U (2002) The SmartKom multimodal corpus at BAS. In: International conference on language resources and evaluation (LREC), Las Palmas, Spain
Schlosberg H (1954) Three dimensions of emotion. Psychol Rev 61(2):81–88
Schmidt G, Haulick T (2006) Signal processing for in-car communication systems. Signal Proces 86(6):1307–1326
Schröder M (2000) Experimental study of affect bursts. In: Proceedings of ISCA workshop on speech and emotion, Belfast, United Kingdom
Schröder M (2004) Speech and emotion research: An overview of research frameworks and a dimensional approach to emotional speech synthesis. PhD thesis, Saarland University
Schröder M, Cowie R (2006) Developing a consistent view on emotion-oriented computing. In: Renals S, Bengio S (eds) MLMI 2005, LNCS 3869. Springer, Heidelberg, Germany, pp 194–205

Schroeder MR, Atal BS, Hall JL (1979) Optimizing digital speech coders by exploiting masking properties of the human ear. J Acoust Soc Am 66(6):1647–1652

Schuller B, Müller R, Lang M, Rigoll G (2005) Speaker independent emotion recognition by early fusion of acoustic and linguistic features within ensembles. In: International conference on speech and language processing (ICSLP), Lisbon, Portugal, pp 805–808

Seneff S, Wang C, Chung G (2004) A flexible mixed-initiative speech interface for restaurant information. In: 5th SIGdial workshop on discourse and dialogue, Cambridge, USA

Shannon CE (1948) A mathematical theory of communication. Bell Syst Tech J 27:379–423, 623–656

Shigeno S (1998) Cultural similarities and differences in the recognition of audio-visual speech stimuli. In: International conference on speech and language processing (ICSLP), vol 2. Sydney, Australia, pp 281–284

Silverman K, Beckman M, Pitrelli J, Ostendorf M, Wightman C, Price P, Pierrehumbert J, Hirschberg J (1992) ToBI: A standard for labeling English prosody. In: International conference on speech and language processing (ICSLP), Banff, Canada, pp 867–870

Sproat R, Hirschberg J, Yarowsky D (1992) A corpus-based synthesizer. In: International conference on speech and language processing (ICSLP), Banff, Canada, pp 563–566

Stallard D (2002) Flexible dialogue management in the Talk'n'Travel system. In: International conference on speech and language processing (ICSLP), Denver, USA, pp 2693–2696

Stevens SS, Volkmann J, Newman EB (1937) A scale for the measurement of the psychological magnitude pitch. J Acoust Soc Am 8(3):185–190

Stibbard R (2000) Automated extraction of ToBI annotation data from the reading/leeds emotional speech corpus. In: Proceedings of ISCA workshop on speech and emotion, Belfast, United Kingdom

Stibbard RM (2001) Vocal expression of emotions in non-laboratory speech: An investigation of the reading/leeds emotion in speech project annotation data. PhD thesis, University of Reading, United Kingdom

Sugimoto T, Ito N, Fujishiro H, Sugeno M (2002) Dialogue management with the semiotic base: A systemic functional linguistic approach. In: Proceedings of the 1st international conference on soft computing and intelligent systems, Tsukuba, Japan

Sun Y, Willett D, Brueckner R, Gruhn R, Bühler D (2006) Experiments on Chinese speech recognition with tonal models and pitch estimation using the Mandarin Speecon data. In: International conference on speech and language processing (ICSLP), Pittsburgh, USA

Sutton S, Cole R, de Villiers J, Schalkwyk J, Vermeulen P, Macon M, Yan Y, Kaiser E, Rundle B, Shobaki K, Hosom P, Kain A, Wouters J, Massaro D, Cohen M (1998) Universal speech tools: The CSLU toolkit. In: International conference on speech and language processing (ICSLP), Sydney, Australia, pp 3221–3224

Swerts M, Krahmer E (2000) On the use of prosody for on-line evaluation of spoken dialogue systems. In: International conference on language resources and evaluation (LREC), Athens, Greece

't Hart J, Cohen A, Collier R (1990) A perceptual study of intonation: An experimental-phonetic approach to speech melody. Cambridge University Press, Cambridge, United Kingdom

Takahash T, Fujii T, Nishi M, Banno H, Irino T, Kawahara H (2005) Voice and emotional expression transformation based on statistics of vowel parameters in an emotional speech database. In: International conference on speech and language processing (ICSLP), Lisbon, Portugal, pp 1853–1856

Tao J (2004) Context based emotion detection from text input. In: International conference on speech and language processing (ICSLP), Jeju, Korea

Tao J, Kang Y, Li A (2006) Prosody conversion from neutral speech to emotional speech. IEEE Trans Audio Speech Lang Proces 14(4):1145–1154

Tato R, Santos R, Kompe R, Pardo JM (2002) Emotional space improves emotion recognition. In: International conference on speech and language processing (ICSLP), Denver, USA, pp 2029–2032

Teixeira A, Silva L, Martinez R, Vaz F (2002) SAPWindows – Towards a versatile modular articulatory synthesizer. In: Proceedings of the 2002 IEEE workshop on speech synthesis, Santa Monica, pp 31–34

Telatar IE (1999) Capacity of multi-antenna gaussian channels. Eur Trans Telecommun 10(6): 585–595

ten Bosch L (2000) Emotions: What is possible in the ASR Framework. In: Proceedings of ISCA workshop on speech and emotion, Belfast, United Kingdom

Tepperman J, Traum D, Narayanan S (2006) "Yeah Right": Sarcasm recognition for spoken dialogue systems. In: International conference on speech and language processing (ICSLP), Pittsburgh, USA, pp 1838–1841

Tesser F, Cosi P, Drioli C, Tisato G (2005) Emotional festival-MBROLA TTS synthesis. In: International conference on speech and language processing (ICSLP), Lisbon, Portugal, pp 505–508

Tolkmitt FJ, Scherer KR (1986) Effect of experimentally induced stress on vocal parameters. J Exper Psychol Hum Percept Perform 12(3):302–313

Tomkins SS (1984) Affect theory. In: Scherer KR, Ekman P (eds) Approaches to emotion. Erlbaum, Hillsdale, USA, pp 163–195

Torres F, Sanchis E, Segarra E (2003) Development of a stochastic dialog manager driven by semantics. In: European conference on speech and language processing (EUROSPEECH), Geneva, Switzerland, pp 605–608

Tsuzuki R, Zen H, Tokuda K, Kitamura T, Bulut M, Narayanan SS (2004) Constructing emotional speech synthesizers with limited speech database. In: International conference on speech and language processing (ICSLP), Jeju, Korea

Turunen M, Hakulinen J (2001) Agent-based adaptive interaction and dialogue management architecture for speech applications. Lecture Note Comput Sci 2166:357–364

Ukelson J, Rodeh M (1989) A dialogue manager for efficient adaptive man-machine dialogues. In: Proceedings of the 13th annual international computer software and applications conference (COMPSAC), Orlando, USA, pp 588–595

Van Bezooijen R. A. MG (1984) Characteristics and recognizability of vocal expressions of emotion. Foris Publictions, Dordrecht, the Netherlands

Väyrynen E, Seppänen T, Toivanen J (2003) An experiment in emotional content classification of spoken Finnish using prosodic features. In: Finnish signal processing symposium, Tampere, Finland, pp 264–267

Veeravalli AG, Pan WD, Adhami R, Cox PG (2005) A tutorial on using hidden Markov models for phoneme recognition. In: Proceedings of the thirty-seventh southeastern symposium on system theory, Tuskegee, USA, pp 154–157

Veldhuijzen van Zanten G (1998) Adaptive mixed-initiative dialogue management. In: Proceedings of the IEEE 4th workshop on interactive voice technology for telecommunications applications (IVTTA), Torino, Italy, pp 65–70

Veldhuijzen van Zanten G (1999) User modelling in adaptive dialogue management. In: European conference on speech and language processing (EUROSPEECH), Budapest, Hungary, pp 1183–1186

Velten E (1968) A laboratory task for induction of mood states. Behav Res Ther 6:473–482

Viterbi AJ (1967) Error bounds for convolutional codes and an asymptotically optimum decoding algorithm. IEEE Trans Inform Theory IT-13:260–269

Vogt T, André E (2006) Improving automatic emotion recognition from speech via gender differentiation. In: International conference on language resources and evaluation (LREC), Genova, Italy, pp 1123–1126

von Neumann J, Morgenstern O (1944) Theory of games and economic behavior. Princeton University Press, Princeton, USA

Wagner T, Dieckmann U (1995) Sensor-fusion for robust identification of persons: A field test. In: Proceedings of the international conference on image processing (ICIP), vol 3. Washington, DC, USA, pp 516–519

Wahlster W (ed) (2006) SmartKom: Foundations of multimodal dialogue systems. Springer, Berlin, Germany

Walker MA, Aberdeen J, Boland J, Bratt E, Garafolo J, Hirschman L, Le A, Lee S, Narayanan S, Papineni K, Pellom B, Polifroni J, Potamianos A, Prabhu P, Rudnicky A, Sanders G, Seneff S, Stallard D, Whittaker S (2001) DARPA Communicator dialog travel planning systems: The June 2000 data collection. In: Dalsgaard P, Lindberg B, Benner H, Tan Z (eds) European conference on speech and language processing (EUROSPEECH), Aalborg, Denmark, pp 1371–1374

Walker MA, Cahn JE, Whittaker SJ (1997a) Improvising linguistic style: social and affective bases of agent personality. In: Johnson WL, Hayes-Roth B (eds) Proceedings of the first international conference on autonomous agents (Agents'97), Marina del Rey, USA, ACM Press, pp 96–105

Walker MA, Litman DJ, abd Alicia Abella C. AK (1998) Evaluating spoken dialogue agents with PARADISE: Two case studies. Comput Speech Lang 12(4):317–348

Walker MA, Litman DJ, Kamm CA, Abella A (1997b) PARADISE: A framework for evaluating spoken dialogue agents. In: Proceedings of the 35th annual meeting of the association of computational linguistics ACL/EACL, Madrid, Spain, pp 271–280

Walker W, Lamere P, Kwok P (2002) FreeTTS – A performance case study. Technical report TR-2002-114, Sun Microsystems, Inc., Burlington, USA

Wallace GK (1991) The JPEG still picture compression standard. Commun ACM 34(4):30–44

Wardhaugh R (1992) An introduction to sociolinguistics. Blackwell textbooks in linguistics. Blackwell, Chichester, United Kingdom

Watson JB (1930) Behaviorism. University of Chicago Press, Chicago, USA

Weiner B, Graham S (1984) An attributional approach to emotional development. In: Izard CE, Kagan J, Zajonc RB (eds) Emotions, cognition and behavior. Cambridge University Press, New York, USA, pp 167–191

Weizenbaum J (1966) ELIZA – A computer program for the study of natural language communication between man and machine. Commun ACM 9(1):36–45

Welch LR (2003) Hidden Markov models and the Baum-Welch algorithm. IEEE Inform Theory Soc Newslett 53(4):1, 10–13

Werner S, Hoffmann R (2007) Spontaneous speech synthesis by pronunciation variant selection – A comparison to natural speech. In: European conference on speech and language processing (EUROSPEECH), Antwerp, Belgium, pp 1781–1784

Whissel CM (1989) The dictionary of affect in language. In: Plutchik R, Kellerman H (eds) Emotion: Theory, research and experience, vol 4. Academic, New York, USA, pp 113–131

Whiteside SP (1998) Simulated emotions: An acoustic study of voice and perturbation measures. In: International conference on speech and language processing (ICSLP), Sydney, Australia, pp 699–703

Wichmann A (2000) The attitudinal effects of prosody, how they relate to emotion. In: Proceedings of ISCA workshop on speech and emotion, Belfast, United Kingdom

Williams JD (2006) Partially observable Markov decision processes for spoken dialogue management. PhD thesis, Cambridge University

Williams JD, Poupart P, Young S (2005) Partially observable Markov decision processes with continuous observations for dialogue management. In: Proceedings of the 6th SIGdial workshop on discourse and dialogue, Lisbon, Portugal

Williams JD, Young S (2007) Partially observable markov decision processes for spoken dialog systems. Comput Speech Lang 21(2):393–422

Wilting J, Krahmer E, Swerts M (2006) Real vs. acted emotional speech. In: International conference on speech and language processing (ICSLP), Pittsburgh, USA, pp 805–808

Witten IH, Frank E (2000) Data mining: Practical machine learning tools and techniques with Java implementations. Morgan Kaufmann, San Francisco, USA

Wolz U (1990) An object oriented approach to content planning for text generation. In Proceedings of the fifth international workshop on natural language generation, Dawson, USA, pp 95–104

Woodland PC, Young SJ (1993) The HTK tied-state continuous speech recogniser. In: European conference on speech and language processing (EUROSPEECH), Berlin, Germany, pp 2207–2210

Wu W, Zeng TF, Xu M-X, Bao H-J (2006) Study on speaker verification on emotional speech. In: International conference on speech and language processing (ICSLP), Pittsburgh, USA, pp 2102–2105

Wu X-J, Zheng F, Wu W-H (2002) A hybrid dialogue management approach for a flight spoken dialogue system. In: Proceedings of the first international conference on machine learning and cybernetics, Beijing, China, pp 824–829

Wu X-J, Zheng F, Xu M (2001) Topic forest: A plan-based dialog management structure. In: International conference on acoustics, speech and signal processing (ICASSP), Salt Lake City, USA

Wundt W (1924) An introduction to psychology. Allen & Unwin, London. (Translated by R. Pintner, original work published in 1912)

Yacoub S, Simske S, Lin X, Burns J (2003) Recognition of emotions in interactive voice response systems. In: European conference on speech and language processing (EUROSPEECH), Geneva, Switzerland, pp 729–732

Yang L, Campbell N (2001) Linking form to meaning: The expression and recognition of emotions through prosody. In: Proceedings of the 4th ISCA workshop on speech synthesis, Perthshire, United Kingdom

Yankelovich N (1996) How do users know what to say? ACM Interact 3(6):32–43

Yildirim S, Bulut M, Lee CM, Kazemzadeh A, Busso C, Deng Z (2004) An acoustic study of emotions expressed in speech. In: International conference on speech and language processing (ICSLP), Jeju, Korea

Young S (1994) The HTK hidden Markov model toolkit: Design and philosophy. Cambridge University Engineering Department, UK, Tech. Rep. CUED/F-INFENG/TR152

Young S (2001) Statistical modelling in continuous speech recognition (CSR). In: Proceedings of the international conference on uncertainty in artificial intelligence, Seattle, USA

Young S, Evermann G, Gales M, Hain T, Kershaw D, Liu X, Moore G, Odell J, Ollason D, Povey D, Valtchev V, Woodland P (2006) The HTK book (for HTK Version 3.4). Cambridge University Engineering Department

Yu C, Aoki PM, Woodruff A (2004) Detecting user engagement in everyday conversations. In: International conference on speech and language processing (ICSLP), Jeju, Korea

Yuan J, Shen L, Chen F (2002) The acoustic realization of anger, fear, joy and sadness in Chinese. In: International conference on speech and language processing (ICSLP), Denver, USA, pp 2025–2028

Yule GU (1927) On a method of investigating periodicities in disturbed series, with special reference to Wolfer's sunspot numbers. Philos Trans Roy Soc Lond 226:267–298

Zajonc RB (1984) On the primacy of affect. Am Psychol 39:124–129

Zell A, Mache N, Sommer T, Korb T (1991) The SNNS neural network simulator. In: Proceedings of 15. Fachtagung für Künstliche Intelligenz, Bonn, Germany, pp 254–263

Zhang S, Ching PC, Kong F (2006) Automatic recognition of speech signal in Mandarin. In: International conference on speech and language processing (ICSLP), Pittsburgh, USA, pp 1810–1813

Zhang T, Hasegawa-Johnson M, Levinson SE (2004) Children's emotion recognition in an intelligent tutoring scenario. In: International conference on speech and language processing (ICSLP), Jeju, Korea

Zhou L, Shi Y, Feng J, Sears A (2005) Data mining for detecting errors in dication speech recognition. IEEE Trans Speech Audio Process 13(5):681–688

Zinn C (2004) Flexible dialogue management in natural-language enhanced tutoring. In: Konvens 2004 workshop on advanced topics in modeling natural language dialog, Vienna, Austria, pp 28–35

Index

LaVergne, TN USA
20 November 2009
164744LV00001B/70/P